A User's Handbook
of Integrated Circuits

A USER'S HANDBOOK
OF INTEGRATED CIRCUITS

EUGENE R. HNATEK

A WILEY-INTERSCIENCE PUBLICATION

JOHN WILEY & SONS, New York · London · Sydney · Toronto

Library of Congress Cataloging in Publication Data:

Hnatek, Eugene R
A user's handbook of integrated circuits.

"A Wiley-Interscience publication."
Bibliography: p.
1. Integrated circuts. I. Title.
TK7874.H54   621.381'73   72-13596
ISBN 0-471-40110-2

Printed in the United States of America

10-9 8 7 6 5 4 3 2

To

JOSEPH, CAROLINE, FRANK, VLASTA,
SUSAN, STEPHEN, and JEFFREY

PREFACE

To really take advantage of integrated circuit techniques, at the practical how-to-do-it level, an electronics designer has to understand something about fabrication. He also must be familiar with the advantages and disadvantages of the various approaches, in order to properly select the one best suited for a particular application.

Integrated circuit selection is not a simple choice among MOS or bipolar or hybrid devices; it involves a continuous investigation of a multitude of combinations of performance, speed, operation, size, and cost. Only when the equipment designer explores all the alternatives available to him, keeping in mind that there are considerable performance–speed–cost trade-offs involved in his choice, is he likely to make the optimum selection.

The main purpose of the book is to give the practicing engineer, or the engineering student, an insight into the basic trade-offs between conventional circuits assembled from discrete components and ICs, and among the different IC families.

Much of the book is devoted to IC processing—not on the level required by a processing specialist, but on the level that a practicing engineer needs to make an informed choice among circuits made by different technologies. Each process has inherent advantages in performance, cost, and reliability. But each also imposes inherent constraints. These trade-offs are far from straightforward, since the combinations of advantages and disadvantages for each technology are unique.

There are many books on semiconductor processing, as well as many detailed design analyses of particular types of ICs. But the author has long felt a need for a handbook that presents a coordinated view of internal IC design processing and IC applications.

The book is organized in six parts:

- Chapter 1 is an introduction to all the major IC technologies and briefly compares them with discrete-component technology.

vii

- Chapter 2 covers the processing of bipolar ICs and explains in more detail why ICs do not have the same performance characteristics as conventional circuits.

- Chapters 3 through 5 cover the design, types, and applications of digital bipolar ICs. Emphasis is placed in the applications chapter on TTL medium-scale integration, since this type of circuit dominates digital system design. This section also includes the first information published in a book on the new tri-state form of TTL and its applications.

- Chapters 6 through 9 present information on MOS arrays. Since processing technology is the main reason for the performance differences among types of MOS ICs, processing of the types is discussed in context with their performance characteristics. This section includes information on how to combine MOS with bipolar ICs to obtain the most benefit from both.

- Chapters 10 through 12 are devoted to linear bipolar ICs, specifically, operational amplifiers and voltage regulators. These represent the great majority of applications.

- Chapters 13 and 14 might be called the "do-it-yourself" section. They tell how to design thin-film and thick-film hybrid circuits, they review the processes, and they explain when each technique should be used.

- Chapter 15 covers both monolithic and hybrid packaging and assembly, including the appropriate bonding methods for use with semiconductor chip devices in a hybrid circuit.

No attempt is made to cover IC applications exhaustively, but applications of the basic types of digital and linear circuits are outlined, along with the basic rules of using them. Particular attention is paid to the "overlap" areas where a design might benefit from using more than one technology — such as MOS in combination with TTL, or monolithic ICs in a hybrid assembly.

The general applications information is designed to provide the reader with insight into the versatility of ICs. Specific devices for particular functions can then be chosen from manufacturers' literature.

Throughout the book, emphasis is placed on comparing performance and, in more general terms, the cost and reliability offered by competing IC technologies. In each of the chapter groups, simple, step-by-step guidelines and tables present the competing technologies. The guidelines and comparative information are those the author has found by experience to be most meaningful to a choice of one technology or another for a particular class of system. Once that choice has been made, the user can go to the literature for the detailed design analyses he needs to complete his design.

I should like to thank my colleagues at National Semiconductor Corp. for permission to use portions of material for the applications sections of this book, specifically, Robert Widlar, Robert Dobkin, and Dale Mrazek. Additionally, I should like to thank my typist Harford Gillaspie for her painstaking hours in preparing the drafts of this manuscript, and Charles Signor for his aid in obtaining the photographs used in this book.

EUGENE R. HNATEK

Santa Clara, California
October 1972

CONTENTS

1 AN OVERVIEW OF INTEGRATED CIRCUIT CHOICES 1

Classification of Integrated Circuits, 1
Monolithic Integrated Circuits, 1
   Bipolar Monolithic Circuits, 2
   Features of Bipolar Integrated Circuits, 4
Metal–Oxide–Silicon (MOS) Integrated Circuits, 8
   MOS Operation, 9
   MOS Characteristics, 9
Hybrid Integrated Circuits, 10
   Thin-Film Hybrid Circuits, 11
   Thick-Film Hybrid Circuits, 12
   Hybrid Versus Monolithic Integrated Circuits, 12
Factors for Selecting an Integrated Circuit Technology, 13

2 BIPOLAR INTEGRATED CIRCUIT PROCESSING 18

Processing Steps, 21
   Substrate Preparation, 22
   Oxidation and Planar Processing, 23
   Oxide Growth, 24
   Oxide Masking, 24
   Photomask Preparation, 24
   Computerized Artwork Generation, 31
   Photoetching Processes, 32
   Diffusion Techniques, 33
   Dopants and Integrated Circuit Characteristics, 39
   Epitaxial Growth, 40

Isolation, 41
Metallization, 44
Vacuum Deposition, 46
Die Passivation, 47
Process Maturity, 48

3 BIPOLAR INTEGRATED CIRCUIT CHARACTERISTICS AND DESIGN 49

Design Rules for Bipolar Integrated Circuits, 50
Monolithic Integrated Circuit Components, 51
Bipolar Integrated Circuit Transistors, 52
Monolithic Diodes, 65
Schottky-Barrier Diodes, 66
Monolithic Resistors, 68
Monolithic Capacitors, 70
Putting the Components Together, 73

4 BIPOLAR LOGIC CIRCUIT FAMILIES 74

Basic Requirements of Digital Circuits, 74
Description of Major Saturated-Logic Lines, 75
Resistor–Transistor Logic, 75
Diode–Transistor Logic, 75
Transistor–Transistor Logic, 76
TTL Output Gate Configurations, 77
Popular TTL Families, 84
The Concept of TRI-STATE TTL Logic, 90
Current-Mode Logic, 92
Emitter-Coupled Logic (ECL), 93
Comparison of ECL with Other Logic Families, 98
Digital Integrated Circuit Nomenclature, 98

5 APPLICATIONS OF TTL AND TRI-STATE MEDIUM-SCALE INTEGRATED CIRCUITS 105

Economic Considerations of MSI, 105
MSI Applications, 107
Counters, 107
High-Speed TTL Adders, 122
TTL MSI Multiplexers and Demultiplexers, 130
Digital Filter, 142
Digital Integrated Circuit Tone Detector, 144
Digital Clock, 144
TRI STATE Logic (TSL) Applications, 148

6 METAL-OXIDE-SEMICONDUCTOR (MOS) INTEGRATED CIRCUITS   161

p-MOS Devices, 162
  p-MOS Load Resistances, 166
  p-MOS Fabrication, 167
The *n*-MOS Structure, 170
Threshold Voltage, 171
Low- Versus High-Threshold Technology, 171
Complementary-Symmetry MOS Circuits, 174
The Silicon-Gate MOS Structure, 177
The Silicon–Nitride MOS Structure, 179
Ion-Implanted MOS Integrated Circuits, 180
Refractory MOS (RMOS) Circuits, 182
The Field-Shield MOS Structure, 183
The Self-Aligned Thick-Oxide (SATO) MOS Structure, 184
Double-Diffused MOS (DMOS), 185
Metal–Alumina–Silicon Integrated Circuits, 186
Silicon-on-Sapphire (SOS) MOS Integrated Circuits, 186
Silicon-on-Spinel MOS Integrated Circuits, 187
Charged-Coupled Device MOS Integrated Circuits, 189

7 CMOS APPLICATIONS   194

Inverter Characteristics, 200
Transmission Gates, 202
4000A-Series CMOS Functions, 205
Interfacing CMOS with Other Logic Forms, 212
4000A-Series CMOS Usage Guidelines, 215

8 APPLICATIONS OF MOS INTEGRATED CIRCUITS   221

Large-Scale Integration (LSI), 221
Static and Dynamic Logic, 222
  Dc or Static Logic, 223
  Ac or Dynamic Logic, 224
Shift Registers, 228
  Dynamic Two-Phase Shift Registers, 229
  Dynamic Four-Phase Shift Registers, 233
  Static Shift Registers, 235
  Frequency and Power Characteristics, 237
  Input Structures and Output Buffers, 238
  Application of MOS Shift Registers, 240
  Dynamic Shift Registers with TRI-STATE Outputs, 242
Read-Only-Memory (ROM), 246
  Character Generation Using ROMs, 247

Random-Access Memories (RAMs), 252
  The Static Ram, 253
  The Dynamic Ram, 254
  Writing Data In, 255
  MOS RAM Main-Memory Application, 257
Computer-Aided Design, 260

9 MOS/BIPOLAR COMBINATIONS AND TRADE-OFFS 265

Comparison of MOS and Bipolar Technologies, 265
  MOS Versus Bipolar, 265
Advantages and Limitations of MOS Devices, 268
MOS–Bipolar Interface, 275

10 LINEAR INTEGRATED CIRCUITS: THE OPERATIONAL AMPLIFIER 276

Integrated Circuit Operational Amplifiers, 276
The Ideal Operational Amplifier, 277
Design Considerations for Frequency Response and Gain, 279
Operational-Amplifier Terminology, 282
Speed and Frequency Response, 284
Factors to Consider in Choosing an Op Amp, 286

11 OPERATIONAL-AMPLIFIER APPLICATIONS 291

Basic Operational Amplifier Connections, 292
  The Inverting Amplifier, 292
  The Noninverting Amplifier, 293
  The Unity-Gain Buffer and Voltage Follower, 294
Operational Amplifier Protection Circuitry, 294
Effects of Error Current, 295
  Bias-Current Compensation, 295
  Offset-Voltage Compensation, 297
  Universal Balancing Techniques, 299
Frequency-Compensation Techniques, 301
Feed-Forward Compensation Techniques, 305
Design Precautions, 306
Typical Operational-Amplifier Applications, 308
  Voltage Comparators, 308
  Ac Coupled High-Input-Impedance Amplifier, 309
  The Integrator, 310
  Sample-and-Hold Circuit, 310
  Active Filters, 311
  Self-Tuned Filter, 312
  One-Shot Multivibrator, 313
  Sine-Wave Oscillators, 314
  Instrumentation Amplifier, 317
  Analog-to-Digital Converter Ladder Network Driver, 319

Integrated Nanoammeter Amplifier Circuit, 321
Electronic Thermometer, 323
Operational-Amplifier Voltage Regulator, 326
Dual-Voltage Regulator, 328

12 MONOLITHIC VOLTAGE REGULATORS   330

Definition of Terms, 331
Basic Regulator Types, 332
Integrated Circuit Regulators, 334
Integrated Circuit Regulator Applications, 340

Constant-Current Source, 340
Temperature Controller, 340
Power Amplifier, 343
High-Voltage Regulator, 344
Light-Intensity Regulator, 345
Tracking Regulators, 346
Negative Regulators, 347
Low-Cost Switching Regulator, 347
Low-r-f-Noise Switching Regulator, 348
Development of a 5-V, 1-A Monolithic Regulator, 351
Precision Regulators, 357

13 THIN-FILM HYBRID CIRCUITS   360

Thin-Film Circuit Design, 360
Thin Versus Thick Films, 362
Film Circuit Substrates, 364
Thin-Film Processes, 365
Vacuum Evaporation, 366
Process Monitoring, 367
Thin-Film Patterning, 367
Thin-Film Materials, 369
Thin-Film Conductor Design, 370
Thin-Film Resistor Properties, 372
Resistor Parasitics, 375
Thin-Film Resistor Design, 376
Thin-Film Resistor Design Example, 378
Thin-Film Capacitors, 378
Thin-Film Inductors, 382

14 THICK-FILM HYBRID CIRCUITS   384

Designing a Thick-Film Circuit, 385
Power Dissipation, 385
Laying Out the Circuit, 386
Thick-Film Conductor Design, 387
Thick-Film Resistor Design, 388
Resistor Power Density, 389

Chip Resistors, 390
Thick-Film Capacitors, 392
Chip Capacitors, 392
Comparison of Thin- and Thick-Film Hybrid Circuits, 394
Thick-Film Processing, 394
Thick-Film Circuit Layout, 396

15 INTEGRATED CIRCUIT PACKAGING AND INTERCONNECTION TECHNIQUES 407

Packaging Materials, 408
Package Types, 409

TO-Style Packages, 409
Flat Packs, 411
Cavity Dual-in-Line Packages (DIP), 412
Noncavity DIPs, 415
Hybrid Packages, 417

LSI (Large-Scale Integration) Package Problems, 417
Die Bonding, 421

The Flip-Chip Process, 422
Beam-Lead Bonding, 423
Flip-Chip and Beam-Lead Comparison, 428
Spider Bonding, 428
Semiconductor on Thermoplastic on Dielectric (STD), 430
Hybrid-Circuit Assembly, 431

Ohmic Contacts and Wire-Bonding Materials, 431
Ware-Bonding Methods, 432

Thermal Compression Bonding, 432
Ultrasonic Bonding, 434
Bond Inspection Criteria, 435

Encapsulation, 440

BIBLIOGRAPHY 441

INDEX 445

Chapter One

AN OVERVIEW OF INTEGRATED CIRCUIT CHOICES

Although there are a great many semiconductor materials in use today, silicon is preferred for integrated circuits (ICs) for several reasons. First, it is chemically simpler than compound semiconductor materials such as gallium arsenide and cadmium sulfide and therefore is not subject to stoichiometric effects of compounds. Second, silicon grows a stable oxide. Third, silicon technology is very highly developed. It is possible to grow large single crystals having very few crystalline imperfections and closely controlled impurity concentrations. Finally, although high costs are encountered in purifying silicon, they are offset by the very small amount of material used in each circuit. With reasonable yield ($> 50\%$), the fully processed semiconductor material represents considerably less than 5% of the total cost. All ICs discussed in this volume use silicon as their semiconductor medium.

CLASSIFICATION OF INTEGRATED CIRCUITS

ICs are divided into two major categories: monolithic and hybrid types as in Figure 1-1.

MONOLITHIC INTEGRATED CIRCUITS

The term "monolithic" describes an IC construction in which all the elements of the circuit are built in or on a single crystal of silicon. Because the monolithic

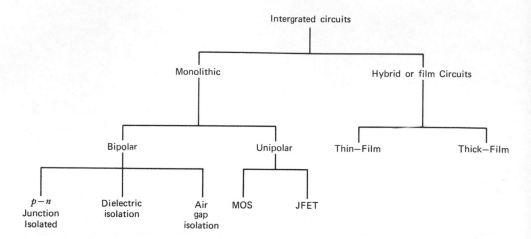

Figure 1-1   Categories of integrated circuits.

circuit layout is produced on the silicon crystal by a photographic process, such devices are easily reduced in size and may be duplicated relatively easily and cheaply. In a hybrid circuit, it is usual to find passive components such as resistors and capacitors deposited on a substrate and active elements added as discrete semiconductor "chips."

Monolithic circuits can be subdivided into bipolar and unipolar types. In bipolar circuits, charge carriers of both positive and negative polarities are required for operation of the active elements; in the unipolar case, either the positively charged "hole" or the electron with its negative charge suffices for the desired electronic function.

Bipolar elements include diodes, which have anodes and cathodes, *npn* or *pnp* transistors, which have emitters and collectors and support current flow in a particular direction, and related devices such as silicon-controlled rectifiers. Unipolar devices include various types of field-effect transistors (FETs), which can support current flow in either direction and which have interchangeable source (positive) and drain (negative) electrodes.

Bipolar Monolithic Circuits

Bipolar monolithic circuits are further subdivided according to the method used to isolate the elements from one another, electrically.

The first and still the most widely used isolation technique is the *p-n* junction method (a reverse-biased *p-n* junction resists the flow of current). However, photocurrent regeneration can break down a reverse-biased junction, making the IC susceptible to malfunction in a radiation environment. Also, the relatively large capacitances associated with isolation junctions adversely affect the speed of digital circuits and the frequency of r-f circuits.

Two other methods are dielectric isolation and air-gap isolation. "Dielectric isolation" ordinarily indicates that the single-crystal elements of an IC are surrounded by a solid dielectric—typically, silicon dioxide. In air-gap isolation, the silicon is etched between the IC components (i.e., in place of the junction isolation). Then an oxide layer is placed over the wafer. These two techniques are generally more complicated to apply and therefore more expensive than the reverse-biased *p-n* junction method. However, they tend to resist radiation better and to give faster performance.

There are many different methods for producing a given bipolar IC. The most important differences result from the following decisions:

1. Technical performance. For example, a diffused silicon resistor has 20% tolerance and a temperature coefficient between 1600 and 2000 ppm/°C; it is not adjustable. Thin-film tantalum resistors are far superior in each respect, and tantalum performance may be vital in certain key areas.
2. The method of achieving isolation between the circuit elements.
3. The choice of the types of transistors (*pnp, npn,* or both).

The foregoing decisions will dictate the type and resistivity of the starting silicon wafer as well as the sequence and details of the processing steps.

Figure 1-2 is schematic cross section of a bipolar IC die with reverse-biased *p-n* junction isolation. The transistor and diffused resistor are produced by diffusing an n^+ buried layer (10) in a *p*-type silicon substrate (1). Then an *n*-type silicon layer (2) is epitaxially grown on the *p*-type silicon substrate (1). Isolation of the components is achieved by a *p*-type diffusion (3) that carries down to the original *p*-type substrate. The resistor is formed by a *p*-type diffusion (4), and the base of the transistor (5) is formed at the same time. A subsequent heavily doped *n* diffusion forms the transistor emitter (6) and the contact region under the collector contact (8). The collector contact (A), the base contact (B), the emitter contact (C), and the resistor contacts (D) are formed by evaporating aluminum, etching away the metal where it is not wanted, and alloying the aluminum into the silicon. Aside from contact regions, the silicon surface is everywhere covered with silicon dioxide (7), which serves as an insulating, passivating diffusion mask. Rather abrupt steps in the oxide, as at (9), must be cut to open windows for diffusions and contacts. Where the aluminum intraconnects pass over these oxide

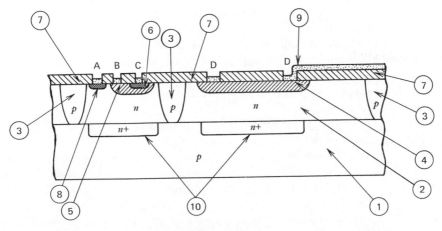

Figure 1-2 Schematic cross section of a bipolar IC die (not to scale).

steps, there is danger that the metal film will thin down and eventually open up. This is one of the weak links in many ICs.

Whereas a hybrid circuit tends to resemble a miniaturized version of a circuit built with discrete devices, the monolithic IC looks very different. This difference in appearance reflects a fundamental difference in design philosophy: It seldom makes sense to take a circuit that has performed well in discrete form and make a one-to-one transformation into an equivalent monolithic circuit. For example, economics usually pushes the discrete designer to use many inexpensive resistors in his circuit and to avoid employing relatively expensive transistors. In designing a monolithic circuit to do an equivalent task, however, the designer would probably avoid resistors and use more transistors. Resistors take up much valuable silicon area. Transistors are far more compact and easier and less expensive to fabricate.

Features of Bipolar Integrated Circuits

Power dissipation of bipolar silicon integrated devices is usually between one and several hundred milliwatts per device, dependent on the function of the device and its design. Both ends of this power spectrum are being pushed by further research and development. For space research and other applications in which power is at a premium, devices that operate at microwatt power levels are needed. At the same time, attempts are being made to expand the high-power end of the spectrum because it is necessary to obtain enough power from silicon integrated devices to perform useful functions, specifically, for high-current monolithic voltage regulators. The upper limit is set by the ability of the package to remove heat from the integrated device. There is a high power level at which the utility of integrated devices is less than that of conventional component circuits. Thin-film circuits have the ability to operate at higher temperatures and thus have more efficient heat transfer for handling more power.

Some of the attributes of bipolar silicon integrated devices are listed in the following paragraphs.

1. *Batch Processing.* One important attribute of silicon integrated devices is that numerous devices are processed as a unit up to the stage where leads are attached and where the devices are encapsulated. This allows a high degree of process control and device uniformity, with relatively low unit cost.

2. *Processing Simplicity.* The number of processes involved in the fabrication of a silicon integrated device is very small when compared with the total number of separate processes required to fabricate the components of the conventional equivalent circuit.

3. *Device Diversity.* A single process can fabricate a variety of integrated devices by variation of the necessary photographic patterns.

4. *Materials.* In the silicon integrated device a small number of different materials are employed. For example, one class of devices employs silicon, silicon oxide, aluminum, and gold; no other materials are necessary. In the other types of integrated devices, additional materials may be employed. They may be other conductors or contacting materials or a resistive metal serving in thin-film resistors. Even with these, the number of different materials involved in the integrated device is small. This tends to promote high reliability.

5. *Area Factor.* The surface area of the single-crystal silicon die on which the integrated device is fabricated is very important. It influences yield and thus cost, allowable power dissipation, required quiescent power, package size, and functional capability. For a given structure, the present lower limit on area may be set by dissipation, current-carrying ability, capacitor and resistor parameters, or resolution limits of the photoengraving process. The latter is most important for low-power circuits, since for a fixed resolution the only trade-offs are between component tolerances and circuit size.

6. *Inverted Economics.* Because of the greater area required for capacitors and resistors on silicon integrated devices, these components add more cost to the integrated devices than do transistors or diodes. This is the reverse of the situation involving circuits designed with vacuum tubes or discrete transistors. The inversion of relative cost of the active and passive circuit components will continue to have significant impact on the design of ICs.

Table 1-1 compares bipolar IC with discrete circuits. Table 1-2 presents a comparative summary of the salient features of bipolar ICs, thin- and thick-film hybrid ICs, and metal–oxide–silicon (MOS) ICs.

Table 1-1 Comparison of Bipolar Integrated Circuits with Discrete Circuits

Bipolar ICs	Discrete Circuits

• Breakdown voltages
   BV_{BEO}: 5–8 V
   BV_{CBO}: 50 V
	• Greatest design flexibility
	• Second sources are easy to obtain
	• Higher total systems cost

• Higher saturation voltage than discrete devices	
• High yield batch process	
• Lower power dissipation capability than discretes	• Capable of high power dissipation
• Limited component values	• Can obtain any desired component values
• Limited high-frequency capability	• Best high-frequency response
• Isolation between components required	• Individual components are isolated inherently by the spacing
• Lowest cost for both small and large quantities	
• Loose component tolerances	• Tight component tolerances
• High reliability	• Lower reliability
• Closely matched components	• Components not inherently matched
• Contains inherent parasitics	
• Available with both *pnp* and *npn* devices	• Available with both *pnp* and *npn* devices
• High initial design cost	• Low initial design cost
• Limited to low breakdown voltage	• High-breakdown-voltage transistors are readily available
• Not suited to very complex devices	
• Packaging complex devices present a problem	• Packaging is not a problem
• Not suited for many devices per chip	
• Use of inductance is prohibited	• All component types are available
• Small size and light weight	Large size and high weight

Table 1-1   Continued

Bipolar ICs	Discrete Circuits
• Highest tooling cost • Active substrate • Nonlinear capacitors • Best suited for large volume production where parasitics, component tolerances, and component temperature variations can be tolerated • Capacitor use should be minimized • Resistor values should be kept below 20 kΩ • Design should utilize active devices wherever possible	• Any values of capacitors and resistors may be used

Table 1-2   Comparison of Classes of Integrated Circuits

Bipolar ICs	Thin-film hybrid ICs
• High reliability	• Require more area than monolithic ICs
• Small and lightweight	
	• Can meet tighter specifications than thick-film ICs
• Good thermal coupling through substrate allows tracking with temperature	
• Highest tooling cost	• Great design flexibility
	• Larger and less reliable than monolithic ICs
• Lowest unit cost for both small and large quantities	• Greater power-handling capability than bipolar
• Typical temperature coefficient of resistance of 2000 ppm/°C	• Low-temperature die attachment
• Typical sheet resistance of 250 Ω/□	• High packing density of components due to fine line widths and conductor bar separation
• Active substrate	• Better stability, precision, temperature coefficient, drift, and noise than thick-film ICs
• Limited range of passive component values	• Must add discrete active elements to passive substrate
• Difficult to achieve close tolerances on passive components	• Second-highest tooling cost
• Limited high-frequency response due to parasitic elements	• Increase interconnections per chip
• Capacitors are nonlinear	• Highest unit cost for both small and large quantities
• Difficult to trim resistors to 1%	

• Basic monolithic structures
  Epitaxial-diffused
  Diffused-collector
  Triple-diffused
	• Typical temperature coefficient of resistance: 50 ppm/°C

METAL–OXIDE–SILICON (MOS) INTEGRATED CIRCUITS

Unipolar monolithic circuits are divided into metal-oxide-semiconductor (MOS) and junction field-effect transistor (JFET) categories, according to the type of FET made. Since almost all monolithic FETs are MOSFETs, only MOSFETs are discussed. In an MOS circuit, the oxide forms a dielectric layer between the semiconductor and the metal overlay. Capacitors and FETs are easily formed by MOS construction.

MOS transistors are smaller than any other type and consume very little power. They may well exceed the bipolar transistor in large-scale integration of both linear and digital circuits in the near future.

To illustrate how technologies proliferate, several varieties of MOS circuits could be listed. The original type of MOS was built on (111)-oriented silicon with thin oxide layers under the metal gates. A thick-oxide approach on (111) MTOS

Thick-film hybrid ICs	MOS
Require greater area than monolithic ICs	• Greatest process repeatability
	• Better reliability than monolithic (bipolar) ICs
Looser specifications	
Great design flexibility	• Increased function complexity possible
Larger size and less reliable than monolithic ICs	• High yield process
Contain only conductors, resistors, and capacitors	• Reduction in number of circuit elements and interconnections per chip
Must add active elements	
	• Complete isolation of input and output
Greater power-handling capability	
	• Single fabrication process No emitter or isolation diffusions Eliminates high-temperature processing
Lowest tooling cost	• Controlling oxide thickness is critical
Increases interconnections per chip	
	• High input impedance, lower power
Second-highest unit cost for both small and large quantities	• Has inherent memory
Greater circuit complexity	• Has bilateral operation
Typical temperature coefficient of resistance: less than 100 ppm/°C	• Typical sheet resistance: 25 kΩ/□
Typical sheet resistance: 2 MΩ/□	• Low transconductance and drive capability
Resistance tolerances without trimming: ±10%	• Tight processing control is required
Permanent drift less than 1%	
	• High-speed limitation

Table 1-2   Continued

Bipolar ICs	Thin-film hybrid ICs
• Best for large-volume production where attendant parasitics, component tolerances, and component temperature variations can be tolerated	• Typical sheet resistance: 5 kΩ/□.
• Good high-frequency capability	• Wider range of passive component values to closer tolerances and with better electrical characteristics
• High reliability	• Thin-film high-frequency limit: 10,000 MHz
• Power dissipation and packaging difficulty with increasing complexity	• Capacitors are linear, temperature coefficients are smaller, and distributed effects are smaller
	• Almost any active component that can be made in discrete form can be utilized in thin-film circuits
	• Minimize effect of intense radiation fields
	• Hybrid circuits are best for low production where low parasitics are required and wide range of component values is necessary, but where smaller size of monolithic is not required
	• Component selection, adjustment, and repair capability
	• Ease of rework

was developed to give superior reliability. To obtain lower threshold voltages, another MOS technology based on (100)-oriented silicon is used.

Silicon nitride combined with oxide is coming into increasing use as a dielectric for gate insulation in MOS devices. One of the latest MOS advances is the silicon-gate approach, in which highly doped polycrystalline silicon replaces aluminum as gate metallization.

MOS Operation

There are four basic types of MOSFET structures: p-channel, n-channel, enhancement, and depletion. The channel can be a p- or n-type silicon, depending on whether the majority carriers are holes or electrons. Also, the mode of operation can be enhancement or depletion, depending on the state of the channel region at zero gate bias. If a conducting channel exists at zero bias, the device is called depletion mode, because current flows unless the channel is depleted by an applied gate field. If a channel must be formed by the gate field before current can flow, the device is termed enhancement mode.

The enhancement mode is attractive in digital circuits because it provides inherent noise immunity—input voltage must exceed a threshold voltage before the

nick-film hybrid ICs	MOS

Capable of high power dissipation	• Four-phase logic
	  Low power
	  Devices all the same size
	  Easier to "lay out"
	  High speed
	  High density
	  Complex clock drive

Easy to adjust to desired tolerance	• Two-phase logic
	  High power dissipation
	  Devices are different sized
	  Dissimilar transistors
	  Simple clock

Flexible technique in which variety of
   patterns can be achieved with little
   process variation
Component selection, adjustment, and
   repair capability	

device turns on. This mode is also suitable for self-biasing circuitry schemes, and it is used in linear ICs.

Depletion-mode devices, on the other hand, conduct at zero gate voltage. They are especially attractive for tuner input stages. The high-impedance gate is simply connected to an antenna coil, and the input signal modulates the conductance between source and drain. Since the depletion-mode device is formed of material with a higher doping level than that in the enhancement-mode device, and since channel mobility is also higher, it can operate at higher frequencies.

MOS Characteristics

The processing of MOS circuit wafers uses essentially the same technology required for the monolithic bipolar devices: mask making, photoetching, diffusion, oxidation, and metals processing. Although MOS technology requires fewer processing steps, tolerance requirements in each part of the process differ significantly. Most critical are surface cleanliness and accurately controlled oxide thickness because the MOS technology is extremely surface sensitive.

MOS technology is inherently less expensive than the double-diffusion or bipolar processes, since it requires only one diffusion as compared with two or three and often four for bipolar ICs. In addition, special isolation provisions need not be made for enhancement-mode MOS devices. Each MOS structure is essentially self-isolating without further processing steps. Steps to control the diffusion are less critical, since the MOS device operation takes place on part of or very near the surface.

A number of MOS device technologies have been developed for use as storage units in semiconductor memories. Besides the conventional p-MOSFET, there are the complementary-symmetry (CMOS), silicon gate, silicon nitride, ion-implanted, metal–aluminum–silicon, silicon-on-sapphire, and silicon-on-spinel transistors, and the charge-coupled device.

HYBRID INTEGRATED CIRCUITS

Most hybrids consist of thin-film or thick-film conductor–resistor networks on an insulating substrate, with semiconductor devices and capacitors added in chip form. However, hybrids are not restricted to this combination. Examples of other combinations include:

- Various types of semiconductor devices, such as MOS and bipolar digital IC chips or various types of transistor IC chips, interconnected with film wiring on a common substrate. These are basically multichip versions of large-scale integrated (LSI) arrays.
- Monolithic ICs with thin-film resistors fabricated on the silicon oxide: This combination, often called "monolithic hybrid," provides resistors more closely toleranced than diffused resistors.
- Microwave ICs containing r-f chip transistors, ferrite-device and thin-film inductors, as well as passive networks. The substrate is an integral part of microwave ICs, being used as a stripline (microminiature wave guide) dielectric. In some cases, resonant cavities for signal sources are formed in the substrate.
- Magnetic thin-film memory planes with attached logic chips.

There are as many reasons for using hybrids in preference to monolithic ICs or discrete-component assemblies as there are possible hybrid combinations. Hybrid ICs often combine specifications that are unequaled by monolithic ICs. Compared with conventional monolithics, they offer:

- Much greater design flexibility
- Lower unit cost for small quantities
- Improved performance, since passive components can be trimmed to precision values and high-frequency transistors or characteristics of other high-performance devices can be optimized individually. Hybrid ICs can combine the best performance features of all IC technologies: MOS, bipolar, thick-film, and thin-film.
- Elimination of "real estate" problems. Hybrid substrates generally run more then 1 in.² which is enormous compared with IC chips, and allows plenty of room for both fabrication and heat conduction from large components. Although high-valued diffused resistors and MOS capacitors are used as individual chips (e.g., in hybrid ICs), they are impractical in monolithic ICs.

Advantages over discrete assemblies are primarily smaller size and lower cost, since the bulk and expense of individually packaging components are saved. In addition, there are sometimes technical advantages. For example, because delays due to wiring lengths and capacitance are minimized, hybrid arrays of digital ICs generally can operate faster than assemblies of packaged digital ICs. In r-f work, hybrids provide a means of reducing package and lead capacitances and inductances, minimizing variations in temperature coefficients, eliminating noisy mechanical connections, and dealing with other factors that are hard to compensate for or limit performance in discrete assemblies.

Thin-film passive elements with semiconductor active elements are receiving widening acceptance. The use of thin films on passive substrates, together with relatively new assembly techniques (e.g., flip-chip), permits a greater range of component values while maintaining some of the economic advantages of monolithics and thick-film circuits.

The development of "flip-chip" bonding for both discrete devices and silicon monolithic ICs provides a method for combining the best of film and silicon technologies. Both thin and thick films can be patterned to accept the flip-chip active devices. The economic gain of flip-chip bonding over conventional wire and die bonds provides a combined technology with economic advantages in most cases. Since wire bonding has been one of the principal IC failure modes, flip-chip technologies may provide improved reliability.

Thin-Film Hybrid Circuits

Hybrid integrated circuits can be divided into at least two categories — thin film and thick film.

The thin-film technology is thus named because conductors, resistors, capacitors, inductors — and, theoretically, transistors — are prepared in the form of films only a few thousand angstroms thick. [An angstrom unit (Å) is 10^{-8} cm.] These thin films are generally put down on insulating substrates such as alumina or glass by a series of high-vacuum vapor-deposition processes. Many materials and processes are involved in the fabrication of thin-film circuits, and all must be compatible with one another for assembly of a complete circuit.

Gold–nichrome is often used for the metallization pattern. Resistors and capacitors may be added in chip form just like the diodes and transistors. Thin-film nichrome resistors can be used if desired.

Thin-film methods may be subtractive or additive or both. In "subtractive" thin-film methods, the metal pattern is made by depositing a continuous film all over the substrate and etching away the unwanted metal after the desired pattern has been defined photolithographically. In contrast, the "additive" methods evaporate successive layers of materials through masks; thus no etching is required.

In one branch of thin-film technology, two-layer metallization improves the interfaces between materials and the compatibility of different materials. For instance, a 5000-Å thick gold film is very good for conducting the electricity around a thin-film hybrid circuit; unfortunately, however, gold does not adhere very well to an alumina substrate. Therefore, we first deposit a few hundred angstroms of nichrome, which adheres well to the alumina, then the thicker gold film is deposited on top.

Once the thin-film metallization pattern is fabricated on the alumina substrate, the rest of the assembly operations are basically the same for thin-film and thick-film circuits. The alumina substrates are different for the two cases, however. The photolithography involved in thin-film hybrids requires a smoother surface finish than is needed for thick-film work.

Thick-Film Hybrid Circuits

Thick-film hybrid circuits are fabricated with screening procedures related to the silk-screen techniques used for years by artisans. No vacuum equipment is required, and the resultant films are on the order of a mil in thickness. A thick-film circuit is formed by printing or silk-screening a pattern onto an insulating substrate, using as the "ink" a slurry of pulverized glass and aluminum or other conductor–nonconductor mixtures. By varying the proportions of the mixture, a wide range of resistor values may be obtained, or a conductive interconnect pattern may be produced. Whatever the network, it is fired in an oven to form a stable circuit.

The films thus produced are typically a fraction of a mil thick or about an order of magnitude thicker than the thin films produced by vacuum techniques. Characteristically, thick-film technology is applied to a rough surface. The most popular substrate has been 96% alumina. The slightly rough surface of alumina permits good adherence but is responsible for limitations of line definition. Since screens are usually made of 200 to 300 per inch mesh stainless steel, they also contribute to the limiting of resolution. The combination of these factors results in patterns coarser than those which can be achieved utilizing the photoresist patterning of deposited thin films. The principal advantage of thick-film technology is that a quantity of substrates can be produced with relatively inexpensive capital equipment.

Hybrid Versus Monolithic Integrated Circuits

Basically, hybrid-circuit fabrication is a matter of miniaturizing and combining in one package what might otherwise be put in several packages. The first requirement is that the active components, such as transistors, be available in chip form. If the components are available only as discrete devices, each in its own package, the hybrid approach is not feasible. More and more components are being made available in chip form, but since some are still unavailable, it is safe to assume that the desired chips can be purchased. When the necessary components are available, the hybrid approach can offer great flexibility. For example, in the same hybrid-circuit package, we can use MOS transistors, JFETs, and bipolar devices.

A schematic cross section of a typical thick-film hybrid circuit is presented in Figure 1-3. The package body (2) and the lid (1) can be made from a variety of materials such as Kovar, glass, or alumina. For discussion's sake, assume that both the package and lid are made of gold-plated Kovar, an alloy having a coefficient of thermal expansion matching that of some glasses. The lead frame (5) is also made of gold-plated Kovar, and the different leads are insulated from the package and from each other by glass (4). The lid (1) is sealed to the package (2) with a gold–tin preform (3).

The hybrid circuit itself is assembled on an insulating substrate, typically 96% alumina. The substrate is fixed to the bottom of the package cavity by epoxy or by

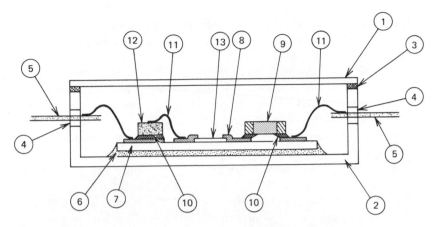

Figure 1-3   Schematic cross section of a thick-film hybrid circuit.

a eutectic gold–germanium preform (6). Thick-film resistors (13) and gold conducting traces (8) are screened and fired onto the substrate. Additional circuit elements such as silicon transistors (12) and capacitors (9) are added to the substrate and secured with conductive epoxy or gold–silicon eutectic preforms (10). Finally, electrical connections are made with fine aluminum or gold wires (11) between circuit elements and from the substrate to the various pins on the lead frame.

This simplified description of a thick-film hybrid circuit points out some of the many interfaces between dissimilar materials that are potential weak points in any electronic system. If the package hermeticity is to be preserved and the physical integrity of the circuit assured, all these interfaces must be able to survive the stresses imposed by the environment in which the circuit has to function.

The eutectic preform and the epoxy assembly procedures for thick-film and thin-film hybrid circuits are chosen because these two approaches can cover the vast majority of end-use requirements while utilizing materials and processes that are compatible, thus assuring reliable performance.

FACTORS FOR SELECTING AN INTEGRATED CIRCUIT TECHNOLOGY

Performance

In most cases, the primary factor in the selection of a basic IC technology is performance. A simple example would be the use of hybrid circuitry rather than monolithic because the power or frequency required could not be achieved in monolithic form. Hybrid circuits can perform functions that are not readily available in completely monolithic circuits. Of course, the hybrid circuit is also more general in that one of its active elements might very well be a monolithic IC chip.

Cost

A second very important factor is cost. At certain production volumes, some technologies are definitely more economical than others. For example, the cost of developing a new monolithic IC can be very substantial. Normally, such an expenditure would be justifiable from a cost standpoint only if many identical circuits were to be produced.

The cost of an IC is probably the prospective user's most important criterion when deciding which manufacturer's device to buy. Thus IC manufacturers are constantly refining processing techniques, technologies, and design practices to improve IC yields and thus reduce finished device costs. This is the "name of the game" in IC manufacturing: keep the manufacturing costs low and thus the cost of the finished device.

The thick-film circuit is usually less expensive for a limited quantity of, typically, fewer than 10,000 circuits, if the circuits are quite simple. On the other hand, if a read-only memory (ROM) containing several thousand transistors of a large-scale logic circuit is required, in quantities of only a few hundred to a few thousand, the development effort may easily pay off in the reduction of assembly labor.

Total systems cost must also be evaluated in determining the choice of an IC technology. In the past, volume buying of universal IC building blocks such as digital-gate ICs and general-purpose IC operational amplifiers provided an effective means of significantly reducing total systems cost.

Today, much of a typical system is made with ICs that provide particular system functions, such as digital counters and shift registers. Some new ICs contain so many circuit functions that they are the equivalent of a subsystem. These reduce the cost per circuit in a system, since the IC is much cheaper than a large number of assembled transistors and other components.

The use of such ICs has drastically reduced the cost leverage per individual circuit or assembly that had been exerted by volume buying of discrete components. Because of the low cost per part or circuit function in IC form, other parts of the system become proportionately more important in terms of total system cost. Their cost may be dropping, too, but they represent a higher percentage of system cost today than they did in the past. These other parts include the printed-circuit (PC) boards, the hardware and packaging, and the power supplies.

In a typical digital system (Table 1-3) the digital ICs comprise about 15% of the total system cost. And of this 15%, only 5% is generally paid for gate circuits bought in large volume. It can be seen that negotiating a penny or so savings per gate unit prices will have a negligible effect on total system cost. Savings can readily be made in other areas of the system by a favorable choice of ICs.

In Table 1-3, although complex ICs account for only 25% of the system IC units, they represent 50% of the IC costs, or almost 10% of total system cost, in the typical system design. To make a significant impact on total system costs, it is necessary to reduce costs of power supplies, PC boards, and so forth, by making

Table 1-3   Logic IC Cost Breakdown

Product Types	Units (%)	% of Logic Cost
Gates	60	35
Flip-flops	15	15
Complex functions	25	50

Table 1-4   Gated System versus MSI System

Variable	System Built with Gates	System Built with MSI
Number of ICs	300	70
IC parts cost	$60	$70
PC board area (in.²)	210	45
PC cost	$80	$16
Connectors	4 @ $7 = $28	1 @ $10 = $10
Power dissipation (W)	22.5	12
PS cost @ $2.50/W	$56	$32
Total system cost	$224	$128

an optimum choice of ICs. As an example, let us compare a system built with gates and one built with complex ICs referred to as medium-scale integrated (MSI) devices (Table 1-4). The table reveals that whereas 300 gate circuits or 70 MSI circuits could be used to build a digital processor, savings in PC board and power costs make the MSI circuits a much better choice.

Volume and Weight

Volume and weight considerations are other factors that affect the selection of a suitable microelectronic technology, particularly for aerospace applications. Not all IC systems are equally efficient in reducing the size of electronic systems. Much of the pressure behind the development of LSI comes from the desirability of reducing size, weight, and cost below the limits that can be achieved with standard IC technologies, and increasing reliability. A comparison of the standard technologies shows that monolithic ICs are generally more compact than thin-film hybrid circuits, which in turn are more compact than thick-film hybrid circuits. In certain weight-sensitive applications (such as satellites) one package may be more attractive than another. For example, a $1/4'' \times 1/8''$ glass-to-metal flat package weighs less than a $1/4'' \times 1/4''$ glass-to-metal flat package. This weight difference is important when several hundred such packages are to be mounted on a PC board. Thus in this situation the package weight (and size) will influence from which manufacturer the ICs will be procured.

Packaging

Another common problem is that of packaging. Packaging ideally provides a hermetic seal because the active elements and the metallization require a completely inert atmosphere. If packaging is not inert, particularly if aluminum is present, a very corrosion-sensitive system may result. Electrolytic action will rapidly eat away the very thin aluminum film and cause open circuits. This happens frequently in nonhermetic packages operated in adverse environments. In terms of packaging, there is not much difference between technologies except that monolithic circuits generally allow the designer to put components into smaller volume and therefore to use smaller packages. This means a smaller perimeter to seal, which makes for higher reliability.

Reliability

Reliability also must be considered. It is a significant advantage of ICs over discrete devices. In fact, the reliability of solid-state components has made possible whole new technologies. The computer industry could not exist without microelectronics. As long as we had to rely on vacuum tubes, large computers were not practical because the failure rate of the tubes was such that no computer could be kept operational. Only the development of solid-state components, with their high reliability and low power dissipation, made possible a mean-time-between-failure that rendered large complex computer systems feasible.

Most common causes of failure in ICs are really common to all these technologies because they are related to wire bonding, metallization, and packaging. All circuits, whether monolithic or hybrid, bipolar or unipolar, involve metallization of one kind or another, as well as packaging. A monolithic system has fewer wire bonds than a hybrid version of the same system, thus reliability should be better. However, metallization is another common cause of failure, and the monolithic circuit is more dependent on metallization for interconnects.

Even where the wire bonding is the same, the reliability may be different depending on the general technology involved. As an example, consider ultrasonically bonded aluminum wire (1-mil diameter). Will the bonds be equally reliable in thick-film hybrid circuits and monolithic circuits? This can become a hotly debated issue, but it might be argued that the two are not equally reliable. There appears to be an advantage in the monolithic case because the surfaces to which wire bonds are made are relatively smooth and reproducible. The hybrid circuit would present similar bonding surfaces at the lead frame and on the aluminized semiconductor chips, but some bonds would have to be made to the rather irregular surface of the thick-film gold metallization. Reliable bonds can be made to such films, but the operator must exercise extreme care in making the bonds. Beam-lead and flip-chip techniques have been developed to increase the reliability of hybrid ICs. These techniques have recently proved to be feasible for monolithic ICs.

In the monolithic case, the utilization of gold wire eliminates problems at the lead frame because the bond is gold on gold. But we have to worry about the interface between the gold wires and the aluminum metallization on the semiconductor chips. In the thick-film hybrid, on the other hand, gold wire in a nonradiation environment is quite attractive, since it features gold-to-gold bonds at the lead frame and on the conducting traces. However, bonding to the active elements must still be contended with.

In a study of the Minuteman system, Autonetics investigators found that about 52% of all the failures were due to packaging, metallization, and bonding-type failures. The failures were not inherent in the semiconductor devices themselves (i.e., the bulk silicon). That gives an idea of the magnitude of these problems which are, at first glance, somewhat peripheral to the fabrication of semiconductor devices. It turns out that they are really the most expensive and least reliable parts of the operation.

Most of the failure mechanisms that have been discussed up to this point have been common to all the IC technologies, or they have related primarily to bipolar circuits. Let us consider briefly the reliability of MOS circuits.

MOS has not been around as long as bipolar, but it is beginning to be evaluated in some significant reliability tests. The evidence to date suggests that MOS technology is as reliable as bipolar technology. MOS is a surface related and sensitive technology. Of course, MOS devices display failure modes different from those of bipolar devices because the former depend on the integrity of a very thin layer of oxide in the gate region and they cannot function if that oxide ruptures. The passivating layer, which is usually silicon oxide, must have a dielectric strength of about 10^6 V/cm. The oxide must also be very pure because any mobile-charged contaminant is going to cause the part to be unreliable because of parameter drift.

MOS processing is simpler in some respects than bipolar, and this property tends to improve reliability. For example, only one diffusion is required to make MOS devices, and that one is not very critical. On the other hand, the whole trend in MOS technology is to crowd components together more closely than is possible in bipolar circuits, and this causes photolithography problems that can affect reliability.

Hybrids are fabricated with active elements that are purchased as uncased chips or dice. It is very difficult to test transistors and diodes in chip form to assure that they are reliable before they are assembled into a hybrid circuit. A transistor in a T0-5 can has convenient external leads that fit into a test socket. Many kinds of ac and dc tests can be performed on the part, as well as environmental tests. When testing is completed, the transistor can be put into a discrete circuit with confidence that it is a reliable part. But if the same basic transistor element is received in the form of an uncased chip, how is it possible to determine whether it is a good component before doing die attach? Since there are no external leads, probes are used on the die for electrical tests; ac testing under these conditions is very limited.

Probably the most costly part of the hybrid fabrication is rigorously testing all the components that are going to be assembled in the circuit. It is likely that this stage also represents the weakest link in the reliability of thick-film and thin-film hybrid circuits. It is imperative, then, that semiconductor device testing be done at the wafer level, just prior to wafer scribing and die sort. The die is then attached to the substrate or package with a minimum of elapsed time and is ready for the electrical preseal test to confirm its acceptability. By this method, any reduction in reliability can be reduced or eliminated.

Thus it is important to realize that the various factors to be considered before deciding to design a system with a particular IC technology are closely interrelated. Also, as in any engineering exercise, trade-offs and compromises must be made. In general, it is impossible to obtain the highest performance and the greatest reliability without also raising the cost of producing the components. Some degradation of performance or reliability may be necessary to bring costs down to an acceptable level.

Chapter Two

BIPOLAR INTEGRATED CIRCUIT PROCESSING

Monolithic integrated circuits require a large number of individual processing steps. Each must be performed with great precision in order to achieve high enough yields of good ICs for economical mass production. In general, about 50% yield of good packaged devices out of the several hundred or thousand made in each silicon wafer can be expected. The yield of good chips must be much higher, to offset losses in packaging and testing (the most expensive phases of IC production).

Generally, a bipolar IC is far more difficult to design than the same circuit in discrete-component assembly or hybrid IC forms. The silicon substrate is an active part of the monolithic IC, and circuit elements such as transistors, diodes, and resistors are capacitively coupled to the substrate. This creates parasitic elements and interactions between components that are not found in conventional circuits, limiting operating frequencies as well. This aspect is discussed in Chapter 3.

On the other hand, monolithic processing offers many advantages, most of them outlined in Chapter 1. To exploit the advantages without running afoul of the drawbacks, it helps to know how ICs are made. At first glance, an IC seems to be a very tiny and fragile thing. It is easily destroyed, if abused by being overstressed electrically or environmentally; but if used with an appreciation of the physical laws and constraints involved in its construction, an IC is as rugged and reliable as a transistor — and, in many cases, less expensive to make than a single transistor despite the many elements (components) in the circuit. That is because the elements are so small and because they are made almost entirely by batch-type photochemical processes.

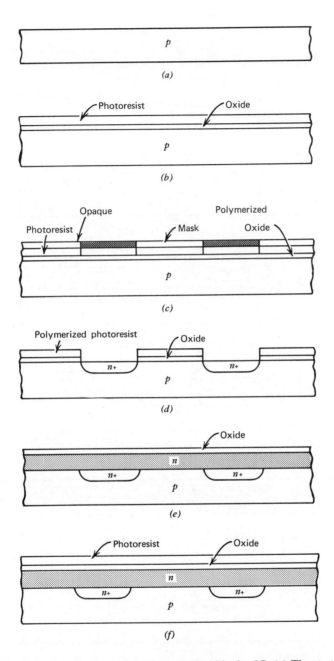

Figure 2-1 Processing steps involved in making a typical bipolar IC. (*a*) The starting material is a wafer of *p*-type silicon between 2 and 3 in. in diameter and a few mils thick. (*b*) An oxide layer is grown to protect the surface of the silicon during the ensuing operations, and the entire surface is covered with a layer of photoresist. (*c*) The mask containing the buried layer pattern is placed on top, and the photoresist is exposed to ultraviolet light. The portion of the photoresist exposed to the light polymerizes, the rest is dissolved. (*d*) The oxide that is not protected by the polymerized photoresist is etched away; *n*+ dopant is diffused through the windows forming the buried layer, and the polymerized photoresist and oxide are washed away. (*e*) A thin layer of *n*-type silicon is grown on the silicon in an epitaxial reactor, and a new oxide layer is grown. (*f*) The entire top surface is again covered with a layer of photoresist.

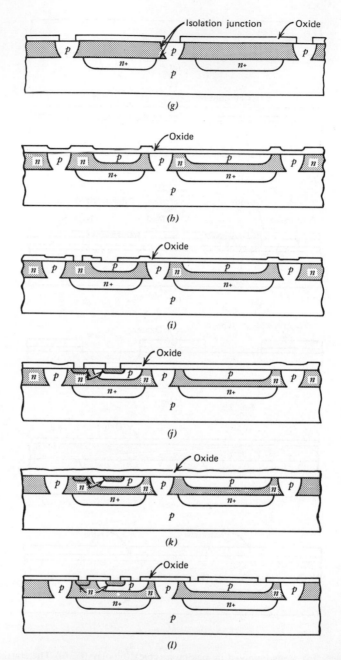

Figure 2-1 (continued) (*g*) *p*-Type dopant is diffused through the windows. The diffused regions connect with the underlying *p* region (the substrate) and form isolation pockets in the epitaxial layer. The edge of the junction is under the oxide. (*h*) *p*-Type dopant is diffused into the unprotected regions, and the wafer is covered with another oxide layer. (*i*) Again the wafer is covered with photoresist. The resist is exposed through the mask, which outlines all shallow *n* regions, and the oxide is etched away in the unpolymerized areas. (*j*) A shallow layer of high *n*-type dopant concentration is diffused into the unprotected areas. (*k*) Another oxide layer is grown. (*l*) The wafer is covered with photoresist for the fifth time. The resist is exposed through a mask in the areas where contact to the devices must be made, and the unprotected oxide is removed.

20

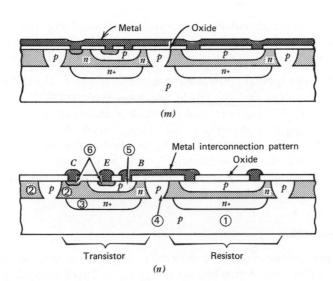

Figure 2-1 (continued) (*m*) The entire wafer is covered with a thin metal film. (*n*) With one more series of photolithographic steps, the portions of the metal layer not needed for interconnection are removed.

PROCESSING STEPS

Before going into the details of monolithic processing, let us briefly run through the major processing steps. These are illustrated in Figure 2-1 for an *npn* transistor and a diffused resistor in a typical bipolar IC. The following steps refer to the diagrams in Figure 2-1:

    a. A starting wafer about 10 mils (0.010 in.) thick is sliced from a single-crystal ingot of *p*-type silicon.

    b. The wafer surface is oxidized and coated with photoresist.

    c. An oxide etching pattern is developed photographically in the resist, and "windows" are etched in the oxide.

    d. *n*+ Impurities are introduced through the windows and diffused into the *p*-type substrate at high temperature. These will become buried layers. The oxide pattern, which served as a barrier or "oxide mask" to the diffusion gas, is removed.

    e. An *n*-type epitaxial layer is grown on the substrate in a reactor, Silicon is deposited, usually by decomposition of a compound containing silicon, and becomes part of the starting single crystal.

    f, g. Oxide masking is repeated, and *p*-type diffusions are made to form *p*–*n* isolation junctions surrounding *n*-type regions.

    h. Again, the oxide masking step is repeated, and a shallower *p*-type diffusion is made for the transistor base and the resistor.

    i, j. After still another oxide masking process, very shallow regions of low-resistivity, *n*-type silicon are formed by a fourth diffusion. This is called the emitter diffusion, which also provides the collector of the transistor in this example.

    k, l. Oxide is grown on the wafer for the fifth time and photoetched to bare the base, emitter, and collector contact regions, and the resistor electrode areas.

m. A thin film of metal, usually aluminum, is deposited over the entire surface.

n. The metal film is etched through a mask of photoresist. Contacts and the interconnections between contacts are left in the final metallization pattern.

Figure 2-2 shows typical vertical dimensions in a wafer. The diffusions actually have a rounded shape and the oxide windows have sloping edges. Diffusants disperse out from a starting point and the oxide etchants attack the upper part of a window for a longer time than the oxide at the bottom. Note in Figure 2-2 that a layer of glass has been deposited over the finished circuit to protect the IC from moisture, contaminants, and scratching of the metallization. This glass will be etched off the bonding pads at the periphery of each IC so that lead wires can be bonded to the metal film during packaging.

Substrate Preparation

Starting wafers or substrates are generally sliced from single-crystal ingots of *p*-type or *n*-type silicon, depending on the process. Ingot diameters are typically about 2 in., giving the wafer enough area for several thousand simple amplifier or logic circuits or about a thousand large-scale digital circuit arrays. More recently, both $2\frac{1}{2}$-in. and 3-in. wafers have been used to increase the number of available dice per wafer (see Figure 2-20).

Slices are much thicker than wafers. The crystal surfaces are damaged and strained by slicing, which degrades resistivity and minority-carrier lifetime. The damage is lapped away by finer and finer polishing compounds until the wafer surface is flat and mirrorlike (Figure 2-3). Chemical etchants are sometimes used, but an etched surface is not as planar (flat) as a lapped surface and is seldom employed when element geometries must be very small and precise.

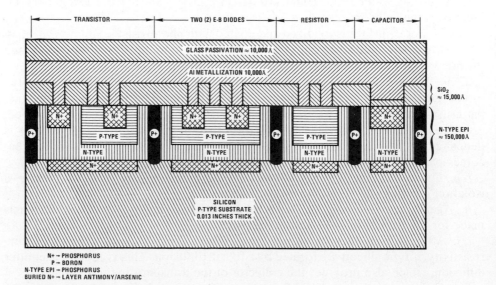

Figure 2-2   Typical vertical dimensions of a wafer.

Figure 2-3   Silicon wafers after polishing.

Since many circuits will be made from one wafer, the overall surface must be highly uniform. Furthermore, the electrical performance of the circuit also depends on the molecular structure of the surface material. If dielectric isolation is to be used, the isolation is formed as part of the substrate processing (see section on isolation). If not, the substrate goes into the junction-isolation process (Figure 2-1).

Oxidation and Planar Processing

Even though IC designers would often prefer to use other semiconductor materials, such as germanium for high-frequency circuits, silicon is the material of all but a negligible percentage of the ICs that have been made to date. One reason is the good high-temperature electrical properties of silicon, but most important, it grows a stable oxide.

Oxidation is the key to low-cost planar processing—the fabrication of electronic devices by photochemical processing of a plane surface. The oxide's three main roles in the process are:

1. Diffusion masking.
2. Sealing the junctions and passivating the silicon surface.
3. Insulating the metal interconnections from the silicon surface and providing a dielectric for MOS capacitors and transistors.

Above a certain thickness, oxide is almost impervious to most silicon diffusants, but thin oxide can be doped by some materials to provide an on-site diffusion source. During diffusion, the $p-n$ junctions move under the oxide, leaving them sealed or passivated. The oxide is stable at diffusion temperatures, which are often in the neighborhood of 1200°C.

The main process constituent is silicon dioxide, which in crystalline form is quartz. Glassy oxide is easily and rapidly grown by passing steam, wet oxygen, or dry oxygen over a heated wafer, and it is easily etched. If properly grown on a clean substrate, the oxide remains an inert barrier to moisture and contaminants. Certain glasses also adhere well to the oxide and can be used as a final passivation layer, sealing the metallization.

Such ideal combinations of materials occur rarely in nature. Depositing acceptable coatings on other semiconductors has so far required very careful selection of materials and very rigorous process controls. Applications have mainly been limited to experimental circuits for microwave systems. Planar germanium logic circuits have also been made in the laboratory.

Oxide Growth

Wafers are usually oxidized in batches in a quartz "boat" that is placed in the quartz tube of a furnace similar to a diffusion furnace. The setup in Figure 2-4 is typical. Nitrogen is a protective environment and, in some processes, a carrier gas. We obtain steam for oxidation by boiling the water in the flask, wet oxygen by bubbling O_2 through water heated to about 95°C, and dry oxygen by venting the oxygen tank into the tube directly.

Forming an oxide layer 10,000 Å thick consumes about 4500 Å of silicon and takes about 4 hr to grow. Color changes with thickness, similar to oil film on smooth water, because of changes in the wavelength of reflected light. The color is used to judge the thickness and uniformity of the layer. Oxide layers are sometimes formed by anodizing or, alternatively, by a deposition process. Deposition gives a thick layer without consuming the substrate.

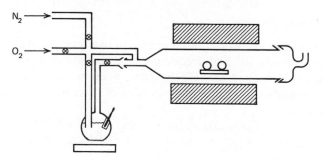

Figure 2-4   Open-tube oxidation–diffusion apparatus.

Oxide Masking

Figure 2-5 shows the oxide masking steps in detail. After the oxide is coated with photoresist, a pattern consisting of a photographic film or a glass plate with an etched pattern of thin metal film is placed as a photomask over the resist (Figure 2-5a). The resist is exposed and developed, and the oxide windows are etched (Figure 2-5b). The resist is removed from the remaining oxide and the desired junctions diffused, in this case the isolation junctions (Figure 2-5c). Figure 2-5d shows the oxide mask for the base diffusion before the resist is stripped.

Photomask Preparation

Relatively few engineers who use, rather than manufacture, ICs participate in the actual processing. However, many systems manufacturers do participate in the

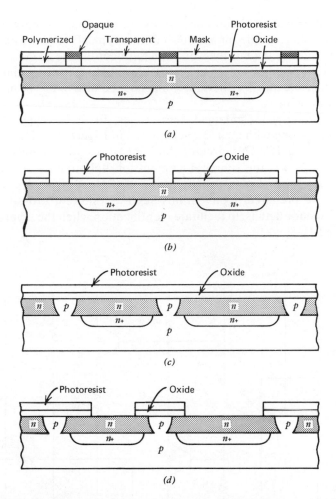

Figure 2-5 Steps in oxide masking. (*a*) The mask containing the isolation pattern is placed on top, and the photoresist is exposed to ultraviolet light. The portions of the photoresist exposed to the light polymerize; the rest can be dissolved. (*b*) The oxide that is not protected by the polymerized photoresist is etched away. (*c*) The diffusion windows are closed with a new oxide layer, and the wafer is again covered with photoresist. (*d*) The photoresist is exposed through the mask, which outlines all shallow *p* regions, and the oxide is again etched away in the unexposed areas.

layout of ICs, particularly complex arrays with specialized system functions. Often, layout design is done with computers, as described in Chapter 6, on MOS ICs.

A large number of circuit designers also prepare the artwork for etching thin-film networks, as outlined in the section on hybrid ICs. The basic procedures for preparing photoresist development patterns or photomasks being quite similar for monolithic and film circuits, these procedures are covered in detail in the following sections. The main differences are that the patterns for monolithic ICs are generally much more complex and are made to much tighter dimensional tolerances.

Mechanical masks, also produced by photoetching, may be used in hybrid IC fabrication. But photo-optical tools are used almost exclusively in monolithic IC work because the finished circuit dimensions are too small for normal mechanical manipulations. For example, an emitter contact typically requires cutting a hole only 0.05 mil wide by 0.4 mil long by 0.08 mil deep in the oxide. And this on a total circuit area ranging from 25 mil² to 0.1 in.². "Huge" monolithic circuits are rarely more than 0.2 in.². Furthermore, a sequence of five or six masks must be registered within microns.

The actual photomasks, enlarged many times, for a 709-type of operational amplifier appear, along with a circuit schematic, in Figure 2-6. This is a relatively simple type of linear IC. This circuit "chip" measures 38 × 38 mils. A composite drawing (Figure 2-7) is the source document for a photomask set. This is a line drawing on stable plastic film, usually Mylar. Different layers are used for each photomask to be prepared. Conventions such as different cross-hatchings and shadings are used for details, to facilitate identification when the layers are stacked and illuminated.

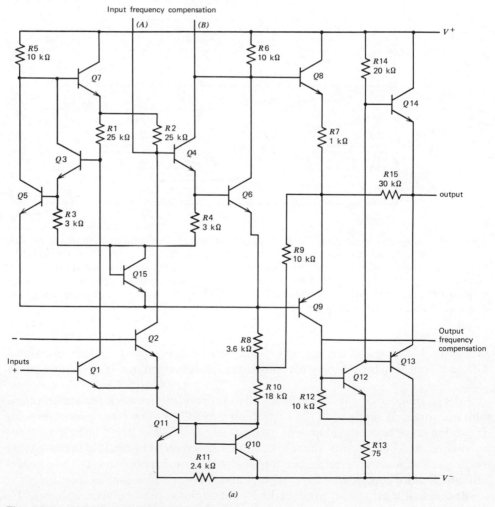

Figure 2-6a 709 schematic diagram.

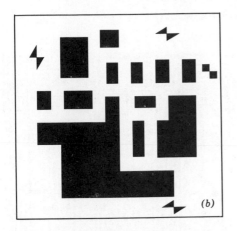

Figure 2-6b Buried layer mask.

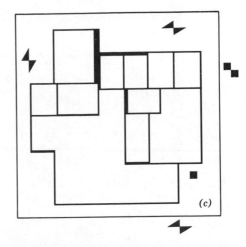

Figure 2-6c Isolation mask.

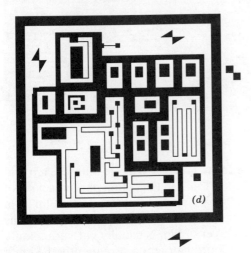

Figure 2-6d Base diffusion mask.

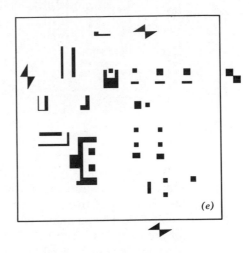

Figure 2-6e Emitter diffusion mask.

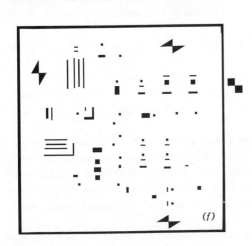

Figure 2-6f Contact window mask.

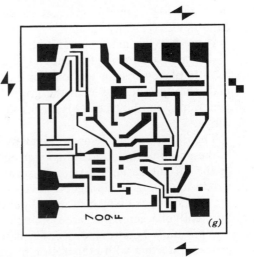

Figure 2-6g Metallization mask.

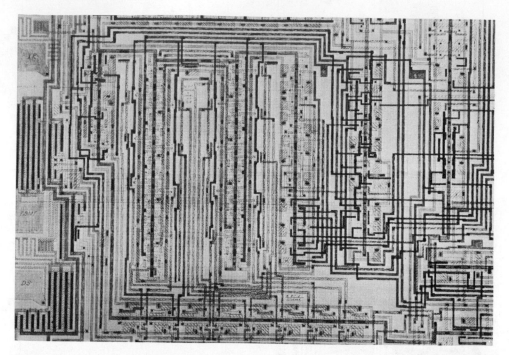

Figure 2-7   Hand-drawn composite.

The composite is generally laid out on a 25/in. grid at a scale of $400X$. Each grid square represents an increment of 100 μin. (0.1 mil) in the actual photomasks. An artwork set is prepared from the composite by stripping a coating of red film from a clear stable film. The laminate most used is trademarked Rubylith, and the films in this set are known as "rubys." The red layer of film is usually cut with a knife mounted on a precision X–Y table called a Coordinatograph. Frequently the knife is positioned under computer control or by a computer-generated tape. The computer is programmed to convert element coordinates to cutting motions. Cutting accuracy is typically 0.1 mil.

If the cuts are made by hand, the composite is placed on a light table under the X–Y traverse carrying the knife. The composite is only a "road map." A sheet of ruby is placed over it and cut for each layer. But the operator uses positioning dials to align the cuts with the grid lines rather than following the drawn lines, which may not be true. This ensures dimensional accuracy and registration of all rubys.

The dimensional tolerances given by the 25/in. grid is consistent with tolerances in the photomasking and diffusion processes. All tolerances are taken into account in the circuit design, of course, since a design based on impractical tolerances would have low yield of good devices. Processes, including artwork preparation, are aimed at achieving guaranteed device specifications, even though much better "typical" characteristics are obtained from many wafers.

What must be stressed is environmental control throughout the photoetching process, starting with preparation of the rubys. Particles or contaminants on the master artwork may eventually show up as pinholes, shorts, or opens in the ICs.

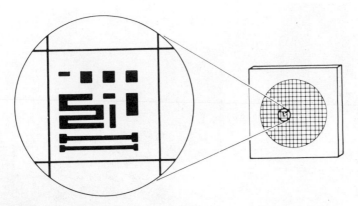

Figure 2-8   Photo mask.

Also, dimensional tolerances can be destroyed by temperature and humidity changes that cause minute changes in the film dimensions.

The rubys are photographed with large cameras to produce photoplates reduced to 10 times the IC size. These are further reduced and replicated to produce the actual photomasks, which are glass plates the size of the silicon wafer (Figures 2-8 and 2-9). The patterns on the plates may be developed in a photographic emulsion or photoetched from a thin film of metal. Metal plates are more expensive than emulsion plates but are more resistant to scratching and wear, which also cause circuit flaws.

A recent development by Bell & Howell is the mass production of chrome masks for use in the IC industry. Chrome masks have some distinct advantages over emulsion masks. Whereas an emulsion mask typically might be used for 10

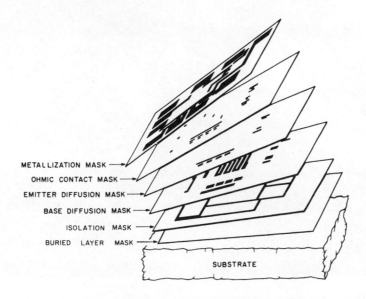

Figure 2-9   Application of photolithography to produce ICs.

Figure 2-10 FEDIS computer-aided IC artwork-generating system.

Figure 2-11 Digitization of composite layout.

exposures, a chrome mask will last for about 150 exposures. In addition, the normal defect level in emulsion masks is 10 to 15%, but in chrome it is below 5%. Frequently encountered defects in masks include pinholes, scratches, and opaque spots. A lower defect level means higher yields and therefore lower costs.

Computerized Artwork Generation

Another means of generating IC artwork is with a computer-aided system. One such system employs Macrodata's Front End Design Information System (FEDIS – Figure 2-10) to ensure quick turnaround time and to minimize human error on complex MOS/LSI design.

First a composite layout is digitized and stored into memory (Figure 2-11). Repetitive cells need be entered only once. The layout is then displayed on an interactive cathode ray tube (Figure 2-12) and checked for accuracy. Changes and the editing of cells may be made on the CRT. In addition, past designs, either standard or custom, can be obtained from storage for present and future use. When the layout is complete, a check print is drawn by a computer-driven plotter.

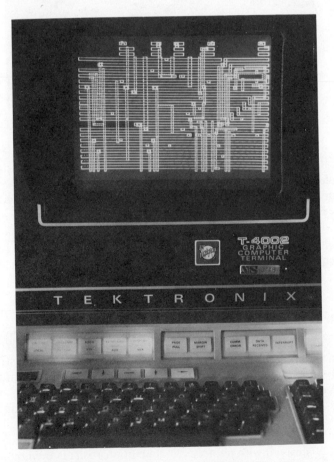

Figure 2-12   Composite layout displayed on interactive CRT.

Figure 2-13   David Mann pattern generator.

A layout verification program is then run to ensure that layout design rules have not been violated.

The output of FEDIS is a magnetic tape containing the entire layout in a digital format. This tape provides the input to a David Mann pattern generator (Figure 2-13) which directly produces an $X10$ reticle. In doing so, "cut-and-peel" errors associated with the Rubylith method are eliminated. Intermediate reductions are no longer required. The $X10$ reticle is reduced to actual size on the six-barrel David Mann step-and-repeat camera.

Photoetching Processes

Photoetching of the oxide masks starts with the application of a thin film of photo-resist to the wafer. Wafers are usually spun while the resist is still liquid (see Figure 2-14). This action spreads the solution in a thin, nearly uniform coat and spins off the excess liquid.

Figure 2-15a depicts an automatic photoresist application/spinning machine. Note that the wafers are handled by means of a special tray. Figure 2-15b is a closeup showing the photoresist being applied. Rougher surfaces, such as ceramic, are usually sprayed with photoresist. The photoresist is then baked to remove the solvents and to make them adhere better to the wafer.

Figure 2-14   Spinning photoresist on a wafer.

The pattern to be etched is developed in the resist by exposing it to ultraviolet light through the photomask. The effect of the light depends on the type of resist. Positive photoresists provide holes where the photomask is opaque (see Figure 2-5a), and negative photoresists leave holes where the photomask is transparent. Figure 2-16 shows an automatic photoresist developing machine.

Two examples of positive resists are Kodak Thin-Film Resist (KTFR) and Kodak Metal-Film Resist (KMFR). The exposed areas polymerize and cannot be dissolved by the developing liquid. Positive resists such as Shipley A2-1350 and AZ-111 become soluble in a water-base developing solution after exposure. Both can be stripped off by solvents after the oxide windows or metallization pattern have been etched.

Alignment machines (Figures 2-17a and b) place the photomasks in close contact with the resist after the operator has lined up the mask with the previous pattern. Patterns such as a set of nested boxes (Figure 2-6), visible through a microscope, are used to align successive photomasks. Microscopes that allow the operator to see the mask and the wafer in the same field of vision are used. The exposed oxide is etched with a hydrofluoric acid solution where it is exposed. This solution does not attack the silicon crystal or the developed resist, although it undercuts the resist somewhat.

Diffusion Techniques

After the resist is dissolved, the oxide-masked wafers are placed in the tube of a diffusion furnace such as that of Figure 2-18. Figure 2-19 depicts a typical IC wafer fabrication line showing the diffusion furnaces. Figure 2-20 shows both 2- and 3-in. wafers being inserted in a diffusion tube. The action of the oxide mask is represented in Figure 2-21. Bared silicon areas are exposed to a gas bearing the diffusion impurity or dopant, such as phosphorus.

Depth, concentration gradients, junction area, and other diffusion parameters

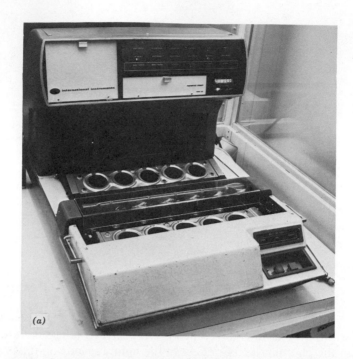

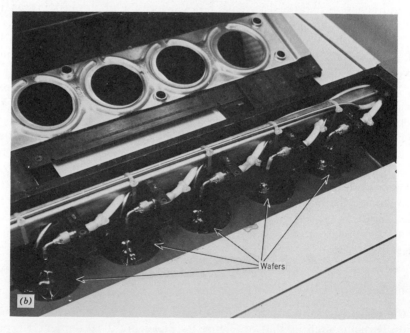

Wafers

Figure 2-15 (*a*) Automatic photoresist application and spinning machine, (*b*) close-up showing photo-resist being applied to wafers.

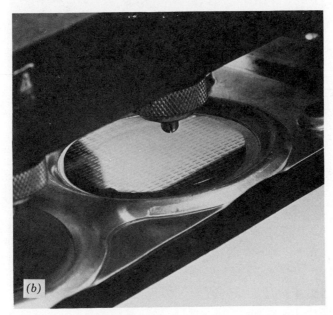

Figure 2-16 (a) Automatic photoresist developer, (b) close-up showing wafer being developed.

can be explained theoretically, but trial and error is used to precisely determine the diffusion temperature–time cycles required to produce particular ICs with particular dopants. A well-developed process can locate a junction within a fraction of a micron of the desired location. Note that the "pull rate" of the diffusion tube from the furnace is probably the most critical wafer fabrication operation in determining the IC's electrical characteristics for wafers in that tube.

Most dopants move through silicon by a substitution process. The gas molecules dissociate when they touch the hot silicon. They enter and move through

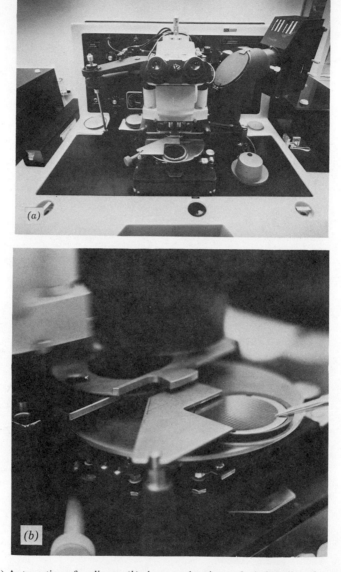

Figure 2-17   (*a*) Automatic wafer aligner, (*b*) close-up showing wafer being aligned.

the crystal by diffusing into vacant atom sites. Some impurities whose atoms are smaller than silicon atoms, such as gold and nickel, move through the silicon lattice between atoms, or interstitially.

The two main diffusion processes go by the names "error function" and "Gaussian." Their impurity distribution profiles are given in Figure 2-22. The first, used for isolation and emitter diffusions, maintains a high surface concentration by a continuous introduction of dopant. The second is used when moderately high sheet resistivity is desired or when multiple diffusions are needed. The surface concentration is diminished by allowing diffusion to continue after the dopant

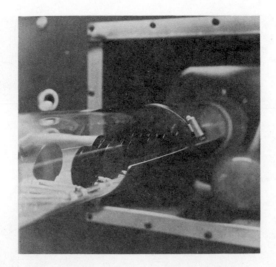

Figure 2-18 Close-up of insertion of wafers into diffusion furnace.

Figure 2-19 Wafer fabrication line showing row of diffusion furnaces.

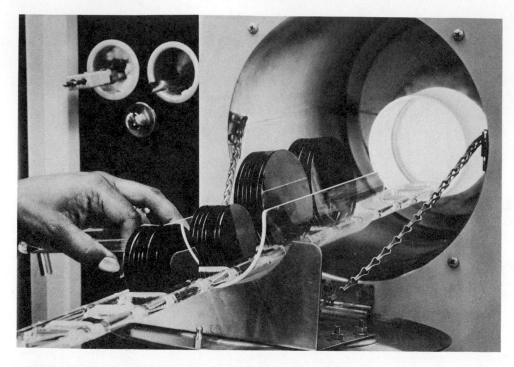

Figure 2-20 Wafers in diffusion tube (2 and 3 in. sizes).

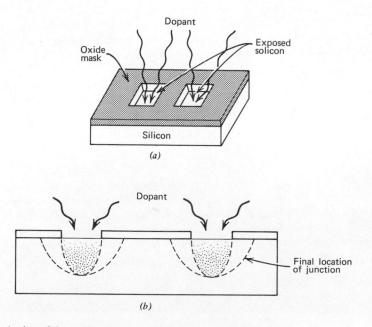

Figure 2-21 Action of the oxide mask: (*a*) etched windows, (*b*) diffusion.

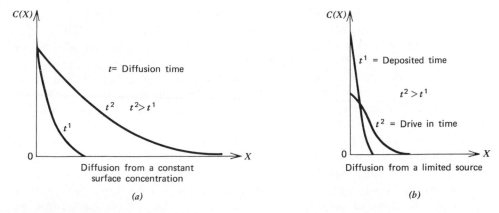

Figure 2-22 Impurity-distribution profiles: (a) one-step (error function) process, (b) two-step (Gaussian) process.

source has been removed. Transistor bases are made by such redistribution diffusions.

Characteristics of commonly used impurities are listed in Table 2-1. Source materials may be solids, liquids, or gases. Boron, for example, may come from boron trioxide or diborane. The solid trioxide would be vaporized in a furnace upstream from the diffusion tube, the liquid carried by an inert gas bubbled through a flask, and the gas metered directly into the tube.

The furnace atmosphere is slightly oxidizing. An oxide skin less than 200 Å thick forms on the exposed silicon. If boron is the dopant, the oxide becomes borosilicate glass, which then acts as the diffusion source. Similarly, a phosphorus diffusion becomes a glassy phosphate. The glass is preferable because it protects the silicon from pitting or evaporating and acts as a "getter" for undesirable impurities in the silicon. It is etched off before a redistribution diffusion is made.

Dopants and Integrated Circuit Characteristics

The effect of the dopant selection on IC characteristics is of considerable interest. Boron and phosphorus are the basic dopants of most ICs. Arsenic and antimony,

Table 2.1   Characteristics of Common Impurities

Element	Maximum solubility (atoms/cm^3)	Temperature for maximum solubility (°C)	Diffusion constant 1300°C (cm^2/sec)	Type
Aluminum	10^{17}–10^{-18}	1150	7.0×10^{-11}	p
Antimony	6×10^{19}	1300	1.5×10^{-12}	n
Arsenic	2×10^{21}	1150	1.5×10^{-12}	n
Boron	5×10^{20}	1200	1.6×10^{-11}	p
Gallium	4×10^{19}	1250	2.2×10^{-11}	p
Gold	10^{17}	1300	2.5×10^{-6}	Ohmic
Phosphorus	1.3×10^{21}	1150	1.6×10^{-11}	n
Silver	2×10^{17}	1350	—	Ohmic

which are highly soluble in silicon and diffuse slowly, are used before epitaxial processing or as a second diffusion. Gold and silver diffuse rapidly. They act as recombination centers and thus reduce carrier lifetime. Gold and silver form ohmic contacts when diffused into silicon.

In the initial development of bipolar ICs, it was found that better *npn* transistors could be made than *pnp* transistors. Since then, the technology has been directed toward improving the *npn*s, and today almost all bipolar ICs are made with processes and geometrics favoring *npn*s.

The various regions in a typical bipolar IC are shown in Figure 2-1*n*. Starting with a polished wafer or substrate of silicon doped with boron to produce a *p*-type material (1 in the drawing), a thin "epitaxial" film or layer of *n*-type silicon (2) whereas the epitaxial layer forms the material into which the transistor structure is diffused. Regions of the "epi" are electrically isolated from each other by isolation diffusions (4) of boron, a *p* dopant which has a medium to low rate of diffusion and well-established properties. If the *n*-type epitaxial region (2) is at a more positive voltage than the *p*-type substrate and isolation regions (1) and (4), the junction between these two types is "reverse-biased" and no current flows across this junction. Therefore, no current will flow between adjacent isolated epi regions when biased in this fashion, and independent transistors can be formed in each isolation region.

The *npn* transistor base material (5) is then diffused into the isolated epi region, again using boron (*p* type). (The same *p* material used for the base serves to generate resistors.) Finally phosphorus, an *n*-type material, is diffused into the *p*-type base to form the emitter of the *npn* (6).

Diffusion is also used to obtain good contact to the collector epi material (2) by providing high-impurity-concentration spots on the surface of the low-impurity-concentration epi.

Also, since the epi concentration is kept low in order to obtain high breakdown voltages, its resistance is high and the voltage drop through it at high currents would be excessive. To alleviate this, a low-resistivity path is provided to help the sidewise movement of current through the collector. This path (3) is an arsenic or antimony layer diffused into the substrate before epitaxial growth. These materials diffuse quite slowly, and the impurities move very little during subsequent process steps.

Where the emphasis is on speed, most digital circuits also have gold diffused throughout the silicon during base diffusion. Gold reduces carrier lifetime in the base region, allowing transistors to come out of saturation quite rapidly. The use of the Schottky-barrier diode technique eliminates the need for gold doping in high-speed circuits. The Schottky diode is discussed in detail in Chapter 3.

It should be noted that at all times a layer of silicon dioxide covers all areas of the chip except those which are exposed to deliberately diffuse impurities.

Epitaxial Growth

Without epitaxy, it would be impossible to make many types of high-performance ICs. In particular, knowledge of the phenomenon has led to the use of buried-layer transistor configurations, with the benefits described in Chapter 3, and improvements in isolation techniques.

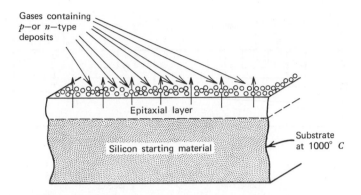

Figure 2-23 Epitaxial growth.

Epitaxy effectively gives the monolithic IC processor two planar surfaces to work with. He can make diffusions in the substrate and in the epitaxial layer. He can therefore have high concentrations of impurities well below the surface without massive diffusions from the surface. More important, a very thin, highly controlled active region is provided, and the starting wafer need only be a "ground plane" and mechanical support. An epitaxial layer is generally best for high-speed logic or high-frequency devices.

A typical process consists of placing the wafers in a furnace, cleaning and etching the surface with hydrogen to reduce any oxides, and adding silicon tetrachloride ($SiCl_4$) to the hydrogen gas. The $SiCl_4$ decomposes into silicon and hydrogen chloride (HCl). At high temperature, the silicon atoms form a single-crystal layer (Figure 2-23). Dopants can be added during deposition to obtain uniform impurity distributions, which is not possible with conventional diffusion.

The improvements that epitaxy has made in transistor characteristics alone have paid off in major improvements in reliability, cost, and performance. These include better frequency characteristics, better linearity, lower base resistance, higher switching speed, higher gain, higher collector–base breakdown voltage and lower $V_{CE(sat)}$.

Isolation

For a circuit to function properly, the elements must be electrically isolated from one another. The method of achieving isolation distinguishes ICs from discrete-component circuits. In the latter, isolation is accomplished by packaging each component separately. In ICs, the various circuits on a particular wafer are isolated by three methods:

- Reverse-biased *pn* junctions (Figure 2-24),
- Electrically inactive dielectric material (Figure 2-25)
- Air–dielectric isolation

All these employ an epitaxial layer for improved performance.

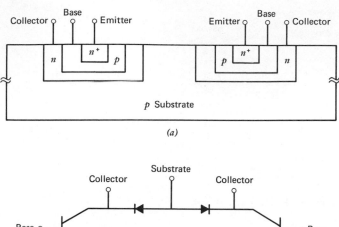

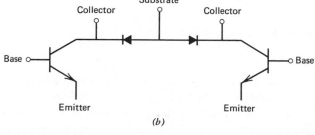

Figure 2-24   *p-n* Junction isolation: (*a*) two transistors on a single wafer, (*b*) equivalent circuit showing junction isolation.

Compared with junction isolation, dielectric isolation is quite complex and expensive. Initially developed as a means of improving the radiation resistance of missile and aerospace circuits, it is now starting to come into use as a high-speed logic and high-frequency circuit technology.

A typical dielectric isolation process is outlined in Figure 2-26. The steps are:

A, B. Moats are etched in the silicon oxide and wafer after the buried layer is diffused.

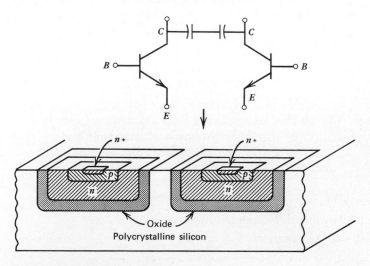

Figure 2-25   Dielectric isolation.

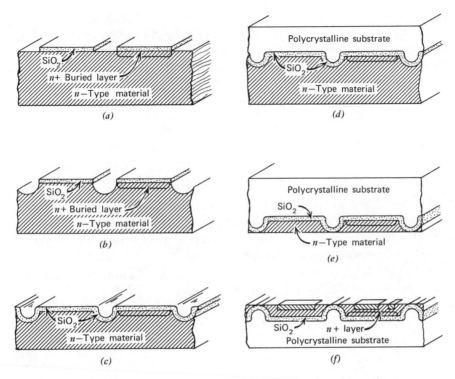

Figure 2-26   The dielectric isolation process: (*a*) oxide etch, (*b*) silicon etch, (*c*) oxide dielectric growth, (*d*) polycrystalline silicon growth, (*e*) *n*-material removal, (*f*) completed diffusion.

C. Oxide is regrown in the moats.

D. A compatible dielectric is thickly deposited over the wafer. Most manufacturers prefer polycrystalline silicon, which can be formed in the same furnace used for epitaxy.

E. The excess, high-resistivity substrate silicon is lapped off.

F. Turned upside-down, the wafer now has silicon "islands" separated by dielectric material. These can now be processed by the methods previously given, except that junction isolation is omitted.

The process decreases leakage currents between collector and substrate by three or four orders of magnitude and capacitance by about one order of magnitude. Current flows and circuit latch-up problems caused by radiation effects in silicon are prevented.

There are many dielectric isolation processes. Some surround the islands with deposited dielectrics, such as oxides. Others have involved experiments with ceramics. Two that are of particular interest because they do not require lapping the substrate are the beam-lead process for bipolar ICs and the silicon-on-sapphire (SOS) process for MOS circuits.

Beam-lead circuits are made with heavy gold leads, built up by plating on a thin-film pattern. When the individual circuit elements are connected by beam leads, the elements such as transistors can be separated by etching. The beam leads hold the islands in place and the air between the islands is the dielectric.

SOS circuits are made in single-crystal silicon epitaxially deposited on a flat plate of sapphire (or on spinel). The sapphire, which is crystalline aluminum oxide, supports the transistors and interconnections after the silicon between the transistors has been etched away. This method has also been used to make large diode arrays for ROMs.

Metallization

Metallization is another process that serves to fabricate both monolithic and thin-film circuits. The metal film is usually deposited by vacuum evaporation and patterned by photoetching. Basic requirements of monolithic IC metallization are:

- It should have the ability to be etched or plated to form high-resolution patterns (i.e., 0.1 mil with 0.1 mil spacing).
- It should make good, low-ohmic contacts to both p-type and n-type silicon without destabilizing the silicon properties.
- It should adhere well to the oxide and also to insulations deposited between layers of multilayer interconnection structures.
- Resistivity should be low initially and should not increase as a result of subsequent processing or operation at high temperatures or power levels.
- The metallization does not form intermetallic compounds that are resistive or weak either at the silicon–metal interface or with the bonded lead wires.
- It does not degrade significantly in atmosphere before hermetic packaging or after packaging in molded plastic.
- It should resist current-induced electromigration and electrochemical corrosion.
- It should resist abrasion.
- Preferably, the metal or metals can be deposited by vacuum deposition.
- The metallization should be metallurgically compatible with the external bonding system.
- It should have controllable alloying characteristics.
- It should have the capability of withstanding high temperatures without bond effect.

Aluminum is the most popular metallization because it adheres well to the oxide, alloys well with silicon for ohmic contacts, and can be applied and patterned with a single deposition and etching process. Good adhesion is promoted by heating the film to about 300°C, which causes the lower surface of the film to oxidize and combine with the silicon oxide. An aluminum metallization process is presented in Figure 2-27. Typically, metallization thickness is on the order of 15,000 Å. This is approximately the ideal thickness to prevent microcracks from occurring at the oxide steps. Aluminum has some drawbacks. If the temperature goes too high (ca. 600°C) during subsequent packaging operations or if a thin or narrow portion is overheated by a current surge during operation, the metal can fuse, go into solution in the oxide, or penetrate through the oxide to the silicon and short the connection. However, adequate process control and testing have made such failures rare in circuits shipped to systems manufacturers.

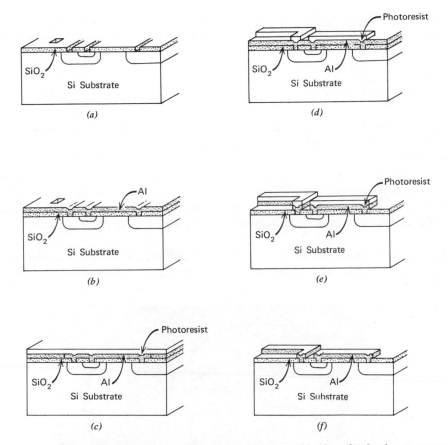

Figure 2-27 Metallization process: (*a*) wafer prepared for metallization, (*b*) aluminum evaporation, (*c*) photoresist coating, (*d*) patterned photoresist, (*e*) etched metallization, (*f*) completed process.

Also, the silicon die is usually mounted in the package by a gold preform or die backing that alloys with the silicon. Gold lead wires have been bonded to the aluminum film bonding pads on the chip, since package leads are usually gold plated. At elevated temperatures, a reaction between the metals of such systems causes the formation of intermetallic compounds, known as the "purple plague" ($AuAl_2$). Purple plague is one of six phases that can occur when gold and aluminum interdiffuse[1]. Voids normally occur in the formation of the $AuAl_2$ phase because of the dissimilar rate at which gold and aluminum diffuse into the intermetallic. The voids may result in weakened bonds, resistive bonds, or catastrophic failure.

The problem is generally solved by using aluminum lead wires, or another metal system, in circuits that will be subjected to elevated temperatures. A popular cure is depositing gold over an underlayer of chromium. The chromium acts as a diffusion barrier to the gold and also adheres well to both oxide and gold—gold has poor adhesion to oxide because it does not oxidize itself. However, the chromium-gold process is comparatively expensive, and it has an uncontrollable

[1]*Intermetallic Formation in Gold-Aluminum Systems*, by Elliot Philofsky, Central Research Laboratories Report 102, Motorola Semiconductor, January 20, 1969.

reaction with silicon during alloying. Two deposition and etching steps are required. Chromium is also used as an underlayer for silver. Nickel and copper have been employed in some processes.

VACUUM DEPOSITION

Films for monolithic ICs are generally deposited by vacuum evaporation. The beam-lead process uses sputtering and electroplating, but these are primarily employed for special hybrid IC materials. In vacuum deposition, the metal is heated in a good vacuum (pressure less than 5×10^{-6} torr) until the vapor pressure of the metal is about $10\,\mu$. Aluminum, for example, reaches this evaporation point at a temperature of 996°C, safely below diffusion temperature, whereas molybdenum does not vaporize below 2533°C. The film deposited is about $1\,\mu$ thick.

The main difference in vacuum evaporation methods lies in the way the metal is heated. Ordinarily, metal clips or wire are wrapped on a resistor–heated fila-

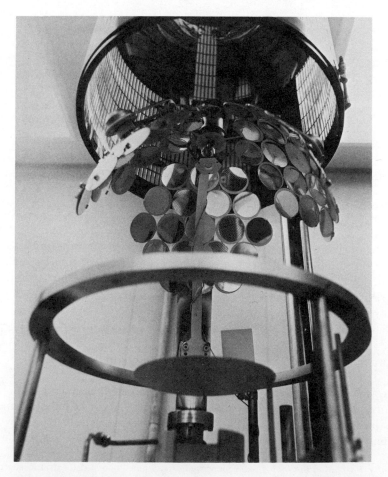

Figure 2-28 Metallization dome.

ment or placed in an electrically heated boat. The filament in the boat is a refractory metal such as tungsten, tantalum, or molybdenum. As the temperature rises, the metal wets the filament or boat and finally becomes hot enough to evaporate.

Another popular vacuum-deposition method is electron-beam evaporation. An electron beam gun with a water-cooled hearth hits the metal with a stream of high-energy electrons. The electrons give up their energy as heat. Figure 2-28 shows wafers in a typical rotating planetary electron-beam vacuum system. The electron-beam system is "clean" as compared with the system that uses a tungsten filament as the heating element. The tungsten boils off and contaminates the wafers being metallized.

Substrates are placed about the evaporation source. They may be heated to between 100 and 300°C before evaporation, in order to improve adhesion or to alter the metal film's properties. Thickness is monitored with a quartz crystal oscillator. A film deposited on an oscillator in the chamber changes its frequency, which is compared with the frequency of a reference crystal outside the chamber. The frequency difference is proportional to the metal thickness. Proportionality constant for the metal used is determined beforehand with an interferometer.

DIE PASSIVATION

As a final process before packaging, most manufacturers coat the wafer with a passivation or "glassivation" film to protect the surface of the wafer from contamination. National Semiconductor Corporation, for instance, deposits a low-temperature glass compatible with oxide made by the reaction of silane with oxygen at around 350 to 450°C (i.e., silicon dioxide lightly doped with phosphorus). Adhesion has to be good and temperature expansion coefficients well matched to preserve structural integrity with the wafer. Typically the thickness of the passivation layer is 10,000 Å.

This added step pays off in protection before and after packaging, in higher yields, and in better reliability. The glass is etched off the thin-film bonding pads and from between the circuits, where the wafer is cut into dice with scribing tools. The rest of the circuit is protected against contaminants and scratching during packaging. Also, passivation gives chips in molded packages much the same environmental protection as hermetic packaging.

Similar processes are used to form insulating layers between the two or three metallization patterns needed on complex circuits such as LSI arrays. Connections between layers are made by etching holes in the interlayer insulation before deposition of the next metal film. An extra layer also allows thin-film components to be formed over the circuit rather than being squeezed between the interconnection traces on the oxide.

Silicon nitride (Si_3N_4) is a passivation material that is finding increasing usage and is employed extensively in the beam-lead process. Although silicon nitride can be sputtered on, it is usually put on in a cold-wall reactor using r-f as a heat source coupled to a graphite susceptor. The reaction can occur between 750 and 1000°C, but it most commonly takes place at 850 to 900°C.

Silicon nitride has many excellent qualities for a surface passivating film. It has high resistivity (10^{16} Ω-cm), high dielectric strength, (ca. 10^7 V/cm), and excellent resistance against chemical and ambient attack; it is an excellent mask against sodium ion migration and, finally, it has superior stability against thermal and electrical stress.

PROCESS MATURITY

Two of the critical requirements for a successful process are freedom from contaminants and maturity. A scrupulously clean process is necessary to avoid wide variations in device parameters and to prevent corrosion of the metallization. Also, minute particles interfere with the processes and cause flaws in the circuits.

Masking and metallization are fairly standard procedures, but diffusion methods and impurity combinations vary from IC family to family. Perfecting a diffusion sequence to the point where device geometries and characteristics can be accurately predicted and used as a basis for circuit design is a long and costly effort, involving the work of many skilled specialists.

This is why "old" IC families produced by well-matured processes in thoroughly debugged equipments are generally a bargain—highly predictable and highly reliable. Once a process is matured, it may be used over again, to produce millions of ICs with very high yields with many different functions. Circuits in a logic family, for example, are generally made by similar processes. The variety comes from changing the numbers, geometries, and arrangement of the elements. In some cases, only an interconnection mask is changed. These are known as masking options. The elements are so small that it is cheaper to change a mask and one process step than a design. A typical case is when a standard MOS shift register is shortened 10 or 20 stages when a large order for a special length is received.

A new IC process is not necessarily expensive or unreliable. Some new IC families, much more complex than the old, are not the result of new processes, but of old processes applied to new designs on much larger dice. Or they represent new combinations of mature processes.

Conservative manufacturers—and most manufacturers have learned to be conservative—generally do not introduce new products for general sale until exhaustive tests or pilot runs have been made.

Chapter Three

BIPOLAR INTEGRATED CIRCUIT CHARACTERISTICS AND DESIGN

Successful IC design can be achieved by any one of the three principal fabrication techniques or a combination of them: film, bipolar, and MOS. Each has specific merits and drawbacks that must be considered when an IC is chosen for use in a system. These considerations are summarized for bipolar monolithic circuits in this chapter.

The main goals of monolithic IC developers are to meet performance specifications in terms of such factors as switching speed, frequency response, and gain, and to reduce fabrication cost. To meet these goals, a number of compromises must be made in design and fabrication.

Economics dictate that many monolithic ICs be made on each silicon wafer and that many wafers be processed in batches with high yield by the steps outlined in Chapter 2. Simultaneous fabrication of all the components of many circuits has disadvantages as well as advantages. The components cannot be sorted for different applications, as is done with discrete transistors. The IC designer must consider process variations in setting the circuit and component tolerances. Usually, tolerances on component values are looser than for discrete components. But integrated elements made simultaneously, side by side, in the same material are very closed matched and are subject to the same junction temperature changes. Thus it is easier to create ICs that stay "in tune" than to select and match discrete components. This matching of elements often offsets the reduced flexibility caused by looser tolerances and the inability to select individual component characteristics.

49

There are many methods for producing an IC. The most important bipolar IC design decisions are (1) isolation method, (2) the types of active elements, such as *npn*, *pnp*, or both *npn* and *pnp* transistors, and (3) technical performance. To date, *npn* transistors predominate because their parameters are easier to control and because they have a better reliability history.

The first bipolar IC amplifiers were actually made by the same processes developed for the early bipolar logic circuits. In recent years, however, configurations and processes more appropriate for linear (analog) circuits have been developed. For instance, the newer IC operational amplifiers now contain FET constant-current sources that are highly temperature stable, as well as differential input transistors with gains exceeding 10,000 for precision operation with very high signal–source impedances.

DESIGN RULES FOR BIPOLAR INTEGRATED CIRCUITS

1. Circuits must be analyzed and tested very thoroughly to be certain that they meet specifications before fabrication is begun. It is much more important that this criterion be met in semiconductor circuitry than in thin-film circuitry, because of the higher initial costs of bipolar IC fabrication. Unfortunately, circuits breadboarded with conventional components do not fully represent the fabricated semiconductor circuit. This is primarily attributable to the interaction between components. Many semiconductor manufacturers fabricate both single-semiconductor components and multiple-semiconductor components on one chip to be used for breadboard purposes. These special components, called kit parts, represent the true fabricated single-semiconductor circuit more realistically than conventional components. Another design technique is analyzing the effect that the interaction between components has on the circuit operation. It is possible to breadboard circuitry in which component interaction is represented by conventional components. In any event, the electronics designer must understand the problems of circuit fabrication, to enable him to help the semiconductor fabricator with his design. The probability of obtaining the desired electrical characteristics in the first effort of circuit design to fabrication, for a simple circuit, is no better than 50%. Since prototype fabrication is expensive, a good understanding of all design processes is necessary to increase the probability of success for subsequent design procedures.

2. Circuits should be designed to employ active components rather than passive components whenever possible, owing to real-estate restrictions, ease of fabrication, and fabrication cost. For example, diode coupling in digital circuitry is preferred to resistor coupling because the active components require very small areas. A diode in an IC is considered to be an active element because it is usually fabricated using one junction of an IC transistor.

3. The system should be designed to use the minimum number of different circuit component types. This will optimize the real estate per chip, increase yields, and thus keep fabricating costs at a minimum.

4. Circuits should be designed to involve ratios of component values rather than the absolute component values whenever possible because ratios are more uniform than are the absolute component values. Thus the absolute accuracy of

resistors should not be required by any design. Also, resistor values should be kept below 20 kΩ.

5. The use of tunable components and inductors should be eliminated from all circuit design.

6. Circuits should be designed to employ direct coupling rather than capacitive coupling. This is especially true with low-frequency amplifiers because of the difficulties of fabricating large capacitors (above 50 pF). Capacitor values should be minimized and their use restricted, where possible, to bypass applications (exceptions include internal frequency compensation of operational amplifiers).

7. Design should be based on matched components. Since similar types of components are fabricated in parallel on the same semiconductor substrate, excellent temperature tracking of both active and passive components can be obtained. This feature permits the designing of balanced drift-free circuitry.

8. Circuits should be designed for minimum power dissipation. Since the IC is small, the circuit power dissipation should not exceed approximately 400 mW for a 55×55 mil die in a metal TO-5 can. A volume power dissipation of 5 W/in.[3] can be obtained with good thermal heat sinks. This consideration is not based on the physical limitations of parts, since diffused components characteristically withstand high power and high temperature without being destroyed. Instead, the main consideration generally affecting maximum power dissipation is the internal heating effect of the circuit, which causes it to drift or fall outside the expected performance range. Such a situation may mean that an effective thermal control element such as a heat sink must be used in direct contact with the circuit package, even in a very low-power circuit.

9. Circuits should be designed to operate on low voltages. Levels above approximately 10 V should be avoided. The power supply should be selected so that the voltage ratings will not be exceeded in any of the circuit parts. The BV_{CEO} rating of the transistor is generally the principal determining criterion. The isolation junction has characteristically the highest voltage breakdown, followed by collector-based junctions and emitter-to-base junctions, in that sequence. In certain cases, these high voltage ratings can be used to advantage if precautions are taken to assure that no operating or transient conditions will allow any voltage rating to be exceeded.

MONOLITHIC INTEGRATED CIRCUIT COMPONENTS

Before discussing active element parameters, let us briefly look at the main elements of an IC—resistors, diodes (which are also capacitors), and a rudimentary IC transistor.

An IC resistor is made by diffusing into the silicon a dopant that converts a filamentary volume to the opposite type. Diffused resistors are typically 20μ wide. Figure 3-1a depicts a resistor of p-type silicon in an n region. The resistor is isolated from the n region by virtue of the p-n junction, provided the resistor is back-biased with respect to the substrate.

The isolation, however, is not complete. A back-biased p-n junction behaves like a nonlinear capacitor. As in the case of a thin-film resistor, the resistance value equals the sheet resistance multiplied by the aspect ratio. The sheet resistance

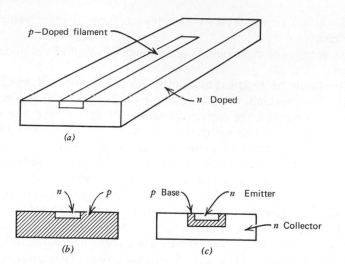

Figure 3-1 Basic IC devices: (*a*) semiconductor resistor, (*b*) diode or capacitor, (*c*) transistor.

depends completely on the distribution of dopant concentration as a function of depth. Sheet resistance is usually 100 to 200 $\Omega/\square$.

A diode or capacitor region, (Figure 3-1*b*) contains a small section of oppositely doped substrate material, producing a *p-n* junction. The element diameter is usually in the 20 to 100 μ range. As previously mentioned, the diode junction acts as a capacitor in the back-biased state. The capacitance depends on the magnitude of the bias voltage. Thus a diode may serve two functions. In practice, the *p-n* junction needed for a diode or capacitor is obtained from a transistor. Depending on the desired characteristics, the fabricator may use the base–collector or the base–emitter junction, or even both in parallel.

Two of the three diffusions required to produce a bipolar monolithic circuit are specifically tailored to produce suitable transistors because the transistor is the basic IC component. A typical transistor structure appears in Figure 3-1*c*. Here the *n*-type silicon substrate is the collector. Some *p* dopant is diffused in to form the base with an initial diameter of perhaps 60 μ and a depth of 10 μ. Finally, *n* dopant is diffused into the *p* region to form the emitter. The emitter diameter may be 20 μ and its depth 9 μ making the base width 1 μ.

BIPOLAR INTEGRATED CIRCUIT TRANSISTORS

In the cross section of a typical integrated transistor with junction isolation (Figure 3-2), the collector is a thin, high-resistivity *n* region on top of a relatively thick (for mechanical strength) *p*-type substrate of even higher resistivity. This creates a *p-n* isolation junction between the collector and the substrate of the monolithic device which is not present in the discrete component.

Note also that the collector contact is made on the top surface of the device rather than on the bottom surface, as for the discrete unit. This top-collector contact is necessary because collector current cannot be permitted to flow to the sub-

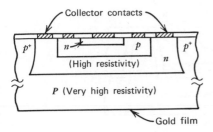

Figure 3-2 Cross section of integrated *npn* transistor using junction isolation.

strate that is common to all other components on the same chip. (Collector current flow to the substrate is prevented by reverse biasing the isolation junction.) Therefore, much of the collector current must flow laterally through the high-resistivity collector region to reach the top contacts. This, of course, results in a voltage drop within the collector resistance. Thus integrated planar transistors have a high collector-series resistance, which degrades both amplifier and switching performance.

A common method of reducing this resistance is by diffusing a very thin layer of low-resistivity *n*-type material between collector and substrate prior to growing the epitaxial layer. Normally referred to as a buried layer, this area is illustrated in Figure 3-3. The buried layer has the effect of shunting the high-resistivity collector region. The process does not affect the collector–base breakdown voltage, which is determined principally by the resistivity of the collector region at the base junction. Likewise, the breakdown voltage and capacitance of the collector–substrate junction remain essentially the same as before, being determined by the resistivity of the substrate.

Junction isolation creates a capacitance between substrate and collector that is usually many times larger than that between the collector terminal and circuit ground of a discrete transistor; this capacitance, moreover, is strongly voltage–dependent. Also, the leakage path between collector and substrate may be significant in low-current applications. Furthermore, junction isolation creates a parasitic *pnp* transistor in the bottom three layers. Under certain voltage conditions, when collector and substrate are negative with respect to the base of the *npn* transistor, this parasitic transistor is in its active region and can therefore conduct significant amounts of current between base and substrate.

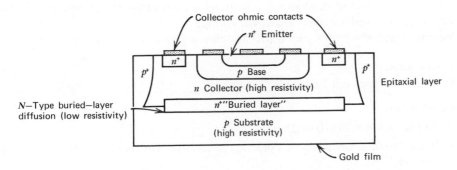

Figure 3-3 Use of a "buried" *n*+ layer to reduce collector-series resistance.

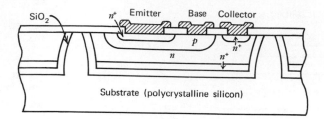

Figure 3-4   Cross section of an integrated transistor using dielectric isolation.

Some junction isolation problems are eliminated by dielectric isolation—in which a nonsemiconductor layer is fabricated between the collector and the substrate, so that there is no *pnp* parasitic transistor (see Figure 3-4). A dielectric layer generally has lower capacitance and leakage current per unit area compared with the reverse-biased junction. But the collector-series resistance must again be reduced, usually with a buried n^+ layer.

Complementary Transistors

In many applications, both *npn* and *pnp* transistors are extremely desirable in a single IC. To date *npn* transistors predominate because their parameters are easier to control and because they have a better reliability history. Recently, however, three methods of fabricating *pnp* transistors have been developed: the lateral *pnp*, the vertical or substrate *pnp*, and the complementary *pnp* with junction isolation. The lateral *pnp* transistor starts with two *p* regions diffused into an isolated *n*-type island (Figure 3-5a) at the same time that the base regions of the *npn* transistors are diffused. The *p* regions are a ring-shaped (circular) collector and a round emitter inside the collector. The *n* region at the surface between the two *p* areas is the base region. Current flows laterally between emitter and collector, thus the name "lateral" base transistor.

One problem with the lateral *pnp* transistor is difficulty in controlling accurately the distance between the two *p*-type regions. A difference in spacing between the emitter and collector so slight that it could not be detected through a laboratory microscope can mean an order of magnitude difference in current transfer ratio (beta) of the transistor. An additional problem with lateral *pnp* transistors is their low base–collector breakdown voltage. This occurs because the base–collector junction is not located deep in the silicon as it is in other structures, and surface effects therefore reduce the breakdown voltage.

Summarizing, the lateral *pnp* transistor characteristics are as follows:

ADVANTAGES
- Collector does not have to be grounded.

DISADVANTAGES
- Thick base width (10 to 25 μ)
- Beta depends on process control; generally, transistors with low beta are obtained.
- Has inherent phase shift because of large base area

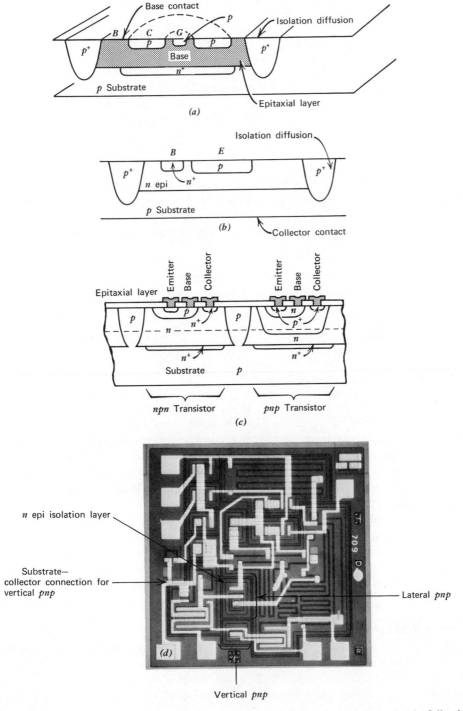

Figure 3-5 Combining *pnp* and *npn* transistors in a single IC can be accomplished using the following structures: (*a*) lateral *pnp*, (*b*) vertical (or collector–substrate contact) *pnp*, (*c*) complementary, (*d*) lateral and vertical (in this photomicrograph of an operational amplifier, the lateral and vertical *pnp* transistors are indicated by arrows).

- Has *pnp* action to substrate—requires use of buried layer to degrade this action.
- Poor frequency response
- Low collector–base breakdown voltage

Control of base width tolerance accounts for most of the difficulty in making the substrate or vertical *pnp* transistor. This structure (Figure 3-5*b*) requires a *p*-type substrate, on which an *n* epitaxial layer is deposited. A *p* region is diffused into the epitaxial layer to form the emitter. The *p* substrate is the collector, and the *n* epitaxial material between it and collector becomes the base.

Since emitter diffusion depth is constant for all transistors on a silicon wafer, the beta of each transistor varies with the epitaxial layer thickness at that point. Even in the best epi material, the variation in thickness causes a wide variation in beta.

Primary characteristics of the vertical *pnp* transistor are as follows:

ADVANTAGES
- High beta
- High base–collector voltage breakdown

DISADVANTAGES
- Applications are limited because all *pnp* collectors are tied together and go to V^-
- Poor frequency response

Truly complementary transistors (Figure 3-5*c*) require additional processing steps, but the performance of the *pnp* transistors is comparable to that of the rest of the circuit. Epitaxial layer formation is interrupted when approximately half the layer is present. Where a *pnp* transistor is to be located, a *p*-type region of high concentration (p^+) is diffused into the first half of the epitaxial layer. The rest of the epitaxial layer is then grown. In a subsequent diffusion, the buried *p*-type impurities diffuse upward to the surface and complete the collector.

The bases and emitters of the two types of transistors are then diffused separately in the normal manner. Compared with the lateral base transistor, the width of the base region is determined by the depth of base and emitter diffusions. As in the *npn* transistor, this depth can be controlled with great precision and the gain of the transistor is therefore in normal range. Figure 3-5*d* depict the actual characteristic geometries of lateral and substrate *pnp* transistors on an IC photomicrograph.

Super Beta Transistors

Transistors with current gains of 2000 to 10,000 at collector currents less than 1 μA can now be made in monolithic circuits. This is more than ten times the gain of present-day discrete transistors. The significance of this breakthrough is greatest for IC operational amplifiers, since lower input bias currents are constantly being sought.

The high-gain transistors have an unusually low breakdown voltage, which precludes their use in standard circuit designs. However, they can be fabricated simultaneously with high-voltage transistors. Circuit techniques are available — namely, bootstrapping and cascade connections — which take advantage of the high current gain of one transistor type and the high breakdown voltage of the second, producing the equivalent of a high-gain, high-voltage device. This may double the number of transistors needed to perform a given function, but it is an economical approach for monolithic ICs because active devices are relatively inexpensive.

In making double-diffused transistors, which are almost universally used in ICs, there is a trade-off between breakdown voltage and current gain. As the emitter is diffused more deeply into the base, reducing the base width, the current gain increases at the expense of breakdown voltage. The process described here uses transistors fabricated with separate emitter diffusions, one being deeply diffused for exceptionally high gain and the other having a normal diffusion for breakdown voltages above 50 V. The results obtained in making transistors are highly dependent on the collector resistivity, as well as the predeposition and diffusion schedules used for the base and emitter.

With a moderate penetration of the emitter into the base, the current gain is low; and the breakdown voltage is essentially equal to the inherent breakdown of the collector–base junction. This is illustrated in Figure 3-6, where a current gain of 10 is obtained for a breakdown voltage of 100 V. If the transistor is returned to a high-temperature diffusion furnace, reducing the base width further, the current gain will go up. However, the breakdown voltage drops and begins to exhibit the familiar negative-resistant BV_{CEO} characteristic of planar transistors. Typical current gains of 250 with a 50-V breakdown can be realized at this point.

Going through another diffusion step gives even more current gain, but the collector–emitter breakdown softens as the device goes into punch-through. This breakdown occurs when the depletion region of the collector–base junction penetrates the base and reaches through to the emitter. Further diffusion proceeds rapidly in the direction of making the transistor a collector–emitter short. Stopped in time, however, it can give current gains above 5000 with breakdown voltages around 5 V.

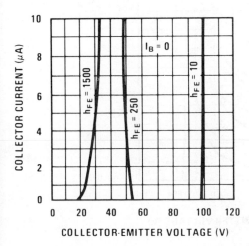

Figure 3-6 Degeneration of the breakdown voltage in a double-diffused transistor as it is diffused for higher current gain.

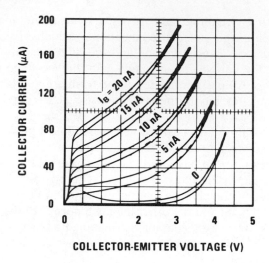

Figure 3-7   Curve tracer display of high-gain transistors.

Figures 3-7 and 3-8 illustrate the characteristics of the two types of transistors that have been fabricated on the same silicon chip. A primary transistor (Figure 3-7) has a breakdown voltage of roughly 4 V and a current gain above 5000. The secondary transistor (Figure 3-8), which is more representative of the ones found in normal ICs, has a breakdown of 70 V and a current gain above 100.

It is interesting to note that transistors that have been driven into punch-through exhibit less falloff of current gain at low collector currents. This probably happens because there is a large difference between the emitter base turn-on voltage in the bulk region near the collector–base junction and the turn-on voltage near the

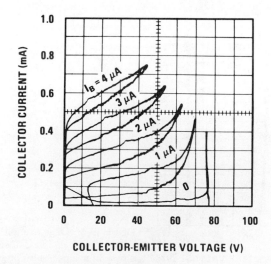

Figure 3-8   Curve tracer display of standard transistors.

surface. A greater difference in these voltages reduces the collector current at which the falloff occurs. Hence, in practice, punch-through transistors can be operated at lower collector currents, giving further reductions in input bias current.

As we mentioned earlier, design techniques are available which take advantage of the high gain of punch-through transistors along with the high-voltage capability of normal transistors. A simple example of this appears in Figure 3-9. A high-gain transistor is operated in the common-emitter configuration, driving a common-base transistor with high breakdown voltage[1]. This cascade connection gives the output voltage capability of the secondary transistor, yet has the current gain of the primary. Breakdown voltage of the primary transistor is not a problem because the transistor is operated at zero collector–base voltage. The circuit uses more components than a single-transistor amplifier, but this poses no problems in a monolithic IC.

Figure 3-10 is a simplified schematic diagram of a voltage follower employing the best characteristics of primary and secondary transistors. Primaries are used for the input stage to get very low input bias current: D_1 is included to operate Q_2 at a near-zero collector–base voltage. In addition, the collector of Q_1 is bootstrapped to the output to reduce the voltage across it. In this circuit, since the only transistor that needs to handle any voltage is Q_3, a high-voltage secondary transistor is used. This example, in fact, employs primary and secondary transistors in a manner that exploits their best characteristics without significantly complicating the design.

Punch-through transistors have advanced the state of the art in the input stage of general purpose operational amplifiers such as the LM108A and LM110. A differential amplifier that can serve in this application is presented in Figure 3-11.

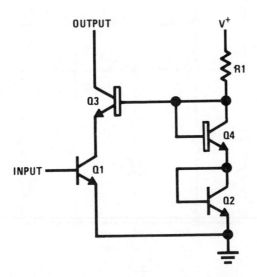

Figure 3-9 Cascade circuit combines high-gain and high-voltage transistors, taking advantage of the best features of each.

[1]In Figure 3-9, the secondary transistors are depicted with a wider base than the primary transistors.

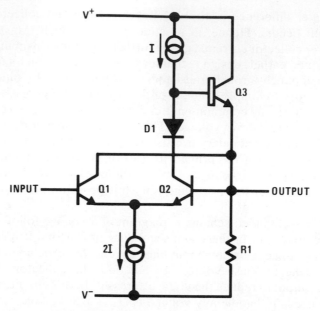

Figure 3-10 Examples of voltage follower using both high-gain and high-voltage transistors.

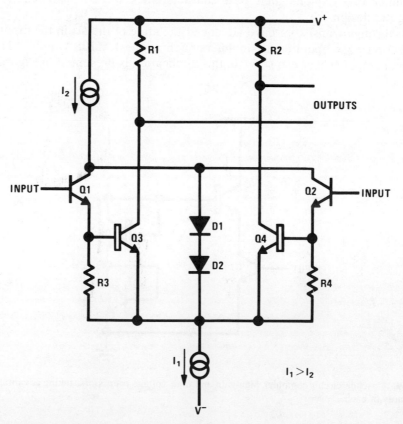

Figure 3-11 Bootstrapped high-gain transistors can be used in a general-purpose operational amplifier.

Again, the primaries are operated at zero collector–base voltage with bootstrapping circuitry. The secondary transistors, on the other hand, stand off the voltage. Thus current gains above 2000 can be obtained with IC transistors by diffusing them into punch-through. This lowers the breakdown voltage to about 5 V – too low for many circuit designs. However, normal transistors can be made on the same chip with the punch-through devices. When used in IC operational amplifiers, this technique can yield input bias currents of less than 1 nA over a −55 to +125°C temperature range. The two types can be combined for better results than can be achieved with state-of-the-art discrete transistors.

Transistor Geometries

Transistor geometries are usually optimized for each application. For example, a small transistor with a short emitter periphery is well suited to low-current and low-noise applications, whereas for high-current applications a long emitter periphery and small access resistance are required. Figure 3-12 is a small-signal transistor with one contact for each region. The rectangular structure is economical in IC area used. The geometry of Figure 3-13 effectively doubles emitter periphery. Two contacts are provided to the base and collector regions so that current flows from the emitter in two directions, rather than one. Also, the collector–series resistance is greatly reduced, since the cross section of the current path is doubled.

A typical IC transistor for very high current applications is shown in Figure 3-14. To obtain as much emitter periphery as possible, the regions are interdigitated. The resistance of the buried-layer diffusion must be low. Figure 3-15 is a photograph of a low-power TTL *J-K* flip-flop. In this particular design, minimum junction areas were desired in order to minimize *p-n* junction capacitances and thus increase the device speed. To accomplish this, the transistor outlines were made smaller and rounded (see arrow in Figure 3-15). Transistor–transistor logic (TTL) transistors are normally square (Figure 3-16) so that more can be fitted in a given chip area, but this geometry enlarges junction areas.

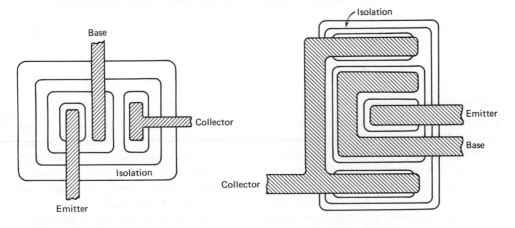

Figure 3-12 Top view of a small-signal transistor.

Figure 3-13 Top view of a high-current transistor.

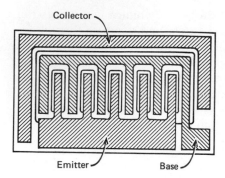

Figure 3-14 Top view of a very-high-current inter-digitated transistor.

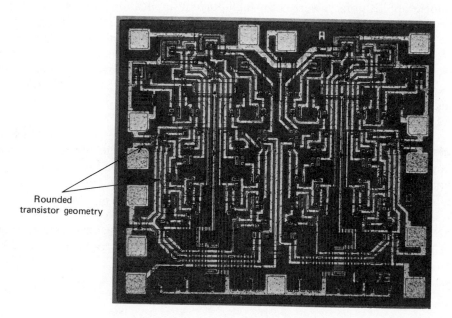

Figure 3-15 Photomicrograph of a low-power TTL *J-K* flip-flop, showing rounded transistor geometries.

Dimensional Tolerances

Practical design allows approximately 4 mA of emitter current for each mil of emitter perimeter. If the current desired is about 42 mA, for example, the emitter dimensions will be about 1.25×4 mils. The other dimensions are scaled to emitter size, as in Figure 3-17. To keep the transistor small (which improves frequency response), base contact widths and spacings and collector contact widths and spacings are each 0.5 mil. These tolerances are compatible with the 0.25 mil tolerances permitted by present manufacturing processes. Figure 3-18 shows typical tolerances for a single-base bipolar IC. Typical IC transistor characteristics for various emitter sizes are listed in Table 3-1.

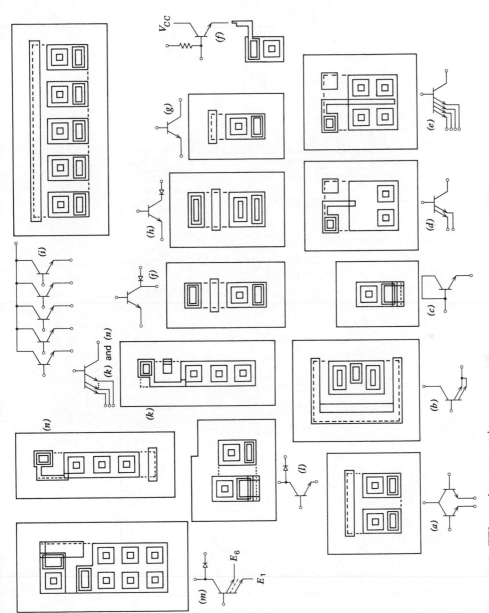

Figure 3-16 TTL transistor geometries.

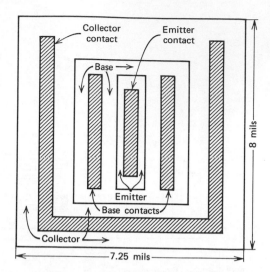

Figure 3-17   Example of integrated circuit transistor with typical dimensions.

How Dopants Determine Performance

As we discussed in Chapter 2, diffusion of impurities into silicon is the primary means of controlling IC characteristics. An impurity is introduced into the surface of a silicon wafer and is subsequently "driven in" at a more elevated temperature. IC transistors are either high-frequency switching types, which are gold-diffused in order to increase their maximum frequency response, or types that are not gold-doped. Gold reduces carrier "lifetime" and therefore permits the transistor to make ON–OFF transitions more rapidly.

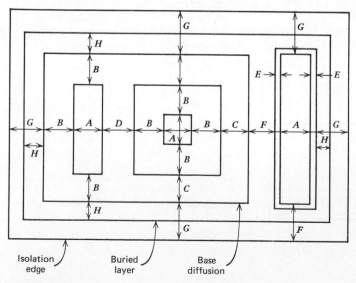

Figure 3-18   Single-base bipolar transistor, showing minimum clearances to achieve practical yields under normal processing variations. Minimum clearances (mils): $A = 0.3$, $B = 0.25$, $C = 0.35$, $D = 0.3$, $E = 0.1$, $F = 0.5$, $G = 2.5$ for dielectric isolation; 1.0 for junction isolation. H is not critical.

Table 3-1 Typical Characteristics of Monolithic Bipolar Transistors

Characteristics	Emitter Size (mils)			
	1×1.5	3×3	2×10	2×10
BV_{CBO}			35 V	
BV_{CEO}			15 V	
BV_{EBO}			7 V	
h_{FE} at I_C	40 at 100 μA	40 at 1 mA	40 at 2.5 mA	25 at 2 mA
h_{FE} at I_C	80 at 5 mA	60 at 10 mA	50 at 50 mA	50 at 5 mA
f_T (MHz)	550	220	—	—
Collector–base C (pF) at 6 V	2	7	10	16.3
Emitter–base C (pF) at 6 V	2	5	10	16
Collector–substrate C (pF) at 6 V	4	10.5	10	17
R_{CS} (Ω)	70	120	35	27

The electrical characteristics of active elements in bipolar ICs are generally determined by the behavior of minority carriers. In contrast, passive circuit components are governed by the flow of majority carriers. Since active elements are the major concern of the designer, the behavior of minority carriers is of considerable interest.

Minority-carrier lifetime depends on impurity concentration and temperature and, most important, on the number of recombination centers in the semiconductor. Introduction of recombination centers (traps) generally reduces lifetime. In silicon, gold acts as a very effective recombination center since its energy level is near the center of the band gap.

Remember from Chapter 2 that impurity concentration is not constant but varies with diffusion depth. There is no abrupt change in impurity concentration in diffused junctions. The junctions are graded, rather than being abrupt steps. Each diffusion must start with an impurity concentration higher than that of the preceding layer; thus the number of successive diffusions in a given area should be minimized—preferably held to two. Even when a diffused area is sealed with oxide, it continues to spread when the wafer is heated during the next operation.

The transistor cannot be symmetrical, because the emitter capacitance is much greater than the collector capacitance. Moreover, parasitic capacitance effects prevail in semiconductor ICs. The predominant parasitics are generally associated with the isolation junction. Through efficient circuit design, the number of isolation areas may be reduced, resulting in a reduction of the undesired capacitance.

Monolithic Diodes

Diodes are widely used in ICs both as nonlinear elements and as isolation elements. Considering diodes mainly from the standpoint of discrete diode functions, the important parameters that can be varied by IC processing include: capacitance, leakage current, breakdown voltage, recovery time, and forward voltage drop.

Since transistors are generally the most critical device from a design standpoint, diode (and resistor) characteristics are dictated by transistor requirements.

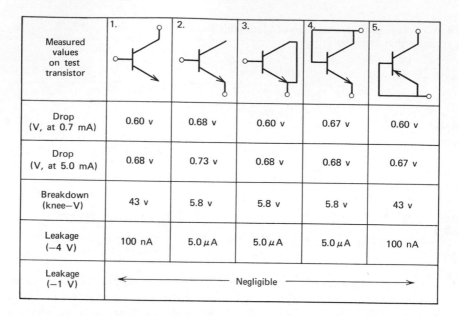

Measured values on test transistor	1.	2.	3.	4.	5.
Drop (V, at 0.7 mA)	0.60 v	0.68 v	0.60 v	0.67 v	0.60 v
Drop (V, at 5.0 mA)	0.68 v	0.73 v	0.68 v	0.68 v	0.67 v
Breakdown (knee—V)	43 v	5.8 v	5.8 v	5.8 v	43 v
Leakage (−4 V)	100 nA	5.0 μA	5.0 μA	5.0 μA	100 nA
Leakage (−1 V)	← Negligible →				

Figure 3-19 Transistor diode connections.

Diodes may be obtained by directly fabricating a *p–n* junction; but in the interest of making the circuit more compact and the element geometries more uniform, one junction of a transistor is often used as a diode. In gold-doped devices, the breakdown of the collector–base junction is determined by optimum gold concentrations and transistor-saturation-resistance requirements. Furthermore, the switching time of the diodes is a function of the amount of gold required to reduce the transistor storage time sufficiently. Such diodes must be either of the base–emitter type or the base–collector type.

Some flexibility in diode switching characteristics can be obtained by connecting the emitter, the base, and the collector of a transistor in various configurations (Figure 3-19). The minority-carrier distribution varies accordingly. Due to the presence of high-impurity concentrations in the emitter and base regions, which makes lifetime short, the switching time of emitter-base diodes is very fast. The various connections also change the coupling capacitance to external regions.

Schottky-Barrier Diodes

A method known as the Schottky process has led to development of high-speed devices. The big advantage of the Schottky diode is that it can be made with existing IC processes. No additional manufacturing steps are required—a very important fact concerning yield.

The principle of operation is relatively simple. When metal is placed in close contact with an *n*-type semiconductor, a voltage barrier is created (the Schottky barrier) because there are many free electrons in the metal, whereas the semiconductor contains relatively few. With a positive voltage applied to the metal, this barrier is overcome and the diode starts conducting. A negative voltage, on the other hand, enlarges the barrier, thus the diode blocks in this direction.

This sounds very much like an ordinary p–n junction diode. But there are some subtle, yet important differences. First, the barrier is only about half as large as that of a junction diode; at low currents, a Schottky diode has a forward voltage drop of only about 0.3 V, close to that of a germanium diode. Second, only majority carriers are involved in the conduction mechanism, which makes the Schottky-diode a very high-speed device (recovery time, < 1 nsec).

The use of Schottky diodes was previously limited by two effects. Silicon has a natural tendency to cover itself with oxide in a very short time. If there is any oxide on the silicon surface before the evaporation of the metal, an almost perfect insulator is obtained instead of a diode. But even if a good contact is established, the metal layer can produce edge effects, causing large leakage currents. Over the last two years, these problems have been gradually overcome. The first one was solved by clean processing. The second was eliminated by layout techniques, as in Figure 3-20. The edge effect is avoided either by extending the metal over the diode area (creating a field plate) or by placing the Schottky diode inside the diffused-base region.

It is important to realize here that the Schottky effect only takes place in relatively high-resistivity semiconductor material. When the semiconductor is heavily doped, a tunneling effect occurs which provides a direct ohmic contact. This is rather fortunate, for otherwise we would not be able to provide good low-impedance contact to the transistor regions.

The first application of the Schottky diode in ICs has been the so-called saturation clamp. When a transistor is allowed to saturate, a large number of carriers accumulate in the collector region. Subsequently, it takes a long time to turn the device off because first all these carriers have to be removed. Placing a Schottky diode between base and collector (as shown in Figure 3-20) cuts off the base current before the transistor can saturate. Now the device can be turned off almost immediately. This configuration is the basis for the Schottky-gate TTL families.

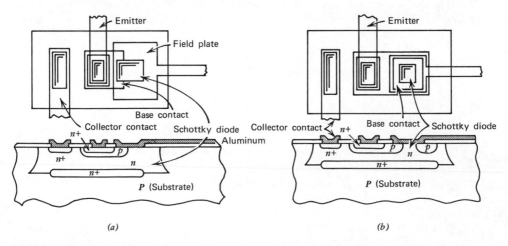

Figure 3-20 Two versions of Schottky diodes: (*a*) with field-plate, (*b*) inside base region. In these configurations, the Schottky diode is connected between base and collector of the *npn* transistor and acts as a saturation clamp.

Monolithic Resistors

The available types of IC resistors are the diffused-base, buried-base, collector–resistor, buried-collector, emitter–resistor, and thin-film types. Of these, the diffused-base type is most frequently used and is thus discussed in greater detail.

The diffused base resistor has a fairly good temperature coefficient, and its processing is well controlled. It has poor absolute accuracy, but excellent relative accuracy and tracking abilities. However, resistivity is determined by transistor requirements. Since diffused-base resistors are usually formed simultaneously with the transistor bases, sheet resistivity is normally optimized for the transistors. Temperature coefficient depends on mobility, another diffusion variable.

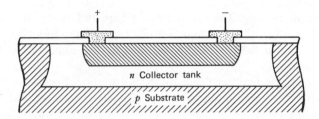

Figure 3-21 Cross section of diffused resistor formed by base diffusion in a previous collector diffusion.

The cross section of a typical diffused resistor appears in Figure 3-21. Note that the resistor is isolated from the substrate by an *n* tank. In either direction, there is one reverse-biased junction. The back-to-back junctions also prevent the current from detouring through the tank. Such monolithic resistors are far from perfect. In the case of Figure 3-21, the triple layer actually forms a parasitic *pnp* transistor (Figure 3-22). Furthermore, any back-biased junction contributes capacitance. In effect, a monolithic IC resistor is a circuit rather than a passive component. Although the "base" thickness of the parasitic *pnp* is relatively great, beta is still high enough (0.5 to 3) to cause a significant shunting of current. The *n* tank should

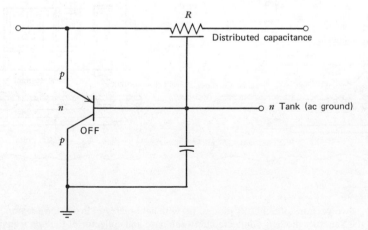

Figure 3-22 Parasitic transistor formed by diffused resistor.

Table 3-2 Typical Tolerances of Diffused Resistors

Operation	Absolute Value (%)	Relative Value (%)
Drawing mask	1	1
Etching	2	1
Diffusion	7	1
Totals	±10%	±3%

be connected to the most positive circuit potential to back-bias (cut off) the parasitic *pnp*.

Typical tolerances of diffused resistor fabrication are listed in Table 3-2 for a standard base diffusion. The minimum practical line and resistor width for volume production is 0.5 to 1 mil—typically 1 mil, and the separation between lines is comparable. A typical resistor requires at least 1 mil²/100 Ω (i.e., the 10-kΩ resistor of Figure 3-23 requires 135 mil², nearly three times the area of a small transistor). Also, 1000-Ω resistors are available that occupy an area of 1 mil². Since monolithic design is dominated by area consideration, the resistor would certainly be discarded if a transistor could serve in its place.

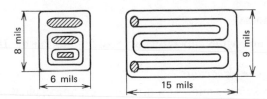

Figure 3-23 Example of diffused resistor.

Sheet Resistivity

Cost and yield factors, as well as the need to reduce parasitic capacitance, make high sheet resistivity important. All depend strongly on the area used. It is usually convenient to specify sheet resistivity R_S (more commonly but inaccurately termed sheet resistance) for the base and emitter diffusion, which are relatively thin and constant in depth.

Sheet resistivity is the resistance value of a thin film with equal width and length (Figure 3-24). The value of this planar resistor is given in ohms per square. If the length is twice the width, the resistance value is equal to twice the sheet resistivity, and so forth. Thickness t must be considered as well. Actually, resistance equals resistivity, ρ times the length of the resistor divided by the cross-sectional area or $R = \rho L/tw = (\rho/t)(L/w)$. A thin film of material of length L and width w has a

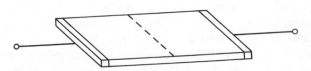

Figure 3-24 Schematic of a monolithic resistor.

cross-sectional area A equal to the thickness times the width ($A = tw$). The technology of ion implantation is used to obtain resistors with a resistivity on the order of 10,000 $\Omega/\square$.

The designer of an IC does not like to worry about anything except the top view of the device — the resistor's length and width — which means that ρ/t (resistivity/ thickness) is a processing rather than layout decision. Resistivity depends on the material and thickness on the processing.

A finite value must be used to define t. For instance, suppose a layer of material 5 μ thick has a bulk resistivity of 10 Ω-cm/$\square$. If we assume that $t = 1$, then

$$R_s = 10\ \Omega\text{-cm}/\square \div (5 \times 10^{-4}\ \text{cm})$$
$$R_s = 2 \times 10^4\ \Omega/\square$$

The total resistance from terminal to terminal is then

$$R = (2 \times 10^4\ \Omega/\square) \times 2\ \square$$
$$R = 40,000\ \Omega$$

Easy enough for thin-film resistors, the calculation is a little more difficult for diffused resistors. If impurities are diffused into the silicon surface, the impurity level drops off as one goes deeper until it reaches zero at the n–p junction interface.

Monolithic Capacitors

Both p–n junction and MOS capacitors are used in monolithic ICs.

Metal–Oxide–Silicon Capacitors

MOS capacitors resemble miniature discrete ceramic capacitors. They consist of a dielectric sandwiched between two layers of highly conducting material. Their capacitance is determined by the dielectric thickness, the dielectric constant, and the area of the insulating layer. The dielectric is usually the oxide grown on the silicon wafers. One terminal is metallization, and the other terminal is silicon. MOS capacitors have very low leakage and are voltage independent. The main problem with MOS capacitors is that thin oxides are needed for reasonably large capacitance values as an alternative to using thick oxides and unreasonably large capacitor areas. The thinner the oxide, the lower its breakdown voltage and the greater the chance for a pinhole in a given area. Useful MOS capacitor values are generally limited to 50 pF and are typically the 30 pF required for internal frequency compensation of operational amplifiers and other designs where one fairly large (for an IC) capacitor with stable characteristics is needed, (see, e.g., Figure 3-25). Note the large physical size of this 30-pF on-chip MOS capacitor as compared with the rest of the amplifiers' circuitry. MOS capacitors are rarely used in logic circuits today.

MOS capacitors are influenced by the silicon doping, which makes silicon a less-than-ideal material for capacitor plates. Surface states and the depletion layer in the silicon electrode can make the MOS capacitance both voltage- and frequency dependent. However, if the emitter doping is heavy and the grown oxide is about 3000 Å thick, MOS capacitors in emitter regions are not voltage dependent at normal IC voltage levels. Since the emitter is usually the last diffusion of a bipolar IC, its surface concentration remains relatively high.

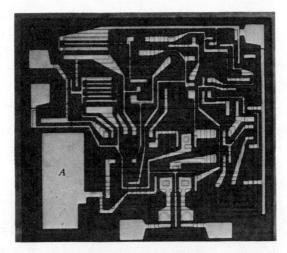

Figure 3-25 Photomicrograph of an LM107 internally compensated operational amplifier showing MOS frequency-compensating capacitor A.

For a given process, the capacitance is calculated in picofarads per square mil or in microfarads per square centimeter, varying from more than 10 pF/mil² at a thickness of 100 Å to less than 0.01 pF/mil² at 10,000 Å when either oxide or a mixture of both oxides is the dielectric.

Junction Capacitors

$p-n$ Junction capacitors are really voltage-variable capacitors, somewhat like varactors, and IC designers must struggle with their tendency to "detune" the circuit except in applications such as voltage-controlled oscillators. The applied voltage must be of a single polarity. The greater the junction gradient, the greater will be the capacitance per unit area. However, the breakdown voltage is also a function of the gradient and thus a limit is placed on the applied voltage.

Any reverse-biased junction has a depletion region that acts as a dielectric between two conductive surfaces. A monolithic capacitor is a back-biased junction, whose capacitance is inversely proportional to the depletion layer width at the junction. This width, in turn, depends on the impurity concentration profile in the vicinity of the junction. Junction capacitors are generally made with emitter diffusions.

For an abrupt transition between p-type and n-type semiconductor material (assuming donor concentration in the n material to be much larger than the acceptor concentration in the p material), the depletion region capacitance is

$$C = A\left(\frac{qKE_0 n_D}{2V}\right)^{1/2}$$

where A = area of the junction
   q = electron charge (1.6×10^{-19} C)
   K = dielectric constant of semiconductor ($K = 12$ for silicon)
   E_0 = permittivity of free space (8.85×10^{-14} F/cm)
   n_D = donor concentration in n-type material
   V = total junction potential

Note that V is the algebraic sum of the applied potential plus the internal contact potential (ca. 1.0 V for silicon). Reverse-bias voltage is considered positive (to be added), whereas forward-bias voltage (considered negative) is subtracted from the contact potential to determine junction potential.

As an example of the magnitude of this capacitance, assume that the resistivity of the n-type material is 0.5 Ω-cm. In this case, $n_D = 1.2 \times 10^{16}$ atoms/cm^3.

$$(qKE_0 n_D)^{1/2} = [(1.6 \times 10^{19})(12)(8.85 \times 10^{-14})(1.2 \times 10^{16})]^{1/2}$$

$$= (20.4 \times 10^{-16})^{1/2}$$

$$\left(\frac{qKE_0 n_D}{2}\right)^{1/2} = 3.2 \times 10^{-8}$$

for
$V = 1$ V (0 bias)	$C = 3.2 \times 10^{-8}$ F/cm^2	or	0.206 pF/mil^2
$V = 6$ V	$C = 1.3 \times 10^{-8}$ F/cm^2	or	0.085 pF/mil^2
$V = 12$ V	$C = 0.92 \times 10^{-8}$ F/cm^2	or	0.059 pF/mil^2

p–n Junctions formed during the transistor-diffusion cycles can be used for this type of capacitor. As indicated, the capacitance of a p–n junction is voltage sensitive, making it useful for special applications, such as voltage-controlled oscillators. However, the voltage-sensitive characteristic is not desirable for all applications, and the oxide capacitor structure is sometimes preferred.

Some of the capacitance of a junction capacitor is necessarily like an MOS

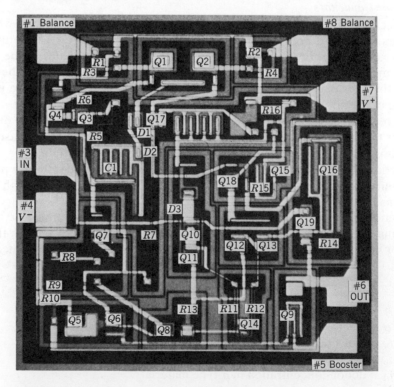

Figure 3-26   Photomicrograph of monolithic LM102 circuit.

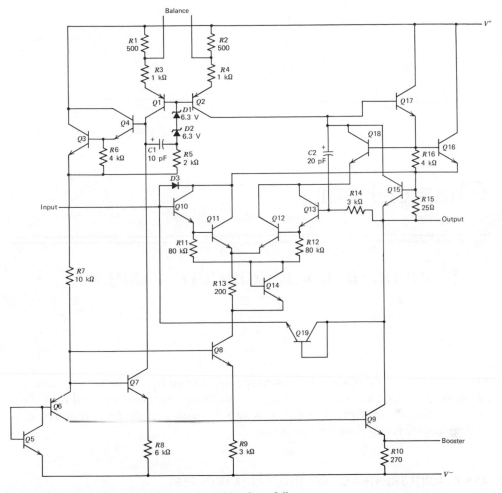

Figure 3-27   Schematic diagram of the LM102 voltage follower.

capacitor, since the terminals lie partly over thermally grown oxide on the silicon wafer. Also there is oxide between the terminal areas, forming what might be called lateral or surface capacitors. These oxides are quite thick and are grown in several successive process steps, however; thus these capacitances have high breakdown voltages and relatively small values. The IC user can let the IC designer worry about them and concern himself mainly with not overstressing the *p–n* junctions.

Putting the Components Together

Figure 3-26 is a photomicrograph of a typical monolithic circuit. Here the various components, in accordance with the schematic diagram (Figure 3-27), are identified, to give an idea of how the components discussed in this chapter are positioned to fit together to form the complete IC circuit layout and thus minimize valuable silicon real estate.

Chapter Four

BIPOLAR LOGIC CIRCUIT FAMILIES

The development of ICs has been strongly motivated by the increasing demand on the performance and economy of digital information-processing machines. This brief examination of digital ICs begins with a discussion of the basic requirements of digital information-handling circuits.

BASIC REQUIREMENTS OF DIGITAL CIRCUITS

A distinct feature of the digital information-processing system is that complex logic functions can be synthesized by the combination and repetitious use of one or a few elementary logic functions. Physically, a giant electronic network that performs complex logic functions can be implemented by interconnecting a large number of elementary digital circuits of one or a few basic forms, commonly called building blocks. The elementary digital circuits must perform the specified elementary logic functions, and they must be interconnectable in the sense that they can be "freely" interconnected with one another in patterns of little regularity without obstructing their basic operations.

DESCRIPTION OF MAJOR SATURATED-LOGIC LINES

Resistor–Transistor Logic

Prior to the development of ICs as we know them today, resistor–transistor logic (RTL) was probably the most popular form of logic in common use. Because it utilized relatively few active components, it was comparatively inexpensive and its performance was quite adequate to meet the computer requirements of the times. Because it was easy to implement and most familiar to logic designers, it was the first line to be integrated. Indeed, the initial basic RTL ICs were direct translations from discrete-component designs.

The simplicity of the RTL design is illustrated in the three-input gate circuit of Figure 4-1. This simplicity imparts some practical advantages to RTL ICs. It permits high-yield production with subsequent low cost. Electrically, the high (logic "1") state noise margin of an RTL gate is independent of fan-out. The same cannot be said of the low (logic "0") state noise margin. The low-state noise immunity is drastically affected by fan-out, and if the operating point were to move too close to the transition region, the gate would become unstable. Thus the fan-out of an RTL circuit is somewhat limited, and the noise margin becomes small as fan-out is increased. RTL circuits are no longer a popular IC logic family. They have been replaced by diode–transistor and transistor–transistor logic ICs.

Diode–Transistor Logic

Where improved noise margins and somewhat higher fan-out capability is required, diode–transistor logic (DTL) circuits are often employed. The basic DTL circuit consists of a diode AND circuit followed by a transistor inverter (Figure 4-2). Since a separate current source is provided for each transistor base, the uneven base current problem is completely eliminated here. The static switching operation of the circuit is almost completely governed by the nonlinear characteristics of the diodes and the transistor, and it is insensitive to the values of the "load" resistor and the power supply. The inputs to the circuit are isolated by the input diodes to allow large fan-in. The fan-out is also large, because the transistor is able to act as a heavy current sink in its saturation region.

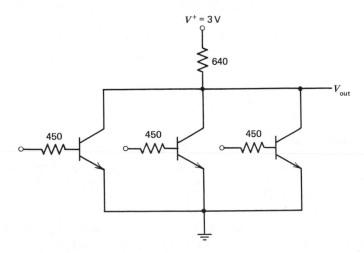

Figure 4-1   Basic RTL circuit.

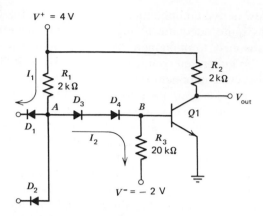

Figure 4-2   Basic DTL gate circuit.

Referring to Figure 4-2, with a low-state voltage, say $V_{CE(sat)} = 0.1$ V, applied to either input diode D_1 or D_2, current I_1 will flow through the diode, causing the voltage at point A to equal one diode drop, or approximately 0.7 V. At the same time, current I_2 will flow through diodes D_3 and D_4, causing the voltage at point B to be two diode drops less than that at point A, or approximately -0.7 V. With a negative voltage thus applied to the base of transistor Q_1, the transistor will be cut off and its collector voltage V_{out} is equal to $V+$, or 4 V (neglecting leakage current).

If a high-state voltage, say 4 V, is applied to both input diodes, they will be reverse-biased and will become nonconductive. The base of Q_1, therefore, "sees" a relatively high positive voltage and the transistor will be driven into saturation. Its output voltage then is $V_{CE(sat)}$. Under no-load conditions, therefore, V_{out} swings from $V_{CE(sat)}$ to nearly $V+$.

The situation does not change a great deal even when the gate is loaded with succeeding identical gates. This is because load gates do not draw current when a positive voltage is applied to the inputs. Therefore, when the output of the driving gate is high, the succeeding gates exhibit virtually no dc loading effect. When the driving gate output is low, the load gates do draw current, which must be supplied by transistor Q_1. This increased collector current tends to raise $V_{CE(sat)}$ of the driving transistor somewhat, although not enough to change the noise margin appreciably. Unlike the situation in RTL circuits, the logic levels and noise margins of DTL gates are relatively independent of fan-out.

DTL ICs are rapidly being replaced by the TTL IC families of devices.

Transistor–Transistor Logic

One of the more recent forms of logic that has gained a high degree of acceptance is the result of another modification of the familiar DTL. Known as TTL, or T²L, transistor–transistor logic is considerably faster than any other form of saturated-mode logic to date.

TTL logic achieves its principal speed advantage over DTL by replacing the diode and gate of the DTL circuit with a multiemitter input coupling transistor, to

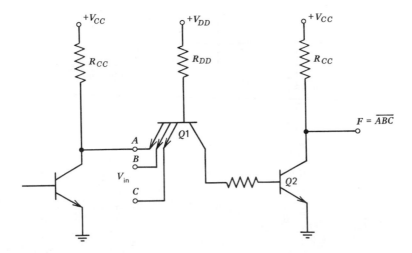

Figure 4-3   A basic TTL integrated circuit.

form the TTL circuit (Figure 4-3). The multiemitter transistor is economically fabricated in monolithic form. A single isolated collector region is diffused, a single base region is diffused and formed in the collector region, and the several emitter regions are diffused as separate areas into the base region.

The multiemitter transistor, driven by heavy base current, is always in saturation. The distribution of the stored charge in the base region of this transistor is rearranged during the switching operation, but the stored charge need not be removed from the transistor; thus the switching speed is high. One requirement of the multiemitter transistor is that its inverse current gain must be low, to ensure isolation between the inputs.

An output stage using an active pull-up transistor is added to the basic logic circuit to give current-gain drive for switching in both directions. This output configuration results in faster switching speed and higher fan-out capability. Since TTL is the predominant digital logic family in use today, an in-depth discussion of the design consideration of the output stage is presented.

TTL Output Gate Configurations[1]

There are a number of ways to build a TTL output gate without altering logic levels. One basic, popular output gate (Figure 4.4a) uses a single transistor for active pull-up. In this configuration, when the input voltage is low (under 0.4 V), Q_1 is in saturation, Q_2 and Q_3 are OFF, and Q_4 is ON, supplying current to the load. The output level is at logic "1", and $V_{out} = V_{CC} - (V_{BE4} + V_D)$.

As the input voltage rises, so does the base voltage of Q_2 and $V_{BE2} = V_{in} + V_{CE(sat)1}$. When the input voltage is 0.5 V, Q_2 turns ON, and its collector voltage drops as its emitter voltage rises. Transistor Q_1 goes from saturation to inverse saturation to

---

[1]Take a Look Inside TTL ICs, by Ury Priel, *Electronic Design*, **8**, April 15, 1971.

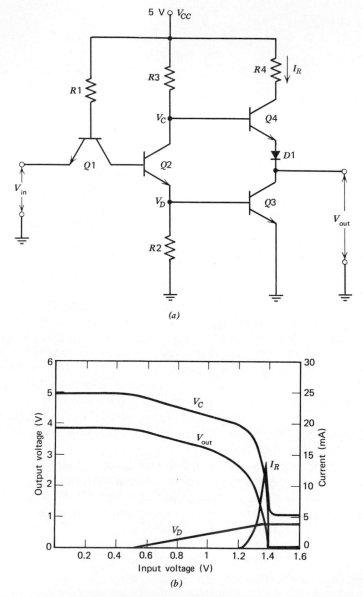

(a)

(b)

Figure 4-4   A popular TTL output gate employs an active pull-up. (*a*) A current spike I_R occurs when the gate switches from logic "0" to logic "1" because a stored charge in Q_2 delays its true turn-off. (*b*) Behavior of the output voltage V_{out} as the gate changes states. The resulting current spike is an ac/dc phenomenon. (From *Electronic Design*, **8**, April 15, 1971.)

an inverse-active mode, as Q_2 goes from OFF to active. The output voltage drops, since the emitter of Q_4 follows its base voltage. The gate is now in the transition region of its transfer curve.

The emitter voltage of Q_2 increases with rising input voltage, until both Q_3 and Q_4 turn ON. When this happens, a current spike I_R (Figure 4-4*b*) occurs. A further increase in input voltage forces Q_2 and Q_3 into saturation, turning Q_4 OFF. The output is now at logic "0", or at $V_{CE(sat)3}$. The current spike is mainly a transient phenomenon. When Q_2 turns OFF, its collector voltage rises quickly and turns Q_4 ON. The stored charge of Q_3 tends to keep it ON for a while, providing a low-impedance path for the emitter current of Q_4 and contributing to the size of the current spike.

Each component of the gate establishes certain performance boundary conditions. For example, R_1 determines the speed at which the voltage at point A will rise, thus influencing gate turn-on time; R_3 affects gate power dissipation when the output is at logic "0." Both R_1 and R_3 influence the fan-in and fan-out properties of the gate and establish input and output load currents. Resistor R_4 protects against short circuits and affects the turn-off delay time when Q_4 is charging a load capacitor. The diode D_1 ensures that Q_4 is OFF when Q_3 is saturated.

When the gate output switches to logic "1", R_2 discharges the input capacitance of Q_3. An important consideration is the ratio of R_3/R_2. Its magnitude affects the size of the current spike as the gate switches from logic "0" to logic "1." A variation of this common gate employs a Darlington configuration (Figure 4-5a) as the active pull-up. The Darlington output structure results in higher gain in the active region, thus lowering gate output impedance and increasing capacitive driving capability.

A variation of the Darlington pull-up itself ties resistor R_5 to the output node rather than to ground as indicated. This modification saves power at the expense of a larger current spike when the gate changes states.

A more recent modification of the Darlington pull-up (Figure 4-5b) includes an active turn-off, which improves circuit transient characteristics and noise immunity by improving the gate's transfer characteristic. This type of turn-off serves as a nonlinear load. Because it is high during the turn-on transient and low during the turn-off transient, it decreases both gate turn-on and turn-off times.

The TRI-STATE® gate[2] which is discussed in detail later in this chapter, is the latest development in TTL output gates (Figure 4-5c). In addition to the low-impedance logic "1" and logic "0" levels typical of TTL, this gate can be switched to a high-impedance state through its control line. In its high-impedance state, the TRI-STATE gate will not supply or sink more than 40 μA when its output voltage is between 0.4 and 2.4 V – the guaranteed output levels of standard TTL. Because of this capability, the TRI-STATE gate is very useful in bus-organized systems. It can be wired-OR like DTL or passive pull-up TTL, but it offers the advantage of active pull-up for good driving performance.

There is a basic functional difference between wired-OR DTL and the TRI-STATE tie. For the DTL gate, the result of two wired-OR functions $f1$ and $f2$ is

$$\overline{f3} = \overline{f1} + \overline{f2}$$

For the TRI-STATE tie, the result is not a Boolean function, but an ability to multiplex many functions economically on a single bus.

A more recent advance in silicon technology has produced yet another clamped circuit, the Schottky-barrier–diode clamp. Figure 4-6a gives the circuit model for

[2]TRI-STATE is a registered trademark of National Semiconductor Corporation.

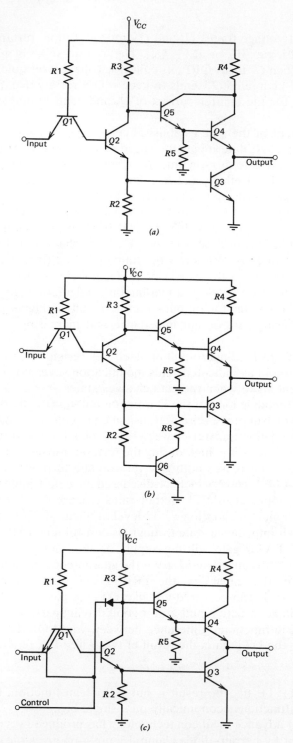

Figure 4-5 A Darlington pull-up on a TTL output gate boosts gain (*a*) to improve gate driving ability. Add an active pull-down (*b*) and turn-off time decreases. The three-state TTL gate (*c*) can be wired-OR because it has a special, controlled high-impedence state in addition to the standard TTL logic "0" and logic "1" states. (From *Electronic Design*, **8**, April 15, 1971.)

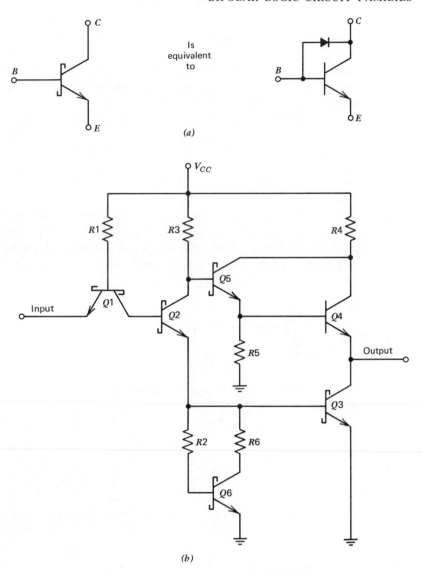

Figure 4-6 (*a*) The Schottky-clamped transistor is formed by shunting the base and collector terminals of a bipolar transistor with a Schottky-barrier diode. (*b*) The Schottky-clamped gate offers better speed performance than the standard TTL gate, but lacks temperature stability because of material properties. (From *Electronic Design*, **8**, April 15, 1971.)

a Schottky-clamped transistor. Since the forward voltage drop of a hot carrier or Schottky diode is approximately 0.3 V, the transistor cannot saturate. The Schottky diode is formed when a metal is used as the anode and the *n*-type silicon of the collector region of the transistor acts as the cathode. The interface between the metal and the semiconductor must form a nonohmic rectifying contact.

An output gate using Schottky-clamped transistors (Figure 4-6*b*) is efficient in terms of the number of components used to stop the transistors from saturating, but it is temperature dependent. As temperature rises, the Schottky clamp becomes less effective, and the transistors tend to saturate.

In addition, the noise immunity of the Schottky-clamped gate is less than that of the other clamped gates because the output logic "0" level of the Schottky-clamped gate is higher. However, the superior speeds possible with Schottky-clamped gates usually outweigh the temperature and noise immunity disadvantages they exhibit compared with other types of TTL gates.

The design of an internal TTL gate is governed by criteria different from those of an output gate. For example, an internal gate may have different logic levels as well as a lower noise immunity, since it is not exposed to external noise. It normally does not require an active pull-up, since parasitic load capacitance is quite low. Power dissipation, the number of components, and operating speed are the most important factors.

Let us look at two commonly used internal gate configurations. In the diode-clamped gate (Figure 4-7a), diode D_1 serves two purposes. When the gate turns OFF, D_1 couples R_3 to the output, thus providing a pull-up. When Q_3 is ON, D_1 acts as a clamp, keeping Q_3 out of deep saturation, since the voltage across the

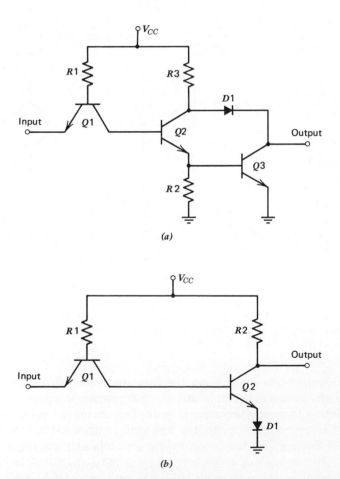

Figure 4-7   (a) A TTL internal gate can be diode-clamped to improve both turn-on and turn-off times. (b) An input-type gate with a higher-than-standard logic "0" level reduces the number of components needed to a minimum. (From *Electronic Design*, **8**, April 15, 1971.)

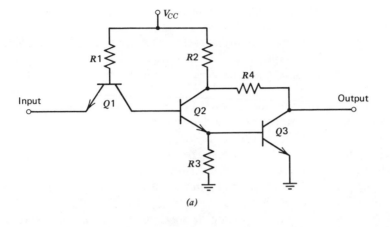

(a)

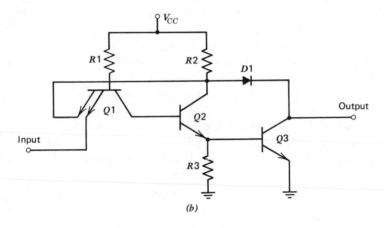

(b)

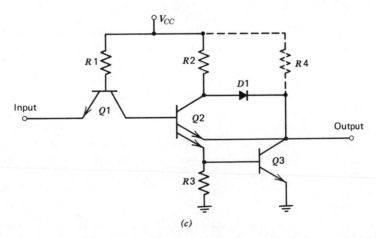

(c)

Figure 4-8 Clamping techniques enhance gate switching speed. Essentially the transistors are kept from saturating and the voltage excursions are reduced. It is possible to use (*a*) a resistor, (*b*) a phase-splitter, or (*c*) an emitter clamp. The dashed lines in (*c*) indicate an alternate pull-up technique for the emitter clamp. (From *Electronic Design*, **8**, April 15, 1971.)

diode is approximately 100 mV less than the V_{BE} of Q_3. The diode's clamping action also enhances gate turn-off delay.

The gate of Figure 4-7b is usually found in the input section of a complex IC. Its threshold is kept the same as that of a standard gate, but its logic "0" level is higher. The main advantage of this configuration is the small number of components needed.

To increase the switching speed of a TTL gate, try to keep the output transistor out of saturation, and limit the voltage excursion between logic "0" and logic "1." There are a number of designs to do this. One method uses a resistor clamp. A resistor-clamped internal gate (Figure 4-8a) uses a low-value resistor ($R_4 \approx 100 \, \Omega$) to prevent the output transistor Q_3 from saturating. The gate's logic "0" level is therefore higher than the $V_{CE(sat)}$ of Q_3. In fact, it is possible to vary the logic "0" level between $V_{CE(sat)}$ for $R_4 = \infty$ and $V_{BE} + V_{CE(sat)}$ for $R_4 = 0$.

Figure 4-8b shows a second clamping method for an internal gate, called a phase-splitter clamp. Here Q_2 is the phase-splitter transistor because its emitter and collector voltages move in opposite directions. In this way Q_2 is clamped out of saturation to $2V_{BE} - V_{\text{offset}}$ by way of the second emitter of Q_1. The collector voltage of Q_3 is also raised, keeping this transistor out of saturation as well.

Another way of clamping an internal gate's logic "0" level is illustrated in Figure 4-8c. The second emitter of Q_2 clamps the collector voltage of Q_3 to V_{BE}. Pull-up can be provided by a diode (D_1) or by a resistor (R_4), as indicated by the dashed lines.

The DTL and TTL IC configurations are presently most popular among the ICs for general digital applications. Because of the less stringent device tolerance requirements, these circuits can be readily batch-fabricated with high yield, low cost, and for relatively high-speed operation.

Popular TTL Families

Standard Series 54/74 TTL

The 54/74 series of TTL was introduced as a standard product line by Texas Instruments in 1964. Today it is the most widely used family of digital logic integrated circuits.

Standard series 54/74 ICs offer a combination of speed and power dissipation best suited to most applications. The basic standard gate circuit (Figure 4-9) features a multiple-emitter input and an active pull-up output configuration. This multiple-emitter input transistor Q_1 offers the most logic for the least physical size, and it is a major contributor to the fast switching speed of TTL. Low output impedance is attained with the active pull-up output transistor Q_3, which also results in improved noise immunity and faster switching.

Series 54/74 Low-Power TTL

Low-power circuits, designated series 54L/74L, give the best speed–power product of all logics available. The basic low-power gate circuit (Figure 4-10) has somewhat the same configuration as the standard 54/74 gate, except that the resistor values are increased and a Darlington booster was added to the TTL "totem pole" output. The increased resistance results in a reduction of power dissipation,

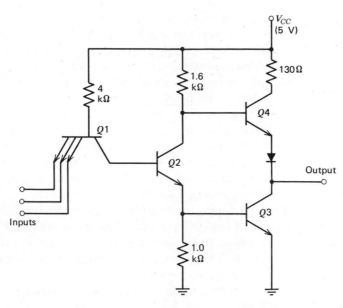

Figure 4-9   Original 54/74 TTL gate.

the power requirements of low power gates are less than one-tenth of those of the standard series TTL gates. The Darlington booster allows the output to source twice as much current as before, and it pulls up about three times faster when driving line capacitances and other loads.

Several recent innovations in low-power design (as compared with standard power design) result in 1 mW per gate power dissipation and operating speeds

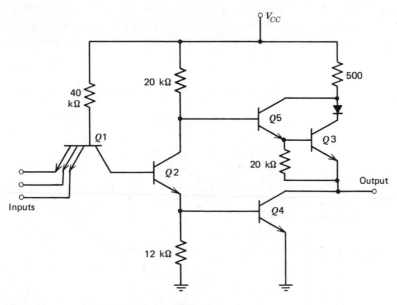

Figure 4-10   Low-power TTL gate (54L/74L).

ranging between 6 and 15 mHz, depending on the function. Aside from the Darlington booster connection, separate diffusions for transistors and resistors, shallow resistor diffusions with very high doping concentration and resistivity, and minimum junction geometries (Chapter 3) to minimize p–n junction capacitances were combined in the design of the 54L/74L series.

The extra diffusion cycle for the resistors allowed very intense doping concentrations without distorting transistor characteristics. The resistors are now made as very narrow, relatively short stripes with high sheet resistivity. Ohms-per-square values are around 260, and the squares are about 50% smaller than usual. Also, the transistor outlines were made smaller and rounded, further reducing junction areas. Since TTL transistors are normally square, more can be fitted in a given chip area; this geometry enlarges junction areas, however.

Another benefit of the high doping concentration in the resistor diffusion is that the temperature coefficient of resistivity dropped to about half that of standard TTL resistors. Input currents and other parameters dependent on resistor values become much more predictable. This factor allowed tight circuit designs and faster switching throughout the operating temperature range.

Series 54L/74L is ideal for applications in which power consumption and heat dissipation are the critical parameters.

Series 54H/74H High-Speed TTL

The circuit configuration of the high-speed gate (Figure 4-11) is basically the same as that of the standard series 54/74 gate. Resistor values are lower, however, and clamping diodes are included on each multiple-emitter input to reduce transmission-line effects that become more apparent with the faster rise and fall times.

The output section consists of a Darlington transistor pair Q_3 and Q_4 (as was the case for the 54L/74L series). This arrangement provides slightly higher speed (6 nsec/gate) than the standard gate because of transistor action and low steady-state output impedance, which is typically $10\,\Omega$ in the unsaturated state and $100\,\Omega$ in the saturated state. However, high-speed TTL circuits, series 54H/74H gates, have the disadvantage of using more power than the standard series 54/74 gates.

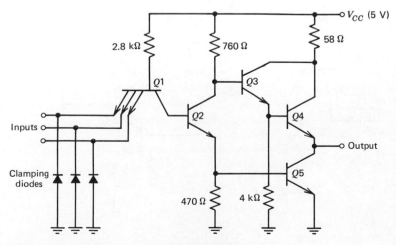

Figure 4-11   High-speed TTL gates (54H/74H).

Series 54S/74S Schottky-Clamped TTL

Series 54S/74S has the highest speed in the 54/74 line. It combines the high speed of unsaturated (emitter-coupled) logic with the relatively low power consumption of TTL. This performance is achieved by using a Schottky-barrier diode as a clamp from base to collector (as in Figure 4-12) of an *npn* transistor. A cross section of the physical implementation of Figure 4-12 appears in Figure 4-13. Since the diode has a lower forward-voltage drop than the base–collector junction, base drive current is diverted through the diode when the transistor is turned on, preventing the transistor from reaching saturation by eliminating the excess charge stored in the base region. As a result, turn-off times are dramatically reduced. For example, the basic 54S/74S gate circuit, (Figure 4-14), achieves 3-nsec propagation delay while maintaining rise and fall times in the 2 to 3 nsec range to minimize system noise considerations.

All input transistors in the basic Schottky TTL gate schematic are clamped by Schottky diodes instead of *p–n* diodes as in conventional TTL. The lower forward-voltage drop of the Schottky diodes gives greater protection against both the negative voltage transients and the positive line reflections possible in high-speed digital systems. And since the reverse input characteristic of a Schottky TTL gate more closely approximates an ideal termination, resistive terminals are usually not required for lines of impedances between 50 and 150 Ω.

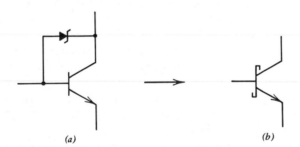

(a)   *(b)*

Figure 4-12 Schottky-clamped TTL (54S/74S): (*a*) transistor and Schottky-barrier diode clamp, (*b*) symbol for transistor with Schottky barrier diode clamp.

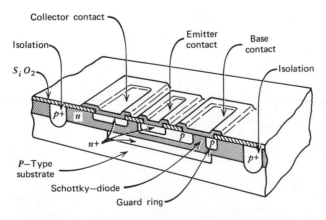

Figure 4-13 Cross-section of typical Schottky-clamp diode.

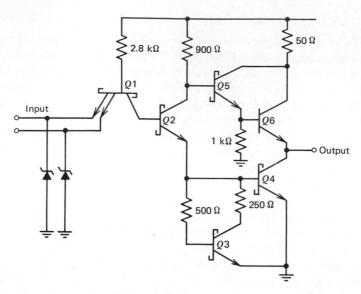

Figure 4-14   Basic 54S/74S series gate.

In addition to the Schottky diode clamps, Series 54S/74S employs other basic design changes to reduce internal circuit-propagation delays. Transistor geometries are approximately one-half the area of conventional TTL transistors. This results in minimized parasitic capacitive effects and thus reduced delays. Although Schottky diodes and advanced-design geometries are used in Schottky TTL, basic processing steps are very similar to those employed in conventional TTL.

Other circuit design techniques minimize switching overlap of the totem-pole output transistor, thus reducing current spikes—about 20% of those in Series 54H/74H high-speed TTL. Thus fewer decoupling capacitors are required with Schottky TTL. Moreover, the lower dynamic power dissipation permits increased complexity and use of smaller power supplies—Schottky TTL circuits of up to 80-gate complexity are now possible.

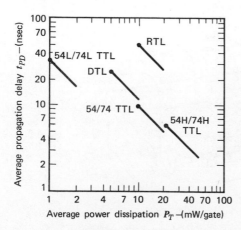

Figure 4-15   Comparison of typical saturated digital logic family speed-power products.

Table 4-1 compares the electrical characteristics of the popular TTL families: standard Series 54/74; Schottky (54S/74S); high-speed (54H/74H); low-power (54L/74L); and TRI-STATE logic (TSL). Figure 4-15 and Table 4-2 compare typical saturated digital logic family speed–power products.

Table 4-1 Comparison of TTL Families

Parameter	54/74	54S/74S	54H/74H	54L/74L	54TSL/74TSL
Logic "1" input voltage, V_{IH} (V)	2	2	2	2	2
Logic "0" input voltage, V_{IL} (V)	0.8	0.8	0.8	0.7	0.8
Logic "1" output voltage, V_{OH} at I_{OH} (V)	2.4	2.5	2.4	2.4	2.4
Logic "1" output current, I_{OH}	$-400\,\mu$A	-1 mA	$-500\,\mu$A	$-200\,\mu$A	-2mA, -5.2 mA
Logic "0" output voltage, V_{OL} at I_{OL} (V)	0.4	0.5	0.4	0.3	0.4
Logic "0" output current, I_{OL} (mA)	16	20	20	2	16
Logic "1" input current, I_{IH} (μA)	40	50	50	10	40
Logic "0" input current, I_{IL}	-1.6 mA	-2 mA	-2 mA	$-180\,\mu$A	-1.6 mA
Short-circuit output current, I_{OS} (mA)	-18 min -55 max	-40 min -100 max	-40 min -100 max	-3 min -15 max	-30 min -70 max
Logic "1" supply current, I_{CCH}	1.8 mA	4 mA	4.2 mA	200 μA	–
Logic "0" supply current, I_{CCL}	5.1 mA	9 mA	10 mA	510 μA	–
Propagation delay time from logic "1" to logic "0", t_{PHL} (nsec)	15	2 min 5 max	10	60	18 (noninvert)
Propagation delay time from logic "0" to logic "1," t_{PLH} (nsec)	22	2 min 4.5 max	10	60	23 (noninvert)
Logic "1" output leakage current, $I_{1(off)}$ (μA)	–	–	–	20	40
Logic "0" output leakage current, $I_{0(off)}$ (μA)	–	–	–	-2	-40

Table 4-2 Further Comparison of TTL Families

Circuit	Logic Type	t_{PD} (nsec)	P_T (mW)	Speed–Power Product	Typical Fan-Out
54L/74L TTL	NAND	33	1	33	10
54/74 TTL	NAND	10	10	100	10
54H/74H TTL	NAND	6	23	138	10
DTL	NAND	25	5	125	8
RTL	NOR	50	10	500	4

The Concept of TRI-STATE TTL Logic

TRI-STATE TTL is the first new concept in digital integrated circuitry in recent years. Because of the third stable state—a high-impedance condition that prevents data transfer—large numbers of three-state TTL circuits can communicate reliably with one another by way of common bus lines, at very high data rates and with excellent noise immunity.

Yet TRI-STATE TTL, or TSL for short, is fully compatible with standard TTL, can also be used with DTL, and does not require digital system designers to change their techniques of controlling bus-organized systems. These advantages have permitted TSL to grow from a novelty to an accepted logic family in a short period of time. At least three manufacturers offer TSL now, and several others are expected to make TSL circuits available soon. Some of the TSL functions being introduced could not have been achieved in practical TTL circuits.

TSL is compatible with TTL and DTL because when TSL is enabled, it has the familiar TTL logic "1" or "0" states. The third, high-impedance state is a disable state. A monolithic equivalent of a virtually open relay contact is provided at the TSL input or output by the third state. The TSL switches perform the functions of input demultiplexers or output multiplexers when groups of TSL devices are operating on a bus line. Logic controls determine which devices receive or transmit the bused data. Standard TTL logic or storage elements are fabricated between the TSL stages to complete the desired IC function.

Breaking away from the TTL mold has brought performance advantages, too. Operating speed is much higher than conventional "wire-OR" logic with passive pull-up—TSL has active pull-up. Noise immunity in the logic "1" state is about 10 times better than TTL. Unlike standard TTL, TSL can drive transmission lines bidirectionally, and TSL drives longer lines than TTL.

The first TSL circuits were introduced by National Semiconductor in 1970. National has trademarked its designs TRI-STATE, leaving TSL, three-state logic, or three-state TTL as the generic names.

TSL Development Goals

TSL was conceived as a high-performance replacement for DTL and open-collector TTL in "wire-OR" designs (it has since found other applications). The goal was a better interface between TTL or DTL and bus lines, rather than a logical wire-AND or "wire-OR." In most cases, "wire-OR" means that several outputs time-share a bus line. All but one output is strobed to prevent data interference.

A TSL output is shown on a TTL gate in Figure 4-16. When the disable input is true, the current switch at the lower left removes output drive current. The output transistors become nonconducting, the output impedance becomes very high, and at most 40 μA of leakage current may flow in the output. In this condition no definable output logic level can be detected. When the current switch is OFF, the output has low impedance and the output state is determined by the gate's state. Thus logical control of the current switch serves the same purpose as strobing "wire-OR" logic.

The Darlington-connected upper stage seen in Figure 4-16 was added to boost the output source current in the enabled logic "1" state. The higher-than-normal

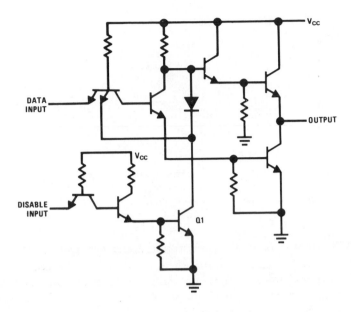

Enabled "0" State > -16 mA at 0.4V
Enabled "1" State > 5.2 mA at 2.4V
Disabled Hi-Z State < 40 µA at 0.4 or 2.4V

Figure 4-16 Typical gate with three output states.

source current is responsible for the exceptional logic "1" state noise immunity of TSL and for its superior line driving. The primary purpose, though, is to permit a driving output to source leakage current for a large number of outputs in the third state on the same bus line. Other design refinements included the use of high-current output transistors and output current limiting. These protect the outputs from being damaged by system malfunctions such as an enabled output being shorted to ground or to V_{CC}. Outputs in the third state are protected by their high impedance.

TSL Bus Connectability

National's TRI-STATE outputs generally source at least 5.2 mA in the enabled logic "1" state at 70°C (derated to 2 mA at 125°C). This is 13 times more than minimum source current, 400 µA, of a standard TTL device with a fan-out of 10. The logic "0" state sink current is 16 mA, the same as standard TTL.

TSL's higher source current capability allows it to supply leakage current into a maximum of 127 third-state outputs ($127 \times 40\ \mu A = 5.08$ mA) while driving at least three standard TTL loads. That is, up to 128 outputs may be bus-connected. Under worst-case temperature conditions of 125°C, at least 40 can be bused, as indicated in Figure 4-17.

Such worst-case leakage demands are rarely found in real designs. First, less than the maximum number may be bus-connected. Second, typical leakage in each TSL output is much less than 40 µA. Consequently, extra source current is available to increase fan-out, drive longer lines faster, or both.

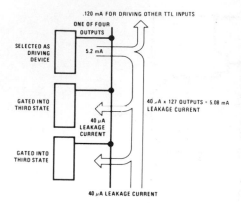

Figure 4-17 Example showing TSL's higher source current capability during worst-case output conditions at 125°C.

TRI-STATE inputs have the same enabled current specifications as TTL— 1.6 mA in the logic "0" state and 40 μA in logic "1." They need drive only the internal TTL stages of the IC. High-impedance leakage current is, again, 40 μA or less. Like TTL, any number of TSL inputs may be active on the same bus line.

TSL inputs have much higher effective fan-in than TTL, which means that bus drivers have much higher effective fan-out. Usually, in a bus-organized system, data are steered or demultiplexed into one or a few of the inputs connected to a data bus. Suppose 50 TSL inputs are bus-connected and up to 5 are simultaneously enabled at any given time. The maximum total logic "1" state current for both enabled and disabled inputs will be 2 mA. The maximum total logic "0" state current will be $(1.6 \text{ mA} \times 5) + (40 \mu\text{A} \times 45) = 9.8 \text{ mA}$. This is well within the capability of either TSL outputs or conventional drivers.

But what if there were 50 standard TTL inputs on the bus? Now the driver must source 2 mA and sink $50 \times 1.6 \text{ mA} = 80 \text{ mA}$. To avoid the excessive logic "0" current requirement, the designer must use demultiplexers, buffers in bus branches, and so on, which complicate the design, add delays to the bus structure, and increase costs.

CURRENT–MODE LOGIC

All the logic forms described so far (with the exception of Schottky-clamped TTL) are saturated-mode logic. As such, they are subject to a delay time which is due to the storage time of a transistor as it is driven into saturation.

One form of logic has been developed that prevents transistor saturation, thereby eliminating storage time as a speed-limiting factor. Called "current-mode" logic, these circuits set the high- and low-voltage states of the transistor at well-defined fixed levels other than zero and $V_{CE(\text{sat})}$. The voltage swing, therefore, is somewhat smaller than for saturated-mode logic, but the circuit speed is considerably improved.

The differential amplifier configuration of the typical current-mode logic circuit of Figure 4-18, employing a pair of matched transistors, provides a sharp and stable switching threshold and a well-defined nonsaturating conduction state of the transistor dictated by the constant-emitter current. The static operation of the

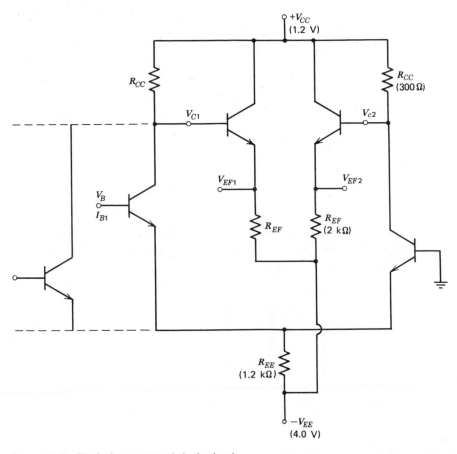

Figure 4-18   Typical current-mode logic circuit.

basic circuit is illustrated in Figure 4-19. The output voltage level of the non-saturating circuit is not compatible with the input level, thus some form of level-translation is required. As shown in Figures 4-18 and 4-20, the level translation is supplied by the emitter followers, which also provide currents for large fan-out. Another noteworthy property of the circuit is the current-limiting feature of the input characteristic (Figure 4-20), which allows parallel fan-out to similar circuits without the uneven base current difficulty. The circuit may have a logic delay of 1 or 2 nsec when operating with a signal voltage excursion of a fraction of a volt and relatively small noise margin. The relatively high-power dissipation of the circuit (usually about 50 mW) is a consideration of high-density IC design.

Emitter-Coupled Logic (ECL)

TTL and MOS circuits may be cheaper than emitter-coupled logic (ECL)—but they are also slower. TTL that incorporates Schottky-barrier diodes may be almost as fast as ECL—but its processing at a given performance level is also more complex. ECL is the most popular form of current-mode logic. It utilizes a pair of input transistors; one is in a conductive state, and the other is nonconduc-

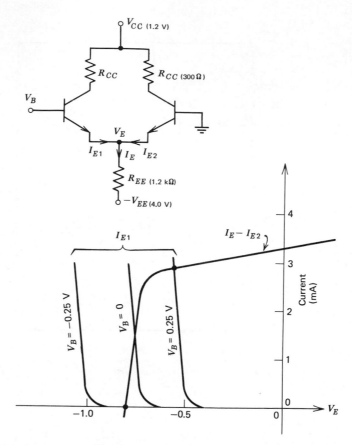

Figure 4-19   Transfer curves illustrating the static operation of the basic current-mode logic circuit.

tive. Switching is accomplished by means of a signal appearing across a common emitter resistor. Hence current-mode logic is often referred to as emitter-coupled logic.

The use of ECL, however, involves certain design trade-offs. Achieving maximum performance often requires the most advanced semiconductor fabrication processes, and this means lower yields. Packaging concepts, thermal considerations, and other user requirements must also be designed to match, or at least not to undercut, the advantages of the circuit family.

In the ECL circuit of Figure 4-21, the three input signals, A, B, C, can be either at -0.8 or at -1.7 V. Only when all three are at -1.7 V can the reference transistor Q4 conduct—at which time its collector voltage drops to about -3.3 V, from near ground. The emitter-follower stage $Q6$ returns these levels to those of the inputs, permitting the circuit to drive another like it.

At the same time, the collectors of the input transistors, electrically common, are complementary to the reference transistor's collector. Another emitter follower provides normal signal levels from this point. Thus the circuit has two outputs that are always complementary except at the moment of transition. At no time is any conducting transistor saturated—which accounts for ECL's speed capability.

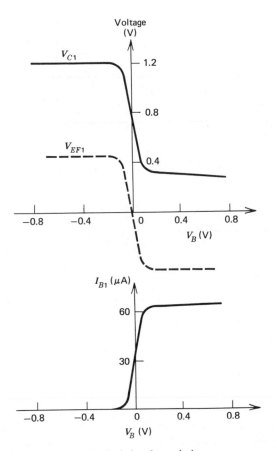

Figure 4-20   Curves depicting current-mode logic level translation.

With the convention that -0.8 V represents a logic "0" and -1.7 V a logic "1," this circuit performs the AND/NAND function—AND from the output at Q_6 and NAND, or NOT-AND from Q_5. The OR function is obtained by connecting the emitter-follower inputs to similar points in another circuit.

No inverters are needed when ECL circuits are used because the logic block that produces any given function also produces its complement. And both emitter-follower outputs can be designed to drive as many similar circuits as desired, reducing the need for special power and line drivers that are often called for by other logic families.

ECL's basic speed advantage is achieved because its circuit configuration avoids saturation; therefore such "process steps" as gold diffusion and Schottky-barrier junctions, which prevent saturation in TTL circuits, are not required. Instead, better performance must be achieved through more sophisticated methods such as reduced component sizes, shallow diffusions, and an optimum choice and placement of diffusion resistivities.

Typical dimensions of ordinary ECL, which switches in 2 to 3 nsec, are smaller than those of circuits like TTL (Figure 4-22). The dimensions of subnanosecond ECL are smaller still. Repeatedly reducing these dimensions, even without

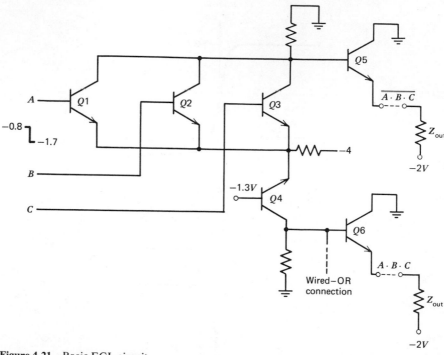

Figure 4-21   Basic ECL circuit.

changing doping concentrations, tends to decrease the capacitance associated with the semiconductor junctions, which in turn decreases the time constant of the transistor and its switching time – up to the point at which photographic technology and the properties of the material make further size reductions impractical.

Beyond this point, the capacitance can be decreased by using a more lightly doped p-type base material and a more lightly doped n-type material for the epitaxial film in which the collector is formed. The base resistance is reduced by adding a p^+ base contact, and the saturation resistance is kept low by an n^+ collector contact that is diffused deep into the n^+ buried layer.

This deep collector diffusion also prevents saturation in a current-source transistor, which sometimes replaces the resistor connected to -4 V in the circuit diagrammed in Figure 4-21. It is also valuable in circuits that are to be used in a high-temperature environment. Most significantly, perhaps, it permits the epitaxial film to be made with a higher resistivity; as a result, although total resistance increases to a degree, the collector–base capacitance decreases to a still greater degree, and overall the time constant also decreases.

Such small dimensions and such refined diffusions, however, entail reduced yield. The small dimensions make photographic alignment, etching, and alloy techniques more difficult and surface cleanliness more critical. Another effect of reduced sizes is metal migration, induced by higher current densities. Near the emitter contact, for example, nearly 10^6 A/cm^2 is quite likely. At densities like this, the electrons in the interconnection metallization are so active that they

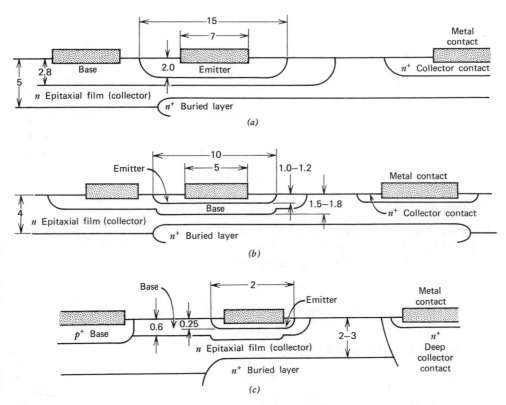

Figure 4-22 Technology comparison (all dimensions in microns): (a) typical saturated logic, (b) typical 2 to 3 nsec ECL, (c) subnanosecond ECL technology. Even the simplest ECL circuits require shallower diffusions and narrower depositions [as in (b)] than do such conventional saturated circuits as (a), which is TTL. Subnanosecond circuits (c) involve even finer control.

dislodge the metallic atoms one by one until the entire metallization path actually slides along the surface of the chip in the direction of electron flow—sometimes far enough to create an open circuit. (There are metallization processes capable of withstanding the high current densities without suffering from metal migration, but they are complicated and expensive.)

The high-speed aspect of ECL tends to create significant voltage and current transients. To overcome them, the designer must lay out his PC boards to include microstrip transmission lines, and he must make sure the circuits are properly grounded–although sometimes he may try to live with the transients or try to control them with simple diode clamps.

The following conclusions can be made regarding ECL logic:

1. When an ECL gate is driven by another ECL gate, it provides two output signals: one identical to the input, the other equal to the inverted input.
2. The fixed logic levels are -0.80 and 1.70 V, resulting in a difference of 900 mV between the ON state and the OFF state.
3. The transistors are never driven into saturation; therefore, storage time is eliminated.
4. The noise margin is approximately 250 mV.

5. The total current flow in the circuit remains relatively constant, thereby maintaining an unvarying drain on the power supply and preventing internally generated noise spikes.

The circuit furthermore has a very high input impedance and a very low output impedance, which makes high fan-out possible. Thus ECL is not only the fastest logic available, it provides other features that make it highly desirable for advanced systems. Its principal detriment is its limited (900-mV) logic swing, which renders it somewhat more sensitive to externally generated noise than some other logic forms.

Comparison of ECL with other Logic Families

Table 4-3 presents a comparative summary of the salient features of RTL, DTL, TTL, and ECL. A comparison of the typical electrical characteristics of the major IC digital logic families is given in Table 4-4.

DIGITAL INTEGRATED CIRCUIT NOMENCLATURE

V_{IH}, ($V_{in(1)}$) High-Level Input Voltage

An input voltage level within the more positive (less negative) of the two ranges of values used to represent the binary variables. A minimum value is specified which is the least-positive (most-negative) value of high-level input voltage for which operation of the logic element within specification limits is guaranteed.

V_{IL}, ($V_{in(0)}$) Low-Level Input Voltage

An input voltage level within the less positive (more negative) of the two ranges of values used to represent the binary variables. A maximum value is specified which is the most-positive (least-negative) value of low-level input voltage for which operation of the logic element within specification limits is guaranteed.

V_{OH}, ($V_{out(1)}$) High-Level Output Voltage

The voltage at an output terminal for a specified output current $I_{OH}(I_{load})$ with input conditions applied which, according to the product specification, will establish a low level at the output.

V_{OL}, ($V_{out(0)}$) Low-Level Output Voltage

The voltage at an output terminal for a specified output current $I_{OL}(I_{sink})$ with input conditions applied which, according to the product specification, will establish a low level at the output.

I_{IH}, ($I_{in(1)}$) **High-Level Input Current**

The current flowing into[4] an input when a specified high-level voltage is applied to that input.

I_{IL}, ($I_{in(0)}$) **Low-Level Input Current**

The current flowing into an input when a specified low-level voltage is applied to that input.

I_{OH}, ($I_{out(1)}$) **High-Level Output Current**

The current flowing into the output with a specified high-level output voltage $V_{OH}(V_{out(1)})$ applied.[5]

$I_{O(off)}$, (I_{off}) **Off-State Output Current**

The current flowing into an output with a specified output voltage applied and input conditions applied which, according to the product specification, will cause the output switching element to be in the off state.[6]

I_{OS}, **Short-Circuit Output Current**

The current which flows into an output when that output is short-circuited to ground (or other specified potential) with input conditions applied to establish the output logic level farthest from ground potential (or other specified potential).

I_{CCH}, ($I_{CC(1)}$) **Supply Current, High-Level Output**

The current flowing into the indicated supply terminal of an IC when the output is (or all outputs are) at a high-level voltage.

I_{CCL}, ($I_{CC(0)}$) **Supply Current, Low-Level Output**

The current flowing into the indicated supply terminal of an IC when the output is (or all outputs are) at a low-level voltage.

Fan-In and Fan-Out

Multiple inputs and outputs are obviously necessary for information manipulation beyond simple transmission. The maximum number of unit loads that may be simultaneously driven under worst-case conditions is termed the fan-out of a

[4]Current flowing out of a terminal is a negative value. This applies to the six following definitions, as well.
[5]This parameter is usually specified for outputs intended to drive devices other than logic circuits.
[6]See note 5.

Table 4-3 Comparison of Bipolar Logic Types

Saturated-Mode Logic			Current-Mode Logic
RTL 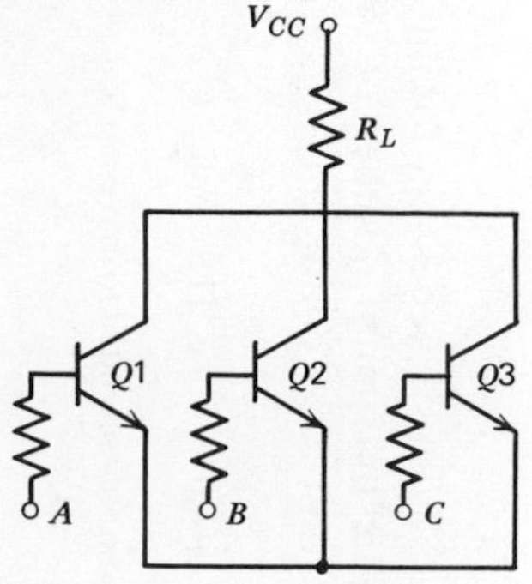	DTL 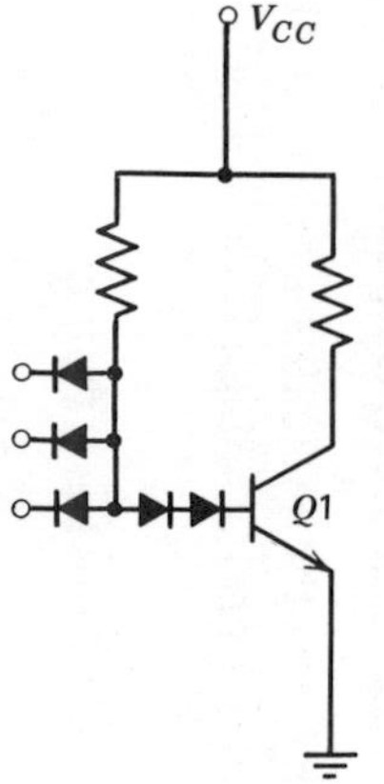	TTL 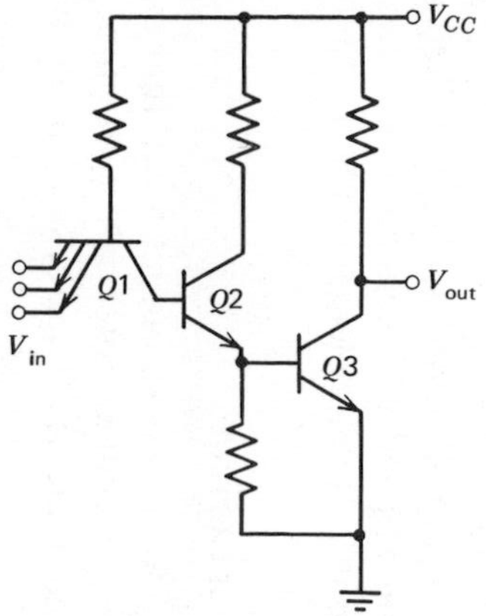	ECL 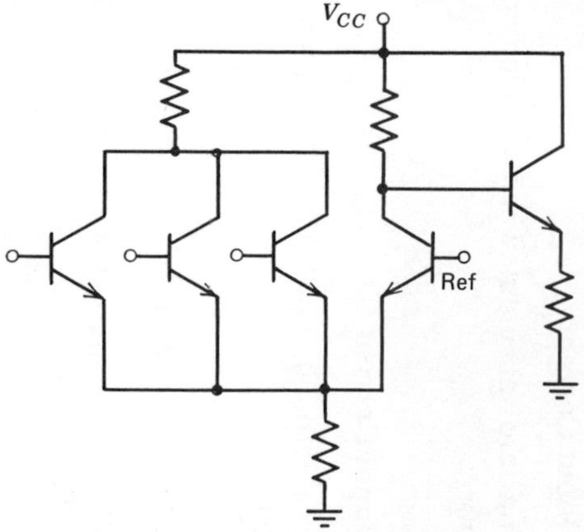

- Oldest circuit
- Good past history
- Simplest circuit
- High-yield production, low-cost circuits
- Facilitates implementation of multifunction designs

- More complex circuits
- Somewhat higher cost circuits
- Faster speed than RTL
- Logic levels and noise immunity independent of fan-out

- More complex circuits
- Somewhat higher cost circuits
- Fastest saturated-mode logic
- Requires gold-diffusion and Schottky-barrier junctions to prevent

- Avoids transistor saturation, thus no storage time
- Reduced turn-off time
- Shallower diffusions and smaller components than TTL
- Costly circuits
- Photographic alignment, etching, alloy-

- Limited fan-out
- Noise margin becomes small as fan-out is increased
- Minimal internal noise
- Approximate noise immunity: 300 mV
- Propagation delay time: 10 to 50 nsec
- Wide range of available functions
- Few manufacturing sources
- Good speed–power product

- Higher fan-out
- Improved noise margin
- Minimal internal noise
- Approximate noise immunity: 1.0 V (less immune to noise)
- Propagation delay time: 10 to 30 nsec
- Low-power dissipation
- TTL compatible
- Wide range of available functions
- Low cost

- saturation
- High fan-out
- Noise spikes generated internally
- Approximate noise immunity: 1.0 V
- Propagation delay time: 6 to 20 nsec
- Inverters needed when using TTL circuits
- Low cost

- ing, and cleanliness extremely important
- Fastest speed logic
- Limited 900-mV logic swing
- Very high fan-out due to high Z_{in} and low Z_{out}
- High speed creates voltage and current transients
- Zero internally generated noise due to constant current
- Approximate noise immunity: 150 to 300 mV (more sensitive)
- Propagation delay time, 5 nsec
- Inverters not required when using ECL circuit
- Excellent speed–power product
- High power dissipation

101

Table 4-4 Comparison Chart of Major Integrated Circuit Logic Families

Logic Families

Parameters	RTL	Low-Power RTL	DTL	HTL[a]	12-nsec TTL	6-nsec TTL	4-nsec ECL	2-nsec ECL	1-nsec ECL
Positive logic function of basic gate	←—— NOR ——X			←—————— NAND ——————→			X	←——— OR/NOR ———→	
Wired positive logic function	←— Implied AND (some functions)		←— Implied AND	Implied AND ←— AND-OR-INVERT (A-O-I) ——→			←— Implied OR (all functions)		
Typical high level Z_{out} (Ω)	640	3600	6000 or 2000	15,000 or 1500	79	10	15	6	6
Typical low level Z_{out}	R_{sat}	R_{sat}	R_{sat}	R_{sat}	R_{sat}	R_{sat}	15 Ω or 2.7 mA	6 Ω or 6.7 mA	6 Ω or 21 mA
Fan-out	5	4	8	10	10	10	25	25 inputs or 50 Ω	10 low Z inputs or 50 Ω
Supply voltage (V)	3.0 ± 10% 3.6 ± 10%	3.0 ± 10% 3.6 ± 10%	5.0 ± 10%	15 ± 1	5.0 ± 10% 5.0 ± 5%	5.0 ± 10% 5.0 ± 5%	−5.2 + 20% −10%	−5.2 + 20% −10%	−5.2 ± 10%
Typical power dissipation per gate (mW)	12	2.5	8 or 12	55	12	22	40	55, plus load	55, plus load
Imunity to external noise	Nominal	Fair	Good	Excellent	Very good	Very good	Good	Good	Good
Noise generation	Medium	Low–medium	Medium	Medium	Medium–high	High	Low	Low–medium	Medium
Propagation delay per gate (nsec)	12	27	30	90	12	6	4	2	1
Typical clock rate for flip-flops (MHz)	8	2.5	12–30	4	15–30	30–60	60–120	200	400

[a] Diode–zener transistor.

102

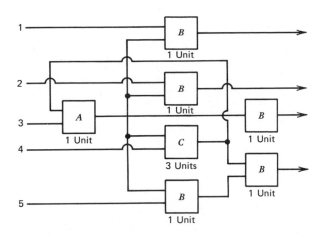

Figure 4-23   Example illustrating fan-out.

package. Packages may be operated into less than the maximum number of loads successfully, but if the fan-out limit is exceeded, the gate will fail under certain (worst-case) conditions.

Example

Every box in the system in Figure 4-23 has been tested and found to be reliable over a temperature range of −40 to +85°C. Yet the system itself is unreliable below −25°C. Why? Assume that boxes A and B have a fan-out rating of 5, and that box C has a fan-out rating of 3. Furthermore, assume that each A or B box input has a one-unit load, and that each C box has an input load one unit greater than the load "seen" by its output. Assume the logic to be correct.

A first glance, no box seems to drive more than five load units. But a careful examination of box C characteristics indicates that box A sees three units looking into the box C − plus four more from the other boxes − or a total of seven. Box A is overloaded. Apparently the fan-out of that particular type of gate rises with temperature; this indicates that the gate works successfully at all but the lowest temperatures, in spite of being overloaded. The concept of fan-out is advantageous for pyramiding logic blocks.

Rise Time and Fall Time

Some computer circuits are activated by logic levels, but others are activated by the rate of change in levels. It is essential that the input signals of the latter circuit type have a sufficiently rapid rate of change; otherwise the circuit will not respond. Rise and fall times are often used as measures of this rate of change.

Referring to Figure 4-24, rise time t_R is the time required for the signal to rise from 10% to 90% of its full amplitude. The time required for the downward passage from the 90% to the 10% point is the fall time t_F. The values of t_R and t_F are not necessarily equal; in transistor circuits having fast rise and fall times, the type of transistors used (i.e., *pnp* or *npn*, will cause t_R and t_F to differ significantly. Usually, t_R and t_F are functions of the amount of loading placed on a gate.

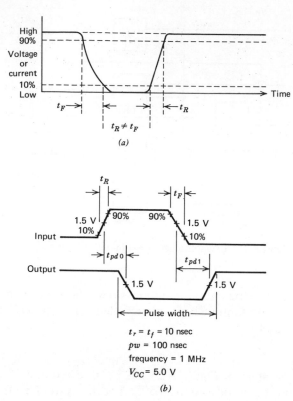

Figure 4-24   (*a*) Digital logic switching waveform illustrating rise and fall times, (*b*) switching-time waveform.

Propagation-Delay Time

The propagation-delay time is a function of the rise and fall times of the gates, therefore it varies with the output load as well as with the loading of the previous stage (driving signal). Propagation delay time is defined as

t_{PD1} = propagation delay from logic "0" state to logic "1" state under loaded conditions

t_{PD0} = propagation delay from logic "1" state to logic "0" state under loaded conditions

Where logic "1" and logic "0" refer to the output states.

Propagation delays for negative- and positive-going outputs are not necessarily the same.

Figure 4-24 depicts a typical switching-time waveform which graphically demonstrates the terms rise time t_R, fall time t_F, and propagation delay times t_{PD0} and t_{PD1}.

Chapter Five

APPLICATIONS OF TTL
AND TRI–STATE MEDIUM–SCALE INTEGRATED
CIRCUITS

ECONOMIC CONSIDERATIONS OF MSI

The economic considerations surrounding the incorporation of MSI circuits in new system designs are considerably more complex and subtle than those considerations facing the IC manufacturer. The systems house must weigh three major areas in its decision to use MSI. The cost of an MSI solution as opposed to a multiple-package solution is the most complex matter the user has to deal with, and the results vary widely from one user to another. Availability of production volumes of an MSI circuit is the user's most important consideration in committing specific pieces of equipment to definite production schedules. An intangible to be born in mind is the possible enhancement of the sales appeal gained by a product incorporating advanced devices.

The cost of the MSI solution versus a multiple-package solution for a given subsystem must be considered in great detail. A check list would include the cost of the piece parts, which for a new MSI circuit is generally higher than for the multiple-package solution. If the MSI piece-part cost is higher than the multiple-package equivalent, other cost considerations may rule. The cost of inserting each package in the PC board can only be determined by the individual user. The actual number of PC boards required can be determined only for each piece of equipment individually. The number and cost of interconnects on a PC board must be considered on a board-by-board basis. The complexity of board layout and the requirement to use high-frequency layout and decoupling techniques to optimize

Figure 5-1   Series 54/74 TTL gate circuit.

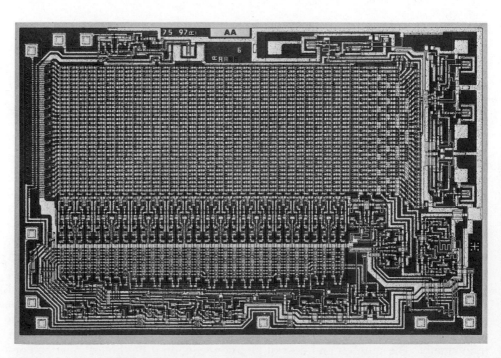

Figure 5-2   1024-Bit TTL ROM employing many of the gates shown in Figure 5-1.

MSI operation add cost that may be eliminated if a slower, noncurrent-spiking logic form such as DTL is used for slower subsystems.

In order to give the reader an appreciation of the degree of complexity of an MSI circuit, the comparison of Figures 5-1 and 5-2 is presented. Figure 5-1 shows a series 54/74 TTL gate circuit; the IC in Figure 5-2 is a TTL 1024-bit ROM circuit. This particular MSI device contains the equivalent of 85 gates and 1024 diodes. Much of the memory-control circuitry is on the chip. There are many gates, similar to those of Figure 5-1, as well as the storage elements, in the memory IC.

With the advent of MSI or complex functions, the cost of the basic gate has been steadily decreasing—so much so that other parts of the system become proportionately more important in terms of total system cost. These costs may be dropping, too, but the parts (which include complex functions, PC boards, hardware, packaging, and power supplies) represent a higher percentage of system costs than they did in the past.

It was indicated in Table 1-3 that although complex ICs are only 25% of the system IC units, they represent 50% of the IC costs, or almost 10% of the total system costs in the typical system design. These are the very circuits that the engineer must choose early in the design cycle, since they have the more specialized functions. And to a very large extent, as a result, they may govern total system cost. Some of these complex circuits may be useful only in one particular system design.

To have a significant impact on total system costs, we must now reduce costs of such items as power supplies, and PC boards by making an optimum choice of ICs.

MSI APPLICATIONS

Counters

A counter is an arrangement of an arbitrary number of flip-flops used to generate an output sequence of binary numbers, directly related to the number of activating (clock) pulses observed at the clock input. Counters are found in almost every kind of digital equipment, and they serve not only for counting but for equipment operation sequencing, frequency division, and mathematical manipulation as well.

Binary Counter

A binary counter is an arrangement of n binaries which provide 2^n separate output codes, each consisting of n binary digits. Each output code is directly related to the number of input pulses observed at the clock terminal. The term "binary counter" is generally intended to mean that the output sequence is comprised of a set of binary digits which are the familiarly weighted binary code (Figure 5-3).

Asynchronous Counter

An asynchronous counter is an arrangement of binaries in which the clock does not activate all the binaries simultaneously with the first one in the counter.

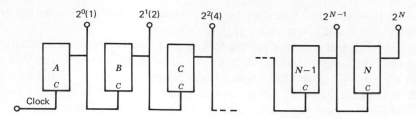

Figure 5-3   A binary counter.

Ripple Counters

The ripple counter is a basic counter commonly implemented with ICs. Of all counters it is the simplest in logic and therefore the easiest to design. In a ripple counter each binary (following the first) is activated, or clocked, by the preceding binary. Thus the nth binary is activated by the output transition of the binary occupying position $n-1$.

The advantages of a ripple counter are that no gating is required between binaries and that lower-power (slower) binaries can be used for each successive stage. The nth binary need only be capable of running at one-half the frequency of the $n-1$ binary, and so on. The ripple counter is limited, however, in its speed of operation. Since the flip-flops in the ripple counter are not under command of a single clock pulse, it is an asynchronous counter.

Figure 5-4 shows a simple 4-bit binary ripple counter. Initially all flip-flops are in the logic "0" state ($Q_A = Q_B = Q_C = Q_D = 0$). A clock pulse is applied to the clock input of flip-flop A causing Q_A to change from logic "0" to logic "1." Flip-flop B does not change state, since it is triggered by the negative-going edge of the clock pulse (i.e., by its clock input changing from logic "1" to logic "0"). With the

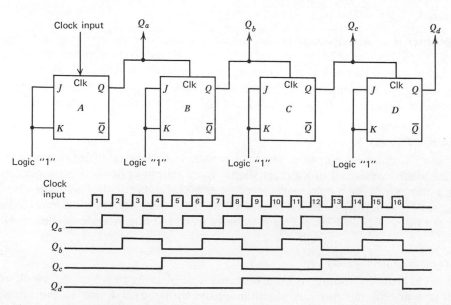

Figure 5-4   A 4-bit binary ripple counter.

arrival of the second clock pulse to flip-flop A, Q_A goes from logic "1" to logic "0". This change of state creates the negative-going pulse edge needed to trigger flip-flop B, and thus Q_B goes from logic "0" to logic "1". Before arrival of the sixteenth clock pulse, all flip-flops are in the logic "1" state. Clock pulse 16 causes Q_A, Q_B, Q_C, and Q_D to go to logic "0" in turn.

In decoding the states of a ripple counter, spikes occur at decode matrix outputs as counter flip-flops change state. The propagation delay of the flip-flops creates these false states for only a short time. Decode spikes are possible in any counter unless all flip-flops change state at exactly the same time, or only one flip-flop changes state for any clock pulse. To eliminate the spikes at decode matrix outputs, a strobe pulse is used (see Figure 5-6). The strobe pulse allows decoding to occur only after all flip-flops in the counter have become stable.

A 4-bit binary counter repeats itself for every 2^n (n = number of flip-flops) clock pulses. This counter sequences in a number system of radix 16 and has 16 discrete states from 0 to $n-1$. The 16 binary states appear in Figure 5-5.

In applications where the binary states of the counter must be converted into discrete outputs, a decoding network is provided. In Figure 5-6 a 3-bit binary counter is presented with decoding of its eight possible states from 0 to 7.

A 4-stage counter can be used to divide by 16 (2^n, n = number of flip-flops). Additional stages can be added if division by a higher power of 2 is required. For division by any integer the following method is used:

1. Find the number n of flip-flops required:

$$2^{n-1} \leq N \leq 2^n$$

where N = counter cycle length. If N is not a power of 2, use the next higher power of 2.

2. Connect all flip-flops as a ripple counter (see Figure 5-7).
3. Find the binary number $N - 1$.

State	Q_D	Q_C	Q_B	Q_A
0	0	0	0	0
1	0	0	0	1
2	0	0	1	0
3	0	0	1	1
4	0	1	0	0
5	0	1	0	1
6	0	1	1	0
7	0	1	1	1
8	1	0	0	0
9	1	0	0	1
10	1	0	1	0
11	1	0	1	1
12	1	1	0	0
13	1	1	0	1
14	1	1	1	0
15	1	1	1	1
0	0	0	0	0

Figure 5-5   State table, 4-bit ripple counter.

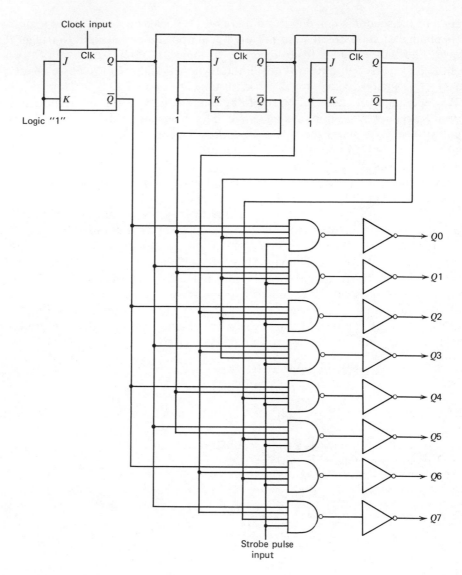

Figure 5-6   A 3-bit ripple counter with decoded outputs.

4. Connect all flip-flop outputs that are 1 at the count $N-1$ as inputs to a NAND gate. Also feed the clock pulse to the NAND gate.

5. Connect the NAND gate output to the preset inputs of all flip-flops for which $Q=0$ at the count $N-1$.

The counter resets in the following manner. At the positive-going edge of the Nth clock pulse all flip-flops are preset to the logic "1" state. On the trailing edge of the same clock pulse, all flip-flops count to the logic "0" state (i.e., the counter recycles).

In many IC counter packages, the preset lines in Figure 5-7 do not exist; only a common clear (or reset) line is available. Figure 5-8 shows a divide-by-12 counter

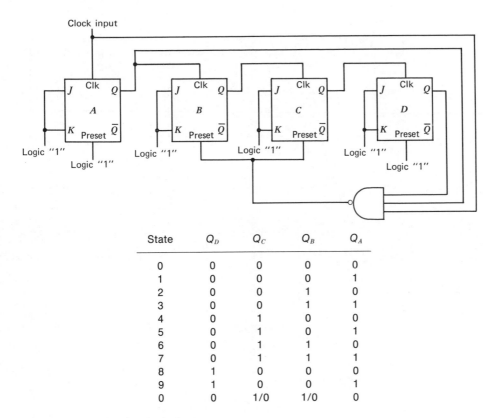

State	Q_D	Q_C	Q_B	Q_A
0	0	0	0	0
1	0	0	0	1
2	0	0	1	0
3	0	0	1	1
4	0	1	0	0
5	0	1	0	1
6	0	1	1	0
7	0	1	1	1
8	1	0	0	0
9	1	0	0	1
0	0	1/0	1/0	0

Figure 5-7   A BCD decade ripple counter.

using a common reset line. Although this is the simplest method for resetting ripple counters, it is not considered the most reliable.

Synchronous Counter

A synchronous counter is an arrangement of binaries and gates in which all binaries are activated (clocked) simultaneously. The advantage is that a synchronous counter does not traverse false states when producing its output sequence. The disadvantage is that maximum gating between binaries is required to produce a binary output sequence.

The synchronous counter eliminates the cumulative flip-flop delays seen in ripple counters. All flip-flops in a synchronous counter are under control of the same clock pulse. The repetition rate is limited only by the delay of any one flip-flop plus delays introduced by control gating. Design of synchronous counters for any number base other than a power of 2 is more difficult than design of a ripple counter, but the design is simplified through the use of the Karnaugh mapping technique.

Figure 5-9 shows a 4-bit synchronous counter with parallel carry. Parallel carry, also known as carry look-ahead, is the faster of the two methods of flip-flop control. According to the state table, flip-flop A is required to change state with the occurrence of each clock pulse. Flip-flop B changes state when $Q_A = $ logic "1";

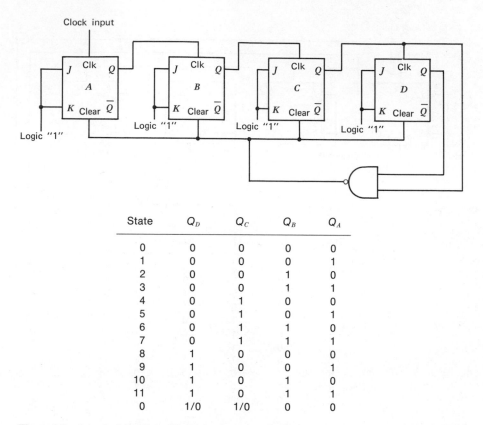

State	Q_D	Q_C	Q_B	Q_A
0	0	0	0	0
1	0	0	0	1
2	0	0	1	0
3	0	0	1	1
4	0	1	0	0
5	0	1	0	1
6	0	1	1	0
7	0	1	1	1
8	1	0	0	0
9	1	0	0	1
10	1	0	1	0
11	1	0	1	1
0	1/0	1/0	0	0

Figure 5-8   A typical divide-by-12 ripple counter.

C changes state when $Q_A = Q_B =$ logic "1"; and D changes state when $Q_A = Q_B = Q_C =$ logic "1." Control of flip-flop A can be accomplished by tieing J_A and K_A to logic "1." Control of flip-flop B is achieved by connecting J_B and K_B to Q_A. Control of flip-flop C can be achieved with the inverted output of a two-input NAND gate whose inputs are Q_A and Q_B. Flip-flop D is controlled as C is, except that the NAND gate inputs are now Q_A, Q_B, and Q_C.

As the number of stages in a synchronous counter with parallel carry increases, the flip-flops must drive an ever-increasing number of NAND gates. Similarly, the number of inputs per control gate increases. Ripple carry (Figure 5-10) eliminates these difficulties, but the clock speed of the counter is reduced. Reduction of clock speed stems from the longer delay through control logic.

BCD Counter

Each flip-flop in a counter like the one in Figure 5-7 has a specific decimal weight assigned to it. Flip-flop A has a weight of 2^0 (or 1) when its output is a logic "1." Flip-flop B has a weight of 2^1 (or 2), C has a weight of 2^2 (or 4), and D has a weight of 2^3 (or 8). The number stored in the counter at any specific time can be determined by summing decimal weights of flip-flops in the logic "1" state. A counter that counts in a standard binary manner and recycles for every 10 clock pulses is called an 8-4-2-1 BCD (binary-coded decimal) counter.

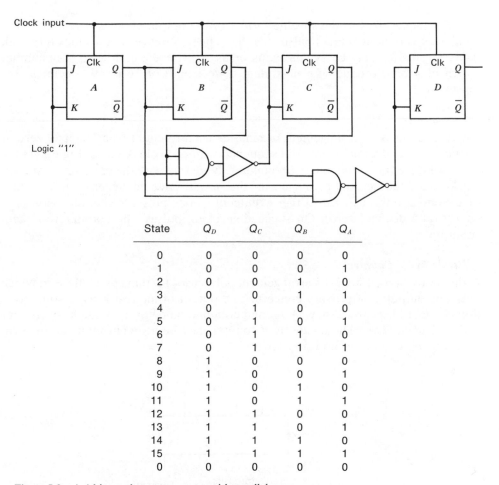

State	Q_D	Q_C	Q_B	Q_A
0	0	0	0	0
1	0	0	0	1
2	0	0	1	0
3	0	0	1	1
4	0	1	0	0
5	0	1	0	1
6	0	1	1	0
7	0	1	1	1
8	1	0	0	0
9	1	0	0	1
10	1	0	1	0
11	1	0	1	1
12	1	1	0	0
13	1	1	0	1
14	1	1	1	0
15	1	1	1	1
0	0	0	0	0

Figure 5-9 A 4-bit synchronous counter with parallel carry.

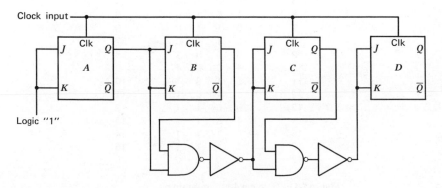

Figure 5-10 A 4-bit synchronous counter with ripple carry.

113

BCD, then, refers to the specific format of the output code as it relates to the number of activating (clock) pulses that have been observed at the clock terminal. A BCD counter may be synchronous or asynchronous, and it has a maximum length of 4 bits, producing a maximum code length of 16 terms (decimal zero through 15).

Up-Counter

An up-counter is an arrangement of binaries used to count from decimal zero to the next higher decimal number in sequence with each clock pulse. Figure 5-11 is the truth table for a 4 bit binary up-counter. Note that with the occurrence of each clock pulse, the decimal equivalent of the output code advances an integer at a time from zero through 15. At the seventeenth clock pulse, the counter returns to binary (and decimal zero). On succeeding clock pulses, the counter will again count-up.

Divide-by-'n' Counter

A divide-by-'n' counter is an arrangement of binaries used in applications in which only one output is of primary concern. The output of interest runs at some fraction of input clock frequency. A second common name for a divide-by-n counter is *frequency divider*. The divide-by-n counter may be a synchronous or an asynchronous up-counter of arbitrary length.

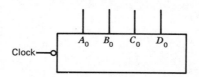

Number of Clock Pulses	Output Bits				Decimal Equivalent
	2^0	2^1	2^2	2^3	
1	0	0	0	0	0
2	1	0	0	0	1
3	0	1	0	0	2
4	1	1	0	0	3
5	0	0	1	0	4
6	1	0	1	0	5
7	0	1	1	0	6
8	1	1	1	0	7
9	0	0	0	1	8
10	1	0	0	1	9
11	0	1	0	1	10
12	1	1	0	1	11
13	0	0	1	1	12
14	1	0	1	1	13
15	0	1	1	1	14
16	1	1	1	1	15
17	0	0	0	0	0

Figure 5-11   Typical 4-bit up-counter.

Decade Counter

A decade counter is an arrangement of binaries specifically designed to provide an output sequence code representing the decimal numbers zero through 9. The decade counter may be synchronous or asynchronous, and the number of binaries may vary depending on the logic implementation. For example, decade counters commonly used incorporate 4, 5, and 10 binary elements. The most common MSI form of decade counter is derived from a 4 bit binary counter ($2^4 = 16$ total coded output states). The binary equivalent (1001) of decimal "9" is internally detected with logic gating and steered to the appropriate binaries, forcing the counter to binary (and decimal) zero at the next clock pulse. Figure 5-12 shows one logic implementation of a decade counter.

Down-Counter

A down-counter is an arrangement of binaries with appropriate gating whose output code runs from the most significant number to the least significant number, in sequence. A BCD decade down-counter and its truth table are set forth in Figure 5-13. A down-counter may be synchronous or asynchronous and of arbitrary length. If a counter is not specifically described as an "up" or a "down" counter, it is generally considered to be an up-counter.

Bidirectional Counter

A bidirectional or up–down counter (Figures 5-14 and 5-15) is an arrangement of binaries used to produce an up-count sequence plus the necessary gating to produce a down-count plus the additional gating required to instruct the desired direction of count while inhibiting the undesired direction of count. A bidirectional counter may be synchronous or asynchronous and of arbitrary length.

Presettable Counter

A presettable counter may have any of the previously described counter configurations having the additional provision for arbitrarily entering a predetermined binary number. The count will start at the predetermined number on the first activating (clock) pulse and will proceed in the normal manner. Presettable counters

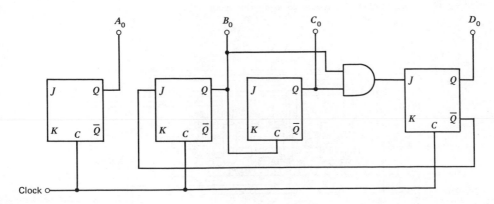

Figure 5-12 Typical decade counter.

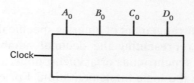

Number of Clock Pulses	Output Bits				Decimal Equivalent
	A_0	B_0	C_0	D_0	
1	0	0	0	0	0
2	1	0	0	1	9
3	0	0	0	1	8
4	1	1	1	0	7
5	0	1	1	0	6
6	1	0	1	0	5
7	0	0	1	0	4
8	1	1	0	0	3
9	0	1	0	0	2
10	1	0	0	0	1
11	0	0	0	0	0

Figure 5-13 A typical decade down-counter.

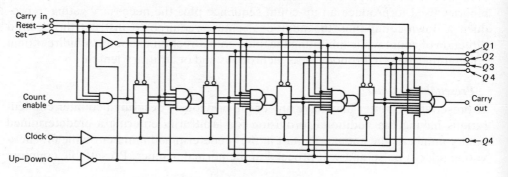

Figure 5-14 Logic diagram of a synchronous up–down counter.

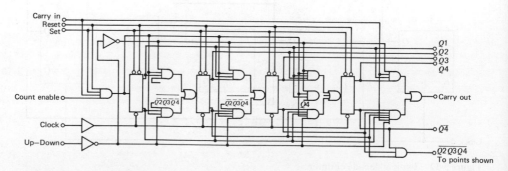

Figure 5-15 Logic diagram of a BCD synchronous up–down counter.

116

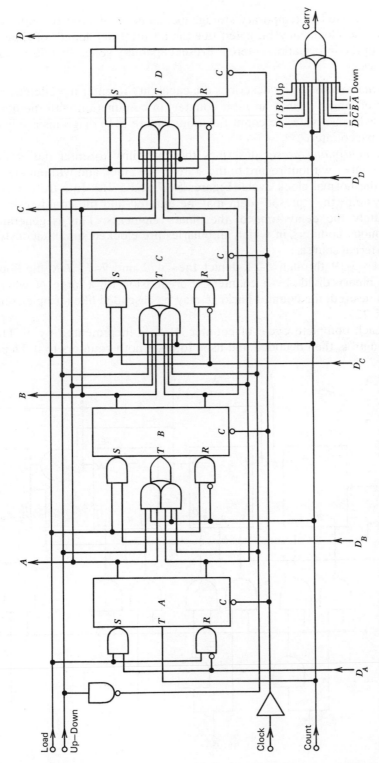

Figure 5-16 A 4-bit up–down binary counter; pin requirement: 15 pins per single quad 56 (with clock amplifier).

can also be used as temporary storage elements since, after the entry of a binary number, the clock may be gated out for an arbitrary length of time, inhibiting change of the information entered. Preset capability generally falls into one of two categories:

1. *Synchronous preset* describes the capability to enter a predetermined binary number of length n (into an n-bit counter) in conjunction with the normal clock cycle used to activate the count function. Figure 5-16 diagrams a synchronously presettable counter.

2. *Asynchronous preset* describes the capability to enter the predetermined binary number without regard to the state of the clock and without specific attention to the normal clock cycle (Figures 5-17 and 5-18). In a truly asynchronous entry system, the "preset" data may be entered and the count restarted in approximately the delay time of the binary elements. This is generally true of synchronous counters in which all binaries are clocked simultaneously from the same external source.

Figures 5-19 through 5-21 depict the 7472 and 7473 *J-K* flip-flops used in various binary divide-by-n counters. For division by a large N when a binary count is desired, the counter package may be cascaded like a ripple counter, as in Figure 5-22.

For each complete cycle of counter A, B is incremented by 1. The desired cycle length is then decoded and used to reset both counters to 0. Depending on

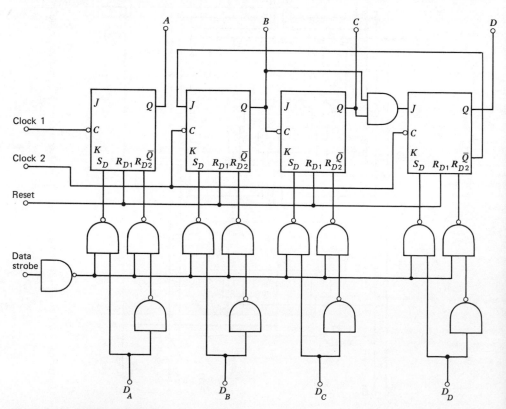

Figure 5-17   Example of asynchronous preset.

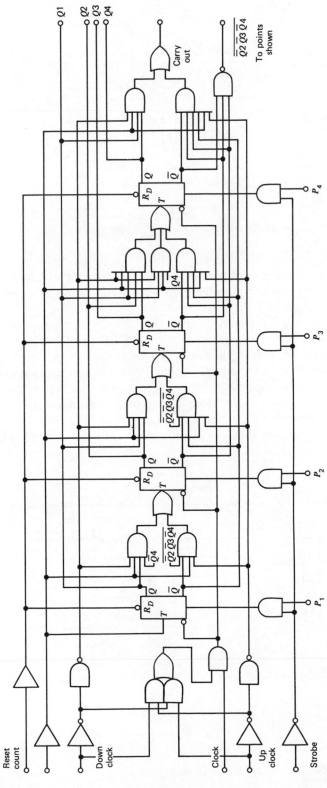

Figure 5-18 A BCD synchronous up–down counter with asynchronous parallel entry.

119

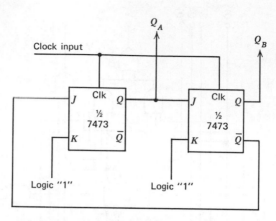

Figure 5-19   A binary divide-by-3 counter.

the cycle length, false outputs may be generated by the NAND gate. It should be strobed when necessary to eliminate false data. Another example using 7493s appears in Figure 5-23.

A common application of decimal decoders is in counter systems. A simple system of this sort if presented in Figure 5-24. Initially the counter is reset to logic "0" and the hold input is low. When the hold input goes to logic "1", the counter is enabled and the cycle begins. If the clock repetition rate is 9 MHz and the hold remains at logic "1" for 100 μsec, the number 900 is stored in the counter at the end of a timing cycle. Hold should then remain low for about 1 sec so that digits can be viewed. A viewing time of 1 sec allows the observer to notice changes in the least significant digits as cycling occurs. Displays used in this example are

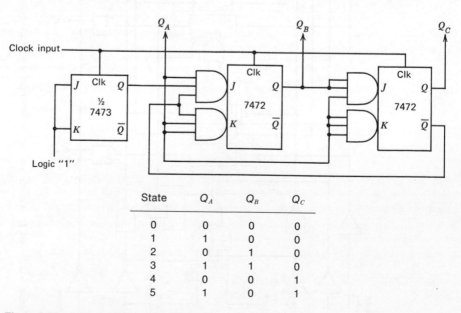

State	Q_A	Q_B	Q_C
0	0	0	0
1	1	0	0
2	0	1	0
3	1	1	0
4	0	0	1
5	1	0	1

Figure 5-20   A binary divide-by-6 counter.

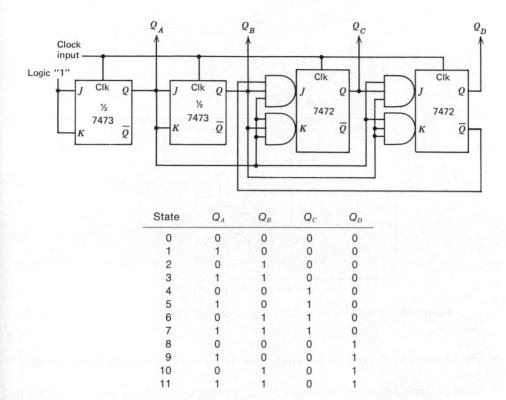

State	Q_A	Q_B	Q_C	Q_D
0	0	0	0	0
1	1	0	0	0
2	0	1	0	0
3	1	1	0	0
4	0	0	1	0
5	1	0	1	0
6	0	1	1	0
7	1	1	1	0
8	0	0	0	1
9	1	0	0	1
10	0	1	0	1
11	1	1	0	1

Figure 5-21   A binary divide-by-12 counter.

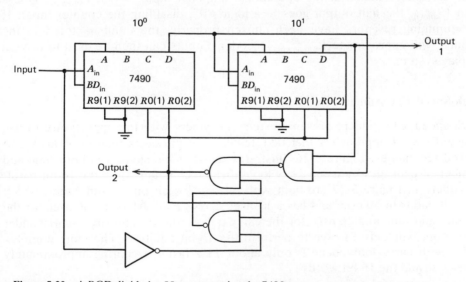

Figure 5-22   A BCD divide-by-88 counter using the 7490.

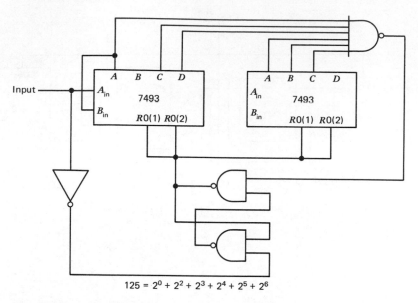

$$125 = 2^0 + 2^2 + 2^3 + 2^4 + 2^5 + 2^6$$

Figure 5-23   A binary divide-by-125 counter using the 7493.

of the filament-bulb type and cannot usually be driven directly from decoder outputs.

Another application for decoders is in analog-to-digital converters, such as that in Figure 5-25. An analog input voltage V_{unknown} is placed on one side of the 709 operational amplifier, and the counter is reset to logic "0". The amplifier output is interrogated to find out whether V_{unknown} is greater than V_{known}. If it is, counting continues. Strobe input allows interrogation of the op amp output only after it has stabilized. The V_{known} is generated in the calibrated resistor matrix, with currents summed at the operational amplifier input. When V_{unknown} is less then V_{known}, the halt output goes to a logic "0", disabling the counter input. If the summing resistors have been chosen properly, the number stored in the counter will approximate V_{unknown}. Decoder outputs can then be used to drive a display system.

High-Speed TTL Adders

High-speed TTL adders present systems designers with the opportunity to use straightforward approaches that take fewer logic packages than previously employed techniques. Figure 5-26 illustrates the possible savings in both time and number of packages by using carry look-ahead techniques. Both the 16-bit parallel adders in Figure 5-27 are built with the 7483 4-bit binary full adder, which typically adds in 30 nsec and has a 12-nsec carry delay. Allowing 30 nsec for the parallel add time and 36 nsec for the carry delays, the straight ripple-carry adder takes approximately 71 nsec to perform the 16-bit addition. The more complex adder with carry look-ahead is only about 11% faster, requiring approximately 63 nsec to add the 16-bit words.

Ripple-carry delays are avoided by the look-ahead design, but the external

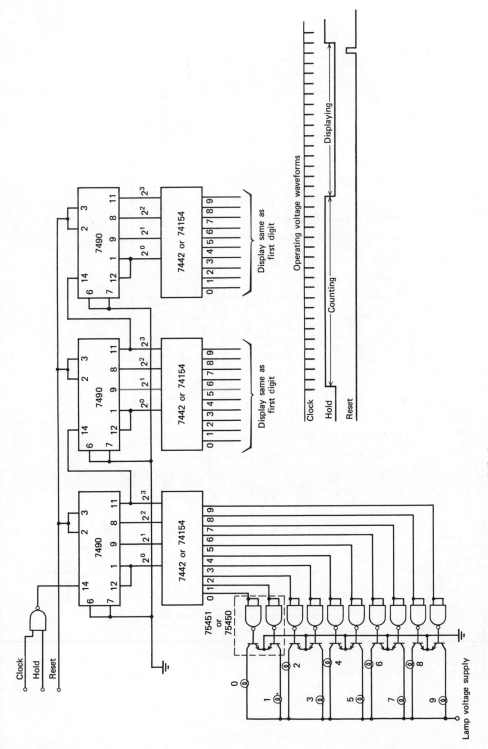

Figure 5-24 Decoders used in a clock-pulse counter and display system.

123

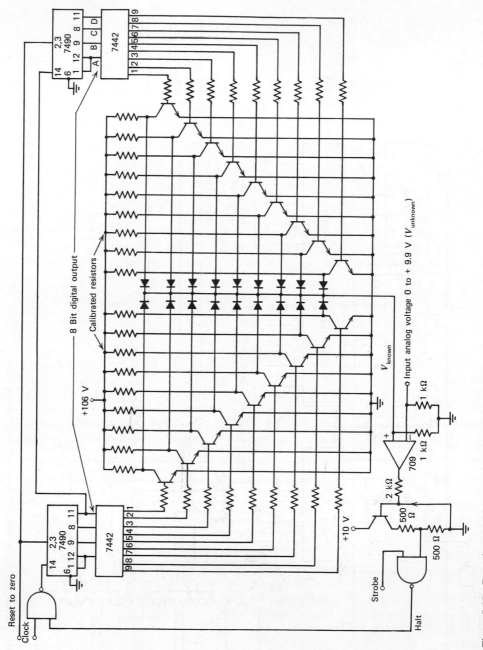

Figure 5-25 Decoders used in an analog-to-digital converter.

124

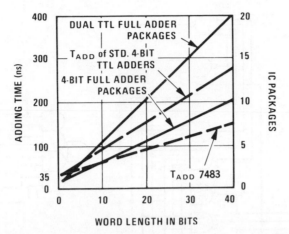

Figure 5-26 Adding times and package counts in ripple-carry parallel adders and subtractors with word lengths to 40 bits.

gates introduce additional system delays. Similar additional delays would be encountered if other gating structures implemented the look-ahead. In view of the extra costs and design problems posed by the gates, it is obvious that a designer should consider look-ahead only when a reduction in time of a few nanoseconds per adder stage is critical to system efficienty or when very long adders are to be built.

In most adder applications, the speed improvement gained by using the 7483 in its normal ripple-carry mode will be significant. Since a full adder is also the basic subtractor device, the data also applies to ripple-carry subtractor and adder–subtractor units.

Binary Arithmetic Units

Connections for ripple-carry parallel addition, subtraction, and addition subtraction units are set out in Figure 5-28. Each design can be extended to any practical number of bits. The false inputs required to operate a full adder as a subtractor are provided by two-thirds of a 7404 hex inverter or one 7486 quad exclusive OR gate.

When single or dual full adders are needed for functions such as obtaining intermediate sums, the 4-bit adder can be logically divided as in Figure 5-28d. The division works as follows.

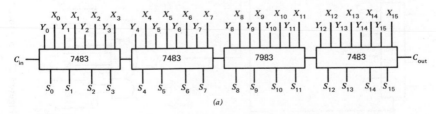

(a)

Figure 5-27 Carry look-ahead slightly improves the speed of a 16-bit parallel adder: (*a*) 16-bit ripple-carry adder.

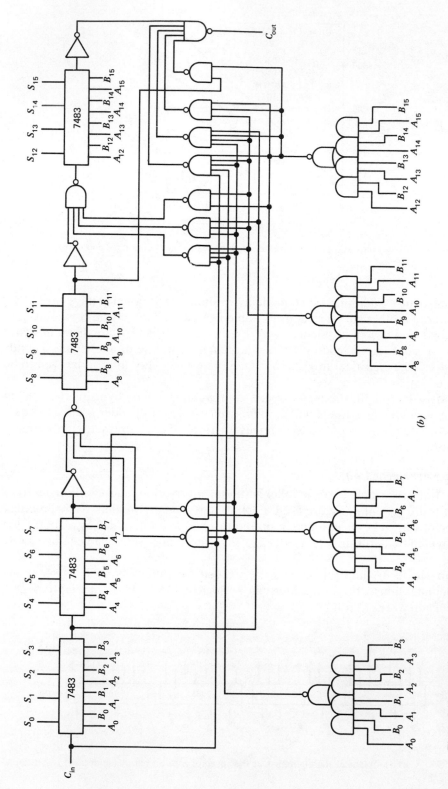

Figure 5-27 (continued) (b) 16-bit adder with carry look ahead.

126

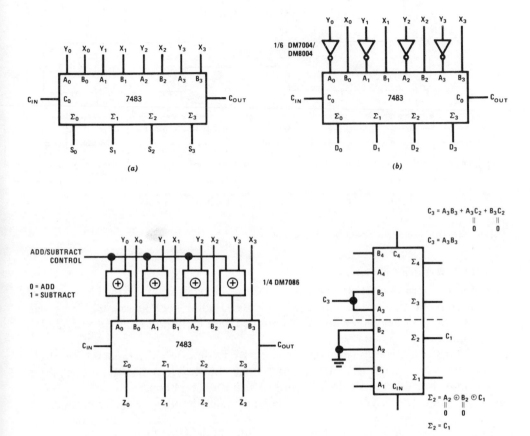

Figure 5-28 Basic connections for ripple-carry adder devices: (a) 4-bit adder, (b) subtractor, (c) adder–subtractor unit, (d) dual full adder.

The first section does not have a conventional carry out. However, $\Sigma_2 = A_2 + B_2 + C_1$, in which C_1 is the internal carry from the first stge. Grounding A_2 and B_2 makes them drop out; thus $\Sigma_2 = C_1$, providing an external carry. The second half lacks an external carry input, but C_3 is the external equivalent. We have $C_3 = A_3B_3 + A_3C_2 + B_3C_2$, where C_2 is the internal carry between the second and third stages. Since A_2 and B_2 are grounded, C_2 becomes logic "0" and $C_3 = A_3B_3$.

The technique is convenient for obtaining functions such as intermediate carries in a system using 4-bit full adders, because conventional dual adders do not have to be stocked and handled by exception. The power dissipation is somewhat larger, due to the additional internal gates used, and carry out of the first half somewhat longer, about 25 nsec. This half can be used where speed is not critical, and the second half where it is. The C_4 carry out takes 12 nsec.

Arithmetic units of any length, for any desired combination of functions are made with modifications of Figure 5-29. If exclusive OR gates are placed at the inputs for adder–subtractor–comparator control, the register outputs would be tied back to the inputs to multiply and divide under clock control.

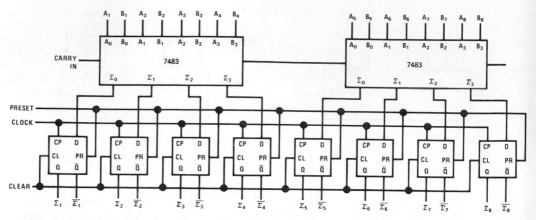

Figure 5-29   An 8-bit section of an arithmetic unit.

BCD Adders and Converters

Arithmetic operations in other number systems, and data conversions, can also be implemented at high speed with the 7483. BCD adder techniques are described, since the BCD codes have wide application. These techniques can be extended to any number of BCD decades.

Two binary adders are needed for each BCD decade. The binary sum of the BCD numbers is obtained with the first stage. In the second stage, a correction factor must be added to the binary sum so that the final binary sum will represent the true sum in the original code. The same principle applies when other codes are used.

Sums larger than 9 are not allowed in any BCD decade. Therefore, larger sums must be sensed and used to generate a carry. The carry is worth more (i.e., 16) in straight binary than in BCD; thus the correction factor is also used to justify the carry.

The 8-4-2-1 BCD adder in Figure 5-30 generates a carry and adds 6 to the first binary sum whenever the first sum exceeds 9. For example, adding 6 and 7 gives the binary sum 13. The output of the 3-input NAND gate goes to logic "1" level, producing the carry and adding 2^1 and 2^2 to the binary sum. We get

$$\begin{array}{r} 1101 \\ 0110 \\ \hline 1\,0011 \end{array}$$

or binary 19 as the final sum. However, this is read out as BCD 3 in the first decade, plus a carry worth $2^0 \times 10^1$ in the second decade, or BCD 13.

The excess-3 BCD addition (Figure 5-30b) takes a correction of 13 if the first binary sum is less than 16 and a correction of 3 if the first sum is 16 or greater. In the first case, since the sum will be in excess by 6 instead of 3, the 3 must be subtracted by adding 13 and ignoring the carry. In the second case, the first four bits of the binary sum will be too low by 3; thus a carry must be assured by adding 3.

For example, add $3 + 3$ in excess 3. The sum is binary 12 and no carry. Now the logic "0" on the carry output is inverted, adding $2^0 + 2^2 + 2^3 = 13$ to the binary sum. The final result is 11001. Since there is no carry, the readout of 9 represents

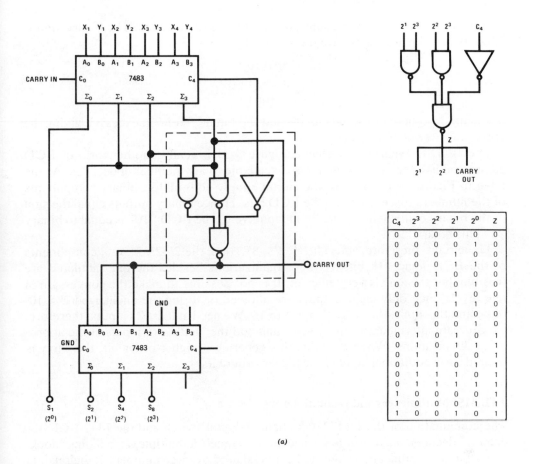

C_4	2^3	2^2	2^1	2^0	Z
0	0	0	0	0	0
0	0	0	0	1	0
0	0	0	1	0	0
0	0	0	1	1	0
0	0	1	0	0	0
0	0	1	0	1	0
0	0	1	1	0	0
0	0	1	1	1	0
0	1	0	0	0	0
0	1	0	0	1	0
0	1	0	1	0	1
0	1	0	1	1	1
0	1	1	0	0	1
0	1	1	0	1	1
0	1	1	1	0	1
0	1	1	1	1	1
1	0	0	0	0	1
1	0	0	0	1	1
1	0	0	1	0	1

(a)

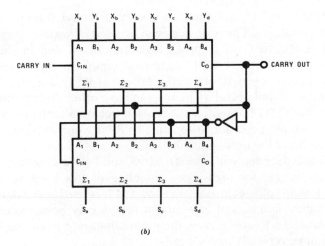

(b)

Figure 5-30 Binary implementation of cascadable BCD adders: (*a*) 8–4–2–1 BCD adder, (*b*) excess-3 BCD adder.

129

6 in the excess-3 code. Now, suppose 8 and 8 are to be added. The binary sum is 22, or binary 6 plus a carry. Adding 3, we get

$$
\begin{array}{r}
10110 \\
0011 \\
\hline
11001
\end{array}
$$

or binary 25. The carry stands for 10 and the binary 9 for 6 in excess-3, giving the correct answer of 16 in excess-3.

The BCD-to-binary converter in Figure 5-31 is really another form of BCD adder, and like the other adders it is cascadable to any number of decades. As the table in Figure 5-31 indicates, the circuit simply adds all like binary components of the numbers represented by the BCD bits. Hence binary outputs equal the sum of the binary equivalents of the BCD inputs (e.g., the BCD "10" is equal to binary "8" plus binary "2").

The circuit, therefore, adds all $2'$s, $4'$s, $8'$s, (i.e., the 2^1, 2^2, 2^3, ... components of the BCD logic "1" bits). Other organizations besides the one tabulated are feasible, but this one is very efficient. It is not possible to make the conversion in one level, since only eight outputs are allowed (ignoring the unchanged $2^0 \times 10^0$ bits), and this limits the parallel inputs to 16. We need at least 23 inputs; therefore, we generate intermediate binary sums and add the "leftover" inputs at the appropriate 2^n positions. When the additions generate a higher-order bit, the carry is treated as an additional input to the higher-order bits.

TTL MSI Multiplexers and Demultiplexers

The combination of the DM7210 8-channel digital switch and the 7442 BCD-to-decimal decoder makes an almost ideal 8-channel demultiplexer building block. It permits communications links to be time-shared by large numbers of digital data channels at rates up to 20 MHz.

The DM7210/7442 combination is very attractive from the standpoint of logic–function versatility. These devices can be used in various ways to generate trans-mission codes, to form complex logic functions, and in the case of the 7442 decoder, to convert any BCD code to decimal. As examples, a keyboard entry encoder, minterm generators, and a universal BCD decoder are illustrated. Both the multiplexer and decoder circuits can have other system functions, as well. For example, the DM7210 switch can be used as a parallel-to-serial converter that can be controlled by a counter to serialize an 8-bit word starting with either the least-significant bit or the most-significant bit.

The 7442 decoder will operate MOS solid-state commutators at rates to the megahertz range, for high-speed sampling of low-level sensor outputs. If the 7441A, a high-voltage decoder like the 5442, is used in the demultiplexer designs, display tubes, lamps, and low-current relays may be actuated directly by trans-mitted data. If a load is driven, the output data rate would be limited by the load-reaction time, typically milliseconds.

DM7210 MSI Multiplexer

The DM7210 multiplexes eight input channels to a single output (see Figure 5-32). It can serve as an 8-bit parallel-to-serial converter, but it differs from a shift reg-

NUMBER	ADDER INPUTS								
	2^1	2^2	2^3	2^4	2^5	2^6	2^7	2^8	2^9
$10^2 \times 2^3 = 800 = 32 + 256 + 512$					X			X	X
$10^2 \times 2^2 = 400 = 16 + 128 + 256$				X			X	X	
$10^2 \times 2^1 = 200 = 8 + 64 + 128$			X			X	X		
$10^2 \times 2^0 = 100 = 4 + 32 + 64$		X			X	X			
$10^1 \times 2^3 = 80 = 16 + 64$				X		X			
$10^1 \times 2^2 = 40 = 8 + 32$			X		X				
$10^1 \times 2^1 = 20 = 4 + 16$		X		X					
$10^1 \times 2^0 = 10 = 2 + 8$	X		X						
$10^0 \times 2^3 = 8 = 8$			X						
$10^0 \times 2^2 = 4 = 4$		X							
$10^0 \times 2^1 = 2 = 2$	X								
$10^0 \times 2^0 = 1 = 1$									

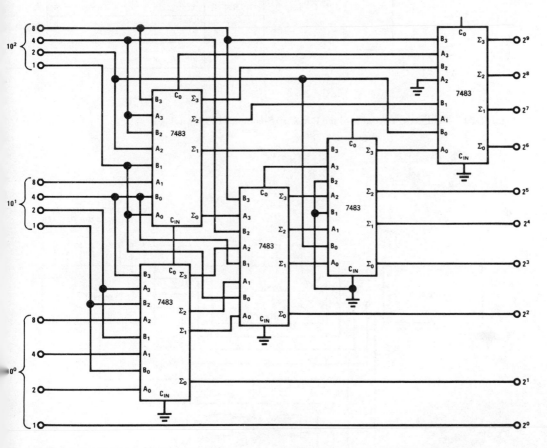

Figure 5-31 Example of BCD-to-binary converter that is cascadable to any number of decades.

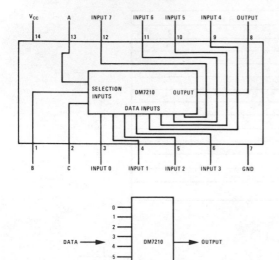

SELECTION INPUTS			DATA INPUTS								OUTPUT
C	B	A	0	1	2	3	4	5	6	7	
0	0	0	0	X	X	X	X	X	X	X	0
0	0	0	1	X	X	X	X	X	X	X	1
0	0	1	X	0	X	X	X	X	X	X	0
0	0	1	X	1	X	X	X	X	X	X	1
0	1	0	X	X	0	X	X	X	X	X	0
0	1	0	X	X	1	X	X	X	X	X	1
0	1	1	X	X	X	0	X	X	X	X	0
0	1	1	X	X	X	1	X	X	X	X	1
1	0	0	X	X	X	X	0	X	X	X	0
1	0	0	X	X	X	X	1	X	X	X	1
1	0	1	X	X	X	X	X	0	X	X	0
1	0	1	X	X	X	X	X	1	X	X	1
1	1	0	X	X	X	X	X	X	0	X	0
1	1	0	X	X	X	X	X	X	1	X	1
1	1	1	X	X	X	X	X	X	X	0	0
1	1	1	X	X	X	X	X	X	X	1	1

X = "Don't Care" Condition

Figure 5-32 Multiplexer operation of the DM7210 8-channel TTL switch.

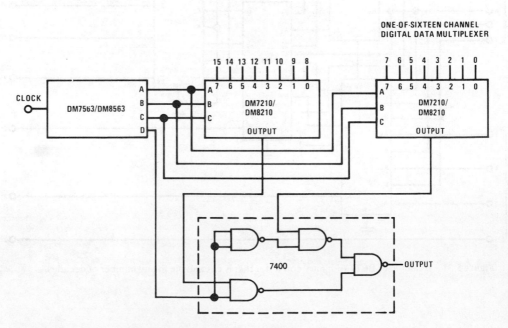

Figure 5-33 Counter controls selection of 16 data channels.

132

Table 5-1 7493 4-Bit Binary Counter Logic Table.

COUNT	OUTPUT			
	D	C	B	A
0	0	0	0	0
1	0	0	0	1
2	0	0	1	0
3	0	0	1	1
4	0	1	0	0
5	0	1	0	1
6	0	1	1	0
7	0	1	1	1
8	1	0	0	0
9	1	0	0	1
10	1	0	1	0
11	1	0	1	1
12	1	1	0	0
13	1	1	0	1
14	1	1	1	0
15	1	1	1	1

ister in that each input channel is selectable with a 3-bit binary number. Channels 0 through 7 can be selected in sequence by the normal outputs of a 3-bit counter.

In Figure 5-33, 16 channels are selected by a 4-bit binary counter. The counter's ABC outputs are used to control the two DM7210s, and the D output controls the quad NAND gate serving as a submultiplexer. If a straight 4-bit counter such as the 7493 is employed, the selection will be sequential, as indicated by the 7493 logic table in Table 5-1. The initial channel and channel-selection order can be made variable by using the 74193 presettable up–down counter.

Channels could be chosen randomly or in nonsequential order with logic or computer commands, or through programs stored in ROMS. Also an MOS ROM may be used when channel-switching rates in the megahertz range are acceptable.

A DM7210 may serve as a submultiplexer to reduce the package counts in large systems, like the 64-channel arrangement in Figure 5-34. This system requires six control bits. Control bits for sequential selection can be provided by 1.5 4-bit binary counters or by a 4-bit counter and a dual flip-flop.

A 7442 MSI Decoder–Demultiplexer

At the receiver, eight channels can be demultiplexed with each 7442 BCD-to-decimal decoder. This device's logic table (in Figure 5-35) shows that outputs 0 through 7 can be selected by addressing only inputs ABC with the 3-bit octal selection code. The selected output will be at the same logic level as input D. Therefore, input D can serve as the data-link terminus. Furthermore, the channel-selection addresses of the 7442 are identical to those of the DM7210. When $ABC = 001$, for example, the DM7210 selects input 4 and the 7442 selects output 4. No reconstruction of data is required in a digital communications system when the channel-selection codes are synchronously applied to the multiplexer and the demultiplexer.

Note in Figure 5-35 that all 7442 outputs not addressed by the normal 4-bit selection code remain high. This condition prevails for channels 0 through 7 when only a 3-bit selection code is employed for demultiplexing. (In normal BCD-to-

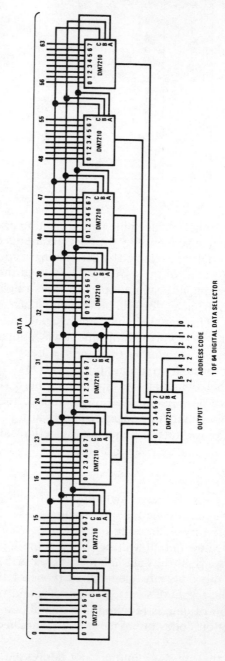

Figure 5-34 DM7210 is a submultiplexer in a 64-channel multiplexer.

134

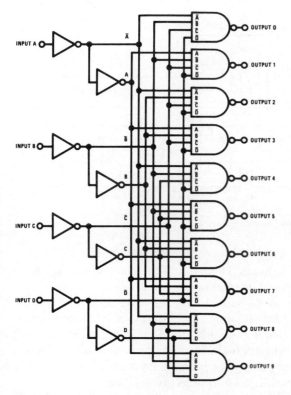

LOGIC TABLE

INPUTS	OUTPUTS					
D C B A	0 1	2 3	4 5	6 7	8 9
0 0 0 0	0 1	1 1	1 1	1 1	1 1
0 0 0 1	1 0	1 1	1 1	1 1	1 1
0 0 1 0	1 1	0 1	1 1	1 1	1 1
0 0 1 1	1 1	1 0	1 1	1 1	1 1
0 1 0 0	1 1	1 1	0 1	1 1	1 1
0 1 0 1	1 1	1 1	1 0	1 1	1 1
0 1 1 0	1 1	1 1	1 1	0 1	1 1
0 1 1 1	1 1	1 1	1 1	1 0	1 1
1 0 0 0	1 1	1 1	1 1	1 1	0 1
1 0 0 1	1 1	1 1	1 1	1 1	1 0
1 0 1 0	1 1	1 1	1 1	1 1	1 1
1 0 1 1	1 1	1 1	1 1	1 1	1 1
1 1 0 0	1 1	1 1	1 1	1 1	1 1
1 1 0 1	1 1	1 1	1 1	1 1	1 1
1 1 1 0	1 1	1 1	1 1	1 1	1 1
1 1 1 1	1 1	1 1	1 1	1 1	1 1

Figure 5-35   7442 BCD-to-decimal decoder logic diagram and table.

decimal decoder applications, the decimal equivalent of the BCD number is determined from the location of the logic "0" bit.)

However, with a 3-bit address, the outputs on 8 and 9 of the 7442 are complements of the outputs on logic "0" and logic "1". The 8 and 9 outputs can be ignored, or they can be used for various dual-control functions. For instance, channels 0 and 1 could carry interlock commands – that is, turn-off or turn-on

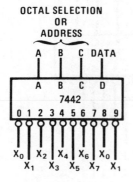

ADDRESS	OUTPUT
A B C	LINE
0 0 0	0
1 0 0	1
0 1 0	2
1 1 0	3
0 0 1	4
1 0 1	5
0 1 1	6
1 1 1	7

Figure 5-36   7442 Demultiplexer with eight channels (last two channels complement first two channels).

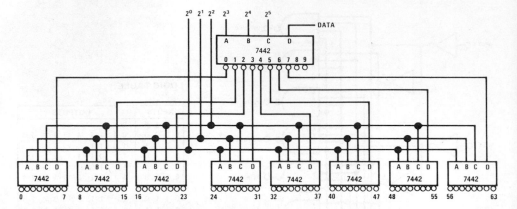

Figure 5-37   One-of-64 decoder or 64-channel demultiplexer.

commands would be received on channels 0 and 1, and the complementary turn-on and turn-off commands would appear on channels 8 and 9.

Figures 5-36 and 5-37 illustrate an 8-channel demultiplexer and an assembly that can be used either as a 64-channel demultiplexer or a 1-of-64 decoder, respectively. The D input may be strobed to hold the 0 through 7 outputs of the input decoder high, thus holding all 64 outputs high. Strobing provides an all-channel inhibit function, since any subsequent data processing or control system would be designed to consider a continuous high-level output as invalid data.

Examples of the 7442's versatility as a decoder appear in Figures 5-38 and 5-39. Figure 5-38 is a 1-of-16 decoder, with complements available as mentioned in the discussion of Figure 5-35. The assembly in Figure 5-39 will decode any BCD

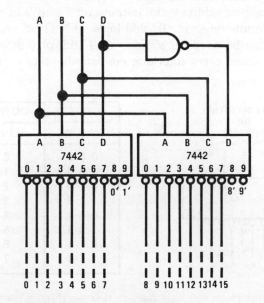

Figure 5-38   One-of-16 decoder.

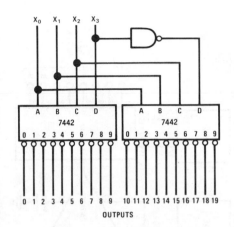

OUTPUT			
BCD CODE			
DECIMAL DIGIT	8421	5421	EXCESS 3	4221
---	---	---	---	---
0	0,18	0,18	3	0,18
1	1,19	1,19	4	1,19
2	2	2	5	2
3	3	3	6	3
4	4	4	7	6
5	5	8,10	8,10	9,11
6	6	9,11	9,11	14
7	7	12	12	15
8	8,10	13	13	16
9	9,11	14	14	17

Decode Any BCD Code
Using two, any 4–Bit BCD Code may be decoded
by selecting outputs as shown in the table.

Figure 5-39   Universal BCD decoder.

code to decimal, when the appropriate output combination is selected with the accompanying table.

Synchronous Data System

Whether counters or other devices control channel selection, some means of synchronously selecting multiplexer inputs and demultiplexer outputs is desirable to minimize data processing. Synch bits (or random-selection control bits) could be worked into the data transmitted and detected at the receiver, but the simplest approach is to encode a synch signal (or random-selection code) in the clock-pulse train, which is sent along with the data on a separate channel.

An example of a two-line sequential-select system is presented in Figure 5-40. The control logic in the multiplexer generates a reset signal and transmits it along with the clock signal to the demultiplexer. Any practical number of inputs (e.g., 16, 32, or 64) in any of the normal digital word lengths can be handled with modifications of the control-counter assembly. As shown, the control logic makes eight 16-bit inputs time-share the data link between resets. At each end, one counter divides the clock by 16 — in this example, to establish the intervals between channel switching. The $f/16$ input to the second counter generates the octal selection sequence.

As the sixteenth bit of the eighth word (channel H) enters the DM7210 watch, the second counter advances to binary 8. The counter's pin 11 goes high, triggering the one-shot in the transmitter. The one-shot resets both counters to zero. The reset pulse also switches the NAND gate, thus clearing the J-K flip-flop and disabling gate A. This prevents one clock pulse from being transmitted to the receiver. Gate B is now enabled, however, and the following clock pulse toggles the flip-flop back to the logic "1" state, enabling gate A and resuming transmission of the clock to the receiver.

At the receiver, the retriggerable one-shot (normally high) detects the missing clock pulse and resets to zero the two counters controlling the demultiplexer. The system is now in synch and begins transmitting and receiving the 16 bits in the

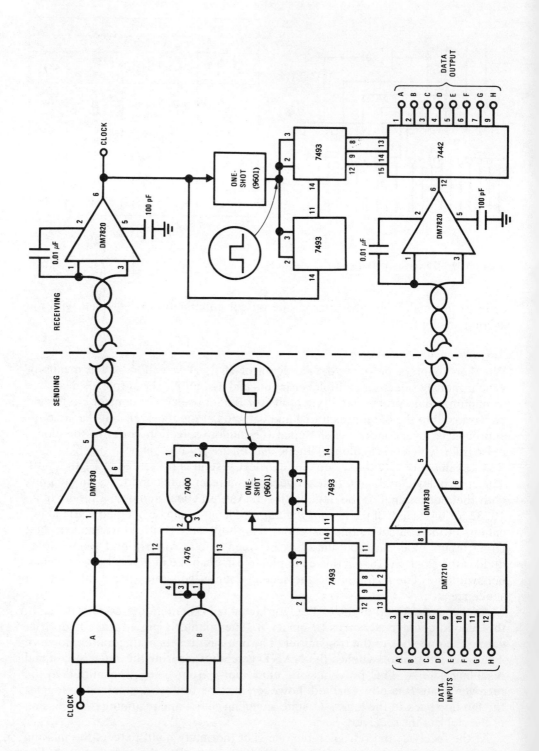

138

Figure 5-40 Two-line digital data-transmission system.

Synchronous Digital Data-Transmission System

This system is suitable for transmitting 8 words of any bit length on a time base multiplex basis. The circuit shown transmits 16-bit words.

The transmitting multiplexer is a DM7210 8-channel switch, and the receiver is a DM7442 BCD to decimal decoder.

In order for the system to function properly, both the 8-channel switch and the BCD to decimal decoder must decode the same channel simultaneously. This is accomplished by synchronizing the four 7493 4-bit counters at the end of each 8-word transmission.

The four counters are synchronized as follows; when the sixteenth bit of the eighth word is sent, the counter connected to the 8-channel switch advances to binary 8, and pin 11 goes high. This triggers the one-shot, which resets both counters on the sending end to zero, and, through the 7400 NAND gate, clears the J-K flip-flop. Clearing of the flip-flop disables gate A so that the next clock pulse will not be transmitted. Gate B is now enabled, so the next clock pulse toggles the flip-flop back to the logic "1" state, enabling gate A again.

On the receiving end, the retriggerable one-shot (normally high) detects the missing clock pulse and resets the two counters on the receiving end to zero. The one-shot should be programmed for a pulse width greater than one clock interval but less than two clock intervals.

The system is now ready to transmit the first word on channel A.

This system could be applied to the pulse code modulation of analog information. First analog-to-digital conversion would be used to obtain the binary equivalent of the analog quantity; the binary word would then be stored in a shift register whose output is connected to an input on the 8-channel switch. When the 8-channel switch is sensing that input, the system clock would transmit the register's contents to the receiver.

To offset the speed difference between the digital transmission system and the analog to digital converter, the converter for channel H could be strobed when channel A is starting to be transmitted.

139

word on channel A. The one-shot is designed for a pulse width greater than one clock interval but less than two clock intervals. Differential line drivers and receivers and twisted pair lines are recommended when transmission links are long or subjected to considerable system noise.

Multiplexing Analog Data

Analog data can be transmitted by the Figure 5-40 system, in a format such as PCM (pulse-code modulation), if analog-to-digital conversion and temporary storage circuits are provided at the DM7210 inputs. For example, the PCM samples may be read into a shift register and read out when the data channel is sensed by the multiplexer. Strobes may be used to offset a difference between the analog-to-digital conversion rate and the digital transmission rate. If a converter on channel H operates at 1 MHz and the system at 4 MHz, for instance, the strobe might be turned on with the channel A select signal, turned off with the channel D select signal, and turned on again with the channel H select signal. During the time channels D, E, F, and G were being sensed, an $f/4$ clock could be applied to the register to load a sample, and the 4-MHz clock applied only during the readout interval. In this simple case, the relationships among the control signals are such that only a few gates need be added to the control logic of Figure 5-40 (e.g., the divide-by-4 output of the first counter could supply the $f/4$ clock).

In Figure 5-41 the 7442 decoder is shown controlling an 8-channel MOS commutator. This design is capable of commutating low-level analog signals at rates in the megahertz range. The low impedance presented by the TTL decoder to the MOSFET gates enhances the MOS switching rates. Channels are selected by the same 3-bit addressing technique employed in the digital demultiplexer.

Keyboard Entry Decoder

Three advantages of the keyboard-entry decoder in Figure 5-42 are (1) the transmitted words are free of logic-circuit noise, (2) the device operates on a demand

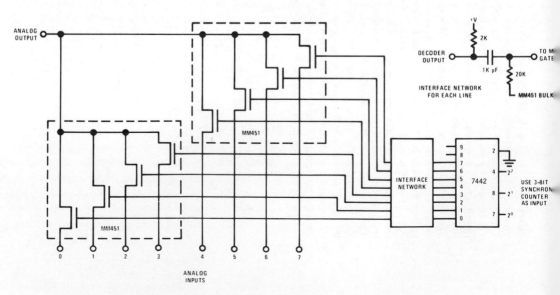

Figure 5-41   TTL decoder controls an 8-channel analog MOS commutator.

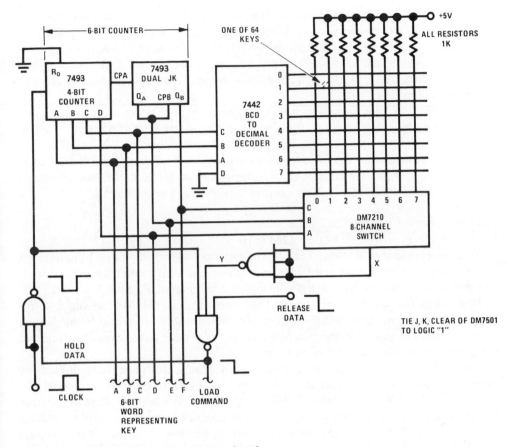

Figure 5-42   A 64-character keyboard entry decoder.

cycle established by the transmission system, and (3) any type of code can be generated by making changes in the 7442, 7493, and DM7210 control or matrix wiring patterns.

Assume that the load-command and release-data lines are high. The line labeled Y is low and the clock signal is inverted from a positive-going to a negative-going pulse by the clock input gate. The clock inhibits the load-command gate and the hold-data line stays high, but the counter continues cycling through its 64 output states.

If the key connecting the number-two lines in the 8×8 matrix is pressed, the counter will produce an output of logic "0" on the decoder's 2 output at a time when the switch's 2 channel is transmitting (within a maximum of 64 clock pulses). All other inputs are held high by the pull-up resistors. Now, switch output X goes low and Y goes high. The release-data line is still high. Therefore, when the clock input to the load-command gate rises, the hold data line goes low and the load-command and clock-input gates latch up.

The counter stops cycling and the word $ABCDEF$ stabilizes on the output to the transmission line. The word in this case would be the decoder and switch select codes 2 and 2, or 010010. After the communications link has accepted the word and the release-data line is made low, a new select cycle begins.

A clock pulse width greater than system settling times is used, to ensure that no noise is present on the output-word lines. This permits typical clock frequencies higher than 5 MHz and clock periods shorter than 200 nsec. Thus the longest it would take to generate an output word is 13 μsec. This is many times shorter than manual keyboard operation, which allows many terminals to time-share the communications link.

At the receiver, the 1-of-64 decoder in Figure 5-37 may be used to actuate a printout. Another alternative is forming an 8×8 matrix with the output lines of two 7442 decoders and sensing the logic "0" coincidences. Bits *ABC* would select one output and bits *DEF* the other output. The 6-bit word may also be used to select a raster-scan or pedestal-scan pattern from an MOS ROM. Such patterns are used to generate CRT displays, control tape printers, and so forth.

Complex Logic Functions

In their basic functioning, both a DM7210 multiplexer and a 7442 decoder resemble programmable ROMs. That is, the channel-selection bits and the data bit on the input can be considered to be an address generating some specific output. Their ROM-like nature is evident in the logic tables given in Figures 5-32 and 5-35. Both devices are considerably faster than conventional ROMs. In suitable applications, each can also replace a gate–logic subassembly of several IC packages, which would not only reduce assembly costs but would automatically eliminate design problems such as critical race condition.

Thus it is possible, and in many cases highly practical, to use these devices as high-speed ROMs in applications such as minterm generators and combinational logic networks. Like conventional ROMs, these devices may be used in cascade to increase the number of combinational logic functions and/or variables handled. The number of functions generated increases by powers of 2 as additional inputs are provided, following the general rules for logic synthesis with ROMs.

The DM7210 makes an efficient minterm generator of up to three variables when used as in Figure 5-43. The minterms are generated when the *L* through *S* data inputs are at the logic "1" level. Likewise, each output of the 7442 can be considered to be a minterm of the input code. As Figure 5-44 illustrates, sums of minterms can be generated with suitable output gating.

Digital Filter

A digital or commutating filter, provides high *Q* bandpass filtering with a few simple low-pass *RC* sections. A four-section commutating filter is schematized in

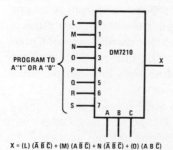

$X = (L) (\overline{A} \, \overline{B} \, \overline{C}) + (M) (A \, \overline{B} \, \overline{C}) + N (\overline{A} \, \overline{B} \, \overline{C}) + (O) (A \, B \, \overline{C})$
$+ (P) (\overline{A} \, \overline{B} \, C) + (Q) (A \, \overline{B} \, C) + (R) (\overline{A} \, B \, C) + S (A \, B \, C)$

Figure 5-43   Minterm generation with DM7210 switch.

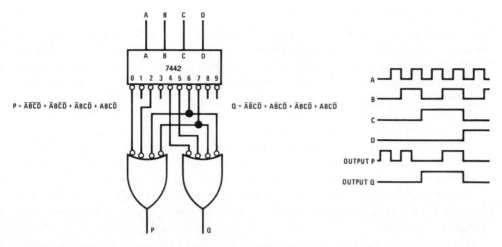

$$P = \overline{A}\,\overline{B}\,\overline{C}\,\overline{D} + \overline{A}\,B\,\overline{C}\,\overline{D} + \overline{A}\,B\,C\,\overline{D} + A\,B\,C\,\overline{D}$$

$$Q = \overline{A}\,\overline{B}\,C\,\overline{D} + A\,\overline{B}\,C\,\overline{D} + \overline{A}\,B\,C\,\overline{D} + A\,B\,C\,\overline{D}$$

Figure 5-44   The 7442 can also be used as a minterm generator.

Figure 5-45. The input to the counter must be $4f$. The bandwidth of this filter will be $1/4\pi RC$. The counter consists of two 7473s connected in the divide-by-4 configuration. The Q of the filter may be increased by adding sections. If there are N filter sections, the clock to the divide-by-N counter must be NF_0. The bandwidth then will be $1/N\pi RC$.

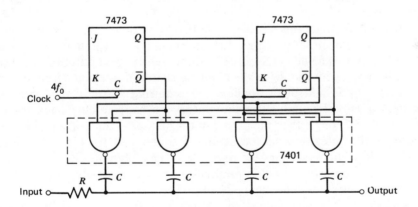

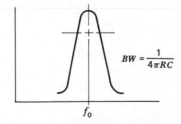

$$BW = \frac{1}{4\pi RC}$$

Figure 5-45   Example of a typical digital filter.

Digital Integrated Circuit Tone Detector[1]

The digital IC equivalent of a tone-activated relay (Fig. 5-46) can quickly discriminate or detect a signal whose frequency falls within its passband. The response is immediate — the output appears within one cycle of the input signal. Applications range from opening garage doors through motor speed control to activating remote machine or telemetry stations.

A sharp, narrow, square-pulse output results when the input frequency is in the circuit's passband. The input signal should be a square wave with a 50% duty cycle. The circuit can be tuned to detect pulse rates from less than 1 pulse per second to several megahertz. Since the upper and lower skirts are individually tunable, both the center frequency and passband are adjustable.

Each half of the detector can be used independently, as high-pass or low-pass frequency detectors. Adjustments are made with the two potentiometers or by changing RC values.

The positive-going edge of the input pulse signal triggers three one-shot pulse generators. Their outputs are a very short reset pulse, an upper-frequency limit pulse and a lower-frequency limit pulse. First, the reset pulse forces the output latches into their OFF state (logic "0" output), which causes the bandpass detector output to be in the untrue state (logic "1" output). The output can go true again only if the input pulse period is within the bounds set by the RC timing components of the upper- and lower-frequency limit generators.

If the first half cycle of the input pulse is greater than the upper-frequency limit pulse width, the upper-limit comparator sets the upper-limit latch. Conversely, the lower-limit latch is set if the incoming pulse width is less than the lower-limit pulse width.

With these two conditions satisfied, the bandpass detector will sense that both limit latches have been set to a logic "1" output and will switch to the true state — a logic "0" output. The detector output will, therefore, be a pulse train within the bandpass frequency for as long as the input signal is at the selected frequency.

The integrator is added for applications requiring a continuous true output, rather than a pulsed output. The integrator is a retriggerable one-shot whose period is made 5% longer than the period of the bandpass center frequency.

As shown, the detector is set for a center frequency of 100 kHz; therefore, the one-shot pulse width is set for 10.5 μsec. A retriggerable one-shot must be used, so that the dc logic level output is reinitiated with each new input pulse without any negative-going steps in the final output.

The circuit components are a 74L04 hex inverter, three 74L00 quad dual-input gates, and a 9601 retriggerable one-shot.

Digital Clock

As shown in Figure 5-47, TTL IC components are the heart of a typical digital clock. The monolithic decade counter (7490) accepts and counts pulses and displays the number of pulses in BCD form on its four outputs. It recycles every 10 pulses. The divide-by-12 counter (7492) connected to divide by 6, allows division

---

[1] Digital IC Tone Detector, by Don Femling, *Electronic Deisgn*, April 15, 1971.

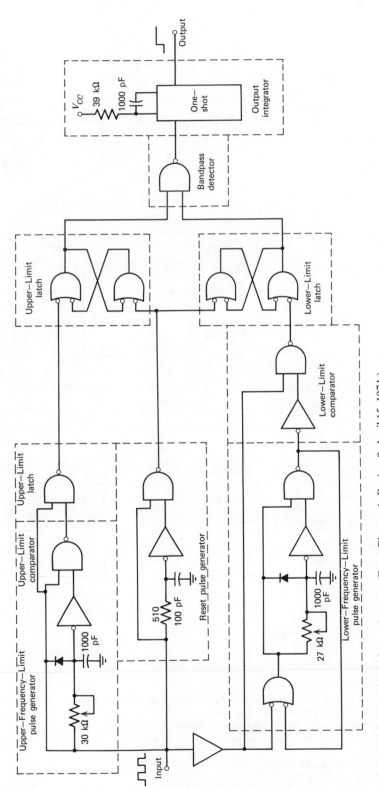

Figure 5-46 A digital IC tone detector. (From *Electronic Design*, **8**, April 15, 1971.)

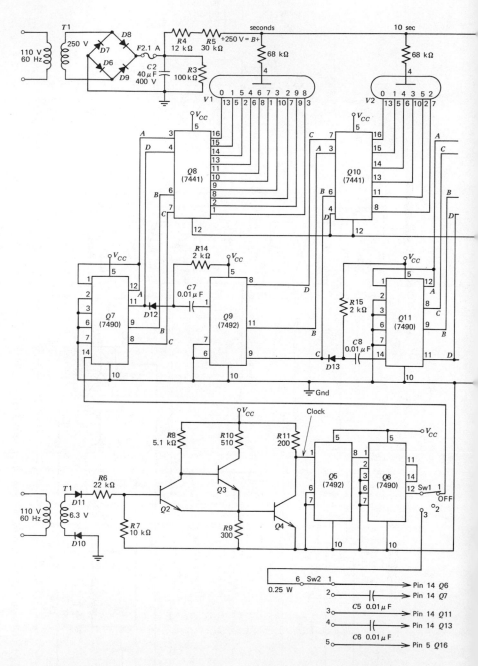

Figure 5-47 A digital clock. A complete digital clock can be easily built using standard TTL components and a display-tube readout. The clock uses the line frequency as a time base, but a crystal oscillator can perform the same function at added cost. By adding comparator circuits (not shown here), the clock can be made to generate command signals at desired times or at desired intervals.
Note:
  1. $D1$ through $D11$ are IN4003's.
  2. $D12$ through $D16$ are INFD100's.

146

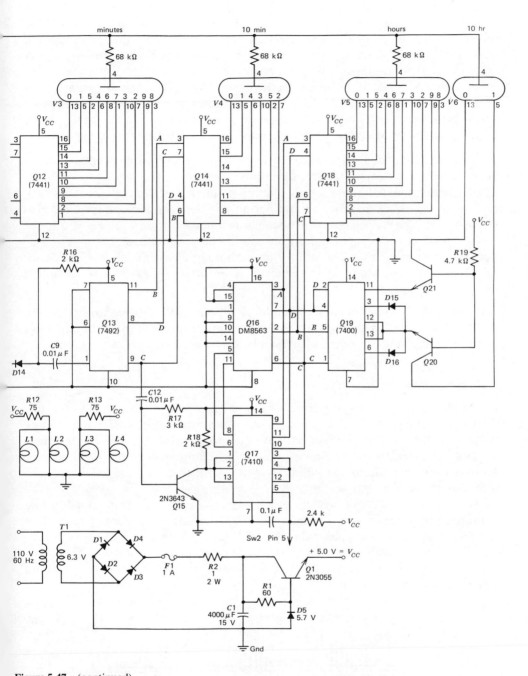

Figure 5-47 (continued)

3. $V1$ through $6NL$ are 840 tubes.
4. All resistors are 0.25 W unless otherwise noted.
5. $Q2$, $Q3$, $Q4$, and $Q15$ are 2N3643.
6. $L1$ through $L4$ are $3 = V$, $14 = mA$ lights.
7. $Q20$ and $Q21$ are 2N718A $BV_{CEO} \leq 70V$.
8. $T1$ is Stancor PS-8416.

147

of the 60-Hz line frequency down to 10 Hz in one step. The display tubes are driven by a decoder-driver (7441) which has the necessary high-voltage drive capability. The up–down counter (8563), which can be programmed to reset and cycle according to states applied at its control inputs, makes possible a counter less complex than would ordinarily be necessary.

TTL circuits typically have a fan-out of 10. This drive capability allows the counter outputs to be tapped to provide logic signals to drive external equipment as well as the decoders. The designer has an almost limitless number of choices for handling and using the counter outputs. A few possible applications are:

1. Making data-acquisition system sample and record experimental data at specific times of day for preset time durations.

2. Controlling apparatus, such as operating electromechanical valves to vary gas flow to ceramic kilns. This should permit any desired temperature–time processing cycle to be carried out with the kiln unattended.

3. Turning on and off applicances, air conditioning, lighting, and other electrical equipment; timing functions in time-lapse photography with automatically controlled camera; and stop-watch applications in sports or event-time recording control.

Anything that can be actuated with a relay or solenoid can be controlled by the clock and suitable TTL detectors. And any action that will make or break an electrical connection can stop the clock or be used to record the time.

TSL Applications

A typical bus-organized system appears in Figure 5-48. Conceptually, the terminal subsystems could be replaced by various control, arithmetic, and storage modules to form a processor or a large multiprocessor system. For structural simplicity and modular expandability, the major requirement is that the bus carry data from any module to any other module. Note that data may flow in both directions on the same bus.

TSL improves effective system-processing rates just as dramatically. The maximum propagation delays in complex TSL devices are generally much less than the delays in simple open-collector TTL gates. A passive pull-up TTL gate delay is 45 nsec maximum, about 30 nsec typical. Even a 40-mA TSL driver offers maximum delays ranging from 12 to 27 nsec, depending on control and data states; 6 to 18 nsec is typical.

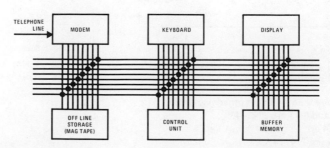

Figure 5-48   Block diagram of typical interactive computer terminal subsystems on data bus.

Furthermore, the open-collector delay is specified with nominal load capacitance of 15 pF and load resistance of 4 kΩ. The pull-up resistor, which must be varied in size with the logic assembly, may be larger than 4 kΩ. Bus capacitance is also higher as a rule. To the actual line capacitance, about 5-pF nodal capacitance must be added at each wire junction. Typical bus delays due to these RC time factors run above 100 nsec unless output buses are buffered, which has the same drawbacks as buffering input buses.

TSL does not use output pull-up resistors because outputs have active pull-up. But delays are typically less than 10 nsec, since the RC time constants are small.

The short TSL bus delays do open the possibility of logic races arising in closed-data processing loops. Any odd number of gates in a closed loop will tend to oscillate when operating nearly in synch. The usual transfer-delaying techniques could be used, but TSL offers a solution that allows high processing rates to be maintained. One simply "opens" a loop by disabling a TSL input when an output is transmitting, or vice versa. Read-in and read-out of a logic block usually are not done simultaneously.

Enable–Disable Controls

TSL devices are controlled by simple techniques, similar to those already controlling "wire-ORed" systems. The first TSL devices use either single-line or NOR-type inputs to control input selection, output selection, or memory chip-enable functions.

One control feature built into all three-state devices is a longer delay in switching from disable to enable states than from enable to disable. The difference prevents data interference by ensuring that a selected device is enabled a few nanoseconds after the previously selected device is disabled. Single changes in system state variables, such as transition of one memory address bit, will control large arrays of TSL devices reliably.

Single-line control is illustrated by the three-state bus buffers in Figure 5-49. The two types have opposite control "sexes" and may be switched in pairs, quads, and so on, by one control line in complementary fashion. Prime application is interfacing conventional logic modules with buses, but these "gates" have many switching uses. A pair of quad buffers, for instance, will multiplex two 4-bit sets of bus lines to a main data bus.

NOR-type inputs must both be low to enable a device which facilitates $X–Y$ selection techniques in large systems. In Figure 5-50, two TTL BCD decoders with active low outputs control up to 100 TSL devices. One input may be grounded or tied to the other if single-line select is desired.

TSL Memories

TTL random-access and read-only memory circuits with TSL outputs are more expandable than memory ICs with open-collector outputs. Up to 128 TSL blocks can be connected to an output data bus, making very large, high-speed memories practical without special output buffering techniques. The TSL circuits are pin-compatible with open-collector TTL memory blocks.

New memory designs also have TSL word-address inputs. The inputs as well as the outputs remain in the high-impedance state until the correct chip-enable code is detected by the control inputs. The 16 ROMs in the expanded array of

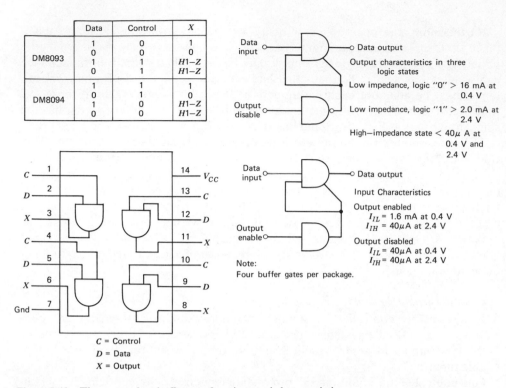

	Data	Control	X
DM8093	1	0	1
	0	0	0
	1	1	$H1-Z$
	0	1	$H1-Z$
DM8094	1	1	1
	0	1	0
	1	0	$H1-Z$
	0	0	$H1-Z$

Data output

Output characteristics in three logic states

Low impedance, logic "0" > 16 mA at 0.4 V

Low impedance, logic "1" > 2.0 mA at 2.4 V

High—impedance state < 40μ A at 0.4 V and 2.4 V

Data output

Input Characteristics

Output enabled
I_{IL} = 1.6 mA at 0.4 V
I_{IH} = 40μA at 2.4 V

Output disabled
I_{IL} = 40μA at 0.4 V
I_{IH} = 40μA at 2.4 V

Note:
Four buffer gates per package.

C = Control
D = Data
X = Output

Figure 5-49 Three-state bus-buffer gate functions and characteristics.

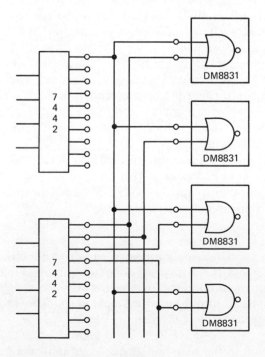

Figure 5-50 TTL decoder control of three-state array.

150

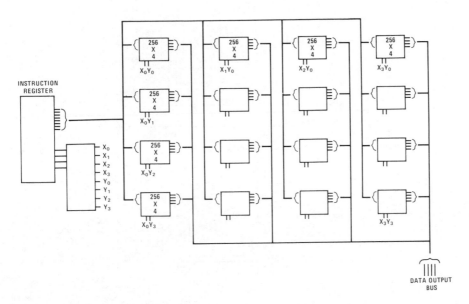

Figure 5-51 Expanded ROM organization.

Figure 5-51 have chip-address decoder inputs. The diagram represents a plane of 4096 4-bit words or word segments (four such planes in parallel, e.g., would store 4096 16-bit words in 64 ROM circuits).

Note that each output of the decoder drives only four chip-enable inputs in the 4×4 plane. Also, only one of the 16 ICs has its word-address inputs (register outputs) enabled at any time. TSL register outputs would easily drive the entire plane directly; in fact, they could drive four such planes at high speed. Conventional TTL ROMs would present the register outputs with more than 100-mA sink current requirement if addressed in groups of 64, whereas by the TSL arithmetic previously cited, TSL inputs require only 8.8 mA to drive the selected input in each of four planes.

This is an excellent example of fan-in and fan-out improvement. Moreover, the TSL input–output bus delays will total only 10 nsec per access, whereas conventional blocks with buffered inputs and unbuffered outputs have about 115 nsec total bus delays. The TSL blocks will have about the same internal delays as TTL blocks, 50 nsec. In other words, the memory cycle time is halved by TSL.

TSL Storage Registers
Another device with TSL inputs and outputs, a quad D flip-flop, is shown in Figure 5-52. This was the first TRI-STATE device produced in quantity. Among its applications are computer registers, variable-delay buffer memories, "store and forward" multiplexers, time-sharing of displays, and busing the outputs of MOS ROMs addressed by high-speed, overlapped addressing techniques. MOS outputs lose speed when they are bus-connected; a TSL interface solves that problem, too.

A unique feature of the flip-flops is that all registers may be synchronously clocked. The clock is gated to a module at the same time the TSL inputs are

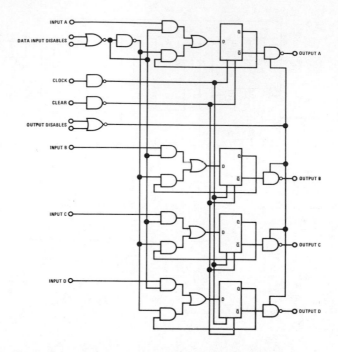

Figure 5-52   The DM7551/DM8551 TRI-STATE Quad-*D* flip-flop.

enabled for data read-in. False-clocking problems are eliminated because the clock does not have to be multiplexed, unlike conventional *D* flip-flops. After the input has been disabled, the stored bit in each binary recirculates without changing state. The data is automatically clocked out when the outputs are enabled.

Figure 5-53 is another unique design. Each latch has a single input–output (I/O) pin, allowing the devices to be hung on system I/O buses like appendixes. They will store data and read them out nondestructively like a scratchpad. They will also reformat data. For instance, shuffling data through a series of these 8-bit latches simulates operation of a push-down stack memory. The bits entered last can be read out first. There is no need to obtain stack operation in the usual way — by transferring data into main memory.

Data-Switching Devices

Data routing is a natural TSL function. The fact that the 2-wide, bidirectional multiplexer of Figure 5-54 uses five fewer gates than a conventional design illustrates the efficiency of TSL in data-transfer applications (the "gates" here are the bus buffers of Figure 5-49).

More complex switching chores are handled by the demultiplexer of Figure 5-55 and the multiplexer in Figure 5-56. The demultiplexer switches either of two inputs to any of its four outputs. Inputs may also be complemented while being switched. Examples of its bus-interchange applications are given in Figure 5-57. Each of the switches could actually be a multichannel assembly of many bus-connected demultiplexers. The multiplexer in Figure 5-56 operates in 8:1 or dual 4:1 modes like a TTL multiplexer. However, the TSL outputs permit multi-

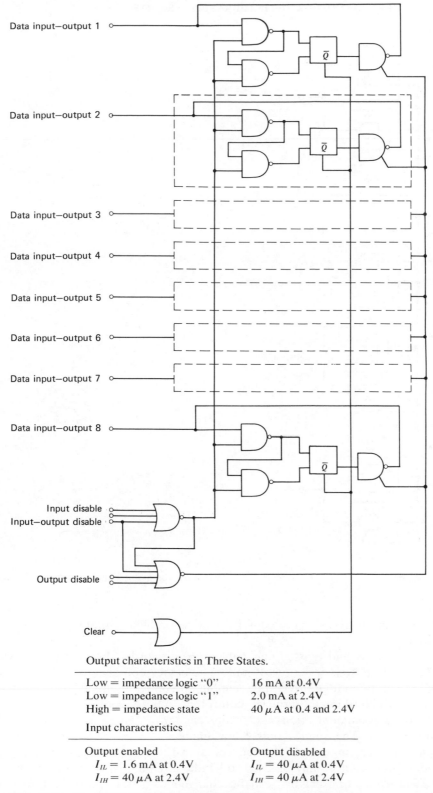

Data input–output 1

Data input–output 2

Data input–output 3

Data input–output 4

Data input–output 5

Data input–output 6

Data input–output 7

Data input–output 8

Input disable
Input–output disable

Output disable

Clear

Output characteristics in Three States.

Low = impedance logic "0"	16 mA at 0.4V
Low = impedance logic "1"	2.0 mA at 2.4V
High = impedance state	40 μA at 0.4 and 2.4V

Input characteristics

Output enabled	Output disabled
I_{IL} = 1.6 mA at 0.4V	I_{IL} = 40 μA at 0.4V
I_{IH} = 40 μA at 2.4V	I_{IH} = 40 μA at 2.4V

Figure 5-53   The DM7553/DM8553 TRI-STATE 8-bit storage element.

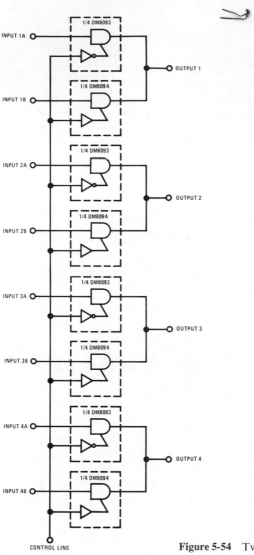

Figure 5-54 Two-wide 4-bit multiplexing with three-state gates.

plexers with as many as 512 channels to be assembled without the usual sub-multiplexers.

Very long buses and twisted-pair cables are driven by the drivers in Figure 5-58. The outputs sink or source 40 mA. The device contains four single-ended drivers or two differential drivers, depending on control inputs to the right-hand NOR. The two NOR inputs at the left control the four TSL outputs.

For two-way communications under logical control, the circuit in Figure 5-59 is recommended. The single control line enables the driver to transmit while disabling the receiving gates, or vice versa. Any number of gate–driver pairs under single-line control may be used to handle any number of bits in the words bused. This data "transceiver" can be changed to a transmitter and an independent receiver simply by having one control line for the drivers and one for the

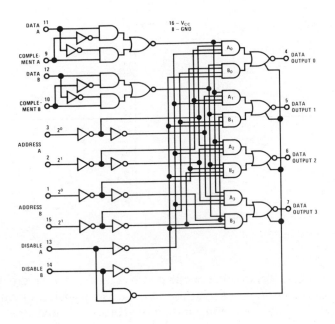

DATA A	COMP. A	DATA B	COMP. B	ADDRESS A 2^1	ADDRESS A 2^0	ADDRESS B 2^1	ADDRESS B 2^0	DIS. A	DIS. B	OUT 0	OUT 1	OUT 2	OUT 3
0	0	X	X	0	0	X	X	0	1	0	1	1	1
0	1	X	X	0	0	X	X	0	1	1	1	1	1
1	0	X	X	0	0	X	X	0	1	1	1	1	1
1	1	X	X	0	0	X	X	0	1	0	1	1	1
0	0	X	X	0	1	X	X	0	1	1	0	1	1
0	1	X	X	0	1	X	X	0	1	1	1	1	1
1	0	X	X	0	1	X	X	0	1	1	1	1	1
1	1	X	X	0	1	X	X	0	1	1	0	1	1
0	0	X	X	1	0	X	X	0	1	1	1	0	1
0	1	X	X	1	0	X	X	0	1	1	1	1	1
1	0	X	X	1	0	X	X	0	1	1	1	1	1
1	1	X	X	1	0	X	X	0	1	1	1	0	1
0	0	X	X	1	1	X	X	0	1	1	1	1	0
0	1	X	X	1	1	X	X	0	1	1	1	1	1
1	0	X	X	1	1	X	X	0	1	1	1	1	1
1	1	X	X	1	1	X	X	0	1	1	1	1	0
X	X	0	0	X	X	0	0	1	0	0	1	1	1
X	X	0	1	X	X	0	0	1	0	1	1	1	1
X	X	1	0	X	X	0	0	1	0	1	1	1	1
X	X	1	1	X	X	0	0	1	0	0	1	1	1
X	X	0	0	X	X	0	1	1	0	1	0	1	1
X	X	0	1	X	X	0	1	1	0	1	1	1	1
X	X	1	0	X	X	0	1	1	0	1	1	1	1
X	X	1	1	X	X	0	1	1	0	1	0	1	1
X	X	0	0	X	X	1	0	1	0	1	1	0	1
X	X	0	1	X	X	1	0	1	0	1	1	1	1
X	X	1	0	X	X	1	0	1	0	1	1	1	1
X	X	1	1	X	X	1	0	1	0	1	1	0	1
X	X	0	0	X	X	1	1	1	0	1	1	1	0
X	X	0	1	X	X	1	1	1	0	1	1	1	1
X	X	1	0	X	X	1	1	1	0	1	1	1	1
X	X	1	1	X	X	1	1	1	0	1	1	1	0
X	X	X	X	X	X	X	X	1	1	Hi-Z	Hi-Z	Hi-Z	Hi-Z

Figure 5-55   The DM7230/DM8230 TRI-STATE demultiplexer.

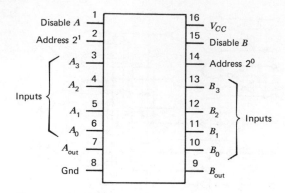

Figure 5-56　DM7214/DM8214 dual 4 : 1 or single 8 : 1 multiplexer.

gates. Large numbers of such independent driver and receiver groups may be placed at the ends of bus branches to accommodate data transfers among peripherals and subsystems, using a common, bidirectional bus structure.

Alternatively, the drivers may be used in a differential transmission system, such as Figure 5-60. Either way, the transmitters and receivers eliminate any need for separate multiplexers and demultiplexers. This should not be tried with conventional drivers — they would destroy one another if their outputs faced one another at these high current levels.

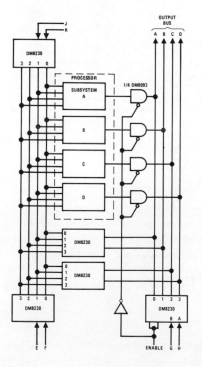

Figure 5-57　Bus-interchange examples.

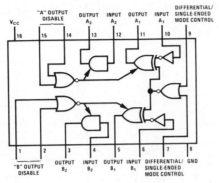

"A" OUTPUT	DISABLE	DIFFERENTIAL/ SINGLE-ENDED MODE CONTROL		INPUT A_1	OUTPUT A_1	INPUT A_2	OUTPUT A_2
0	0	0	0	Logical "1" or Logical "0"	Same as Input A_1	Logical "1" or Logical "0"	Same as Input A_2
0	0	X 1	1 X	Logical "1" or Logical "0"	Opposite of Input A_1	Logical "1" or Logical "0"	Same as Input A_2
1 X	X 1	X	X	X	High impedance state	X	High impedance state

X = Don't Care

Figure 5-58 The DM7821/DM8831 TRI-STATE line driver.

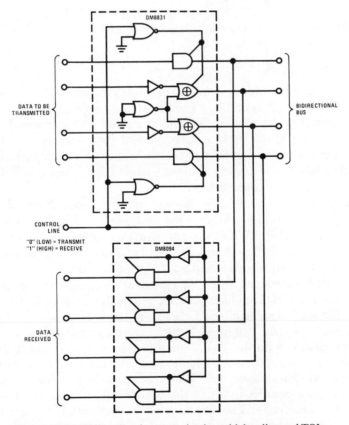

Figure 5-59 Bidirectional communication with bus line and TSL.

157

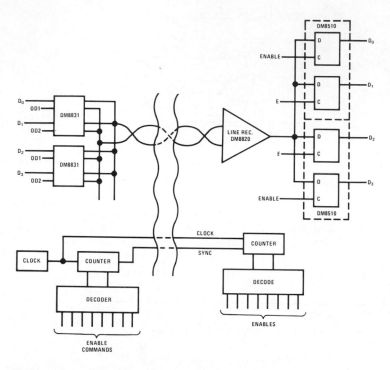

Figure 5-60   Multiplexing with TRI-STATE drivers.

Unusual Applications

The device in Figure 5-61 does not fit into any single category. It is a 4-bit counter, latch and multiplexer all in one IC. The counters cascade to 8 bits, 12 bits, and so on. The three-state outputs of the latches do the multiplexing. Uses include time-sharing of peripherals such as displays and data loggers. For instance, in Figure 5-62, the counters sample control events, transfer their contents on command to their latches, and resume sampling while the latches await their turn to use the single-indicator decoder-driver.

Many recent applications of TSL do not involve buses. Some designers are more interested in the other features. One designer is using TSL to improve the fan-out of low-power (LP) TTL in systems using a combination of LPTTL, TTL, and MOS. The best LPTTL has only a fan-out of 2 at low temperature into standard TTL, and a fan-out of one standard TTL load is all that the specifications guarantee at high temperature. But LPTTL can drive extra TSL inputs when only one input is enabled at a time. Therefore, he puts a device with TSL inputs where storage or switching is needed after a LPTTL function.

Functions such as data "slipping," clock complementing, and clock division may be accomplished with the TSL demultiplexer or driver. These devices perform many ordinary—and unusual—logic functions when data, complement, and control inputs are cross-connected. Some connections provide an output pulse only when one data input is true and the other false, and some when both data inputs have the same state. This implies that control-interlock signals could be generated and distributed by the same device. It would not make sense to use

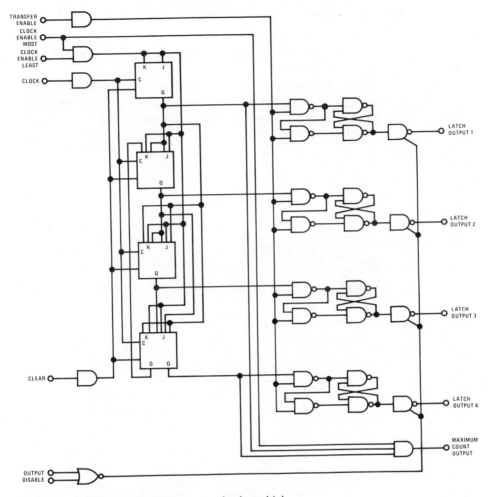

Figure 5-61 DM7552/DM8552 counter–latch–multiplexer.

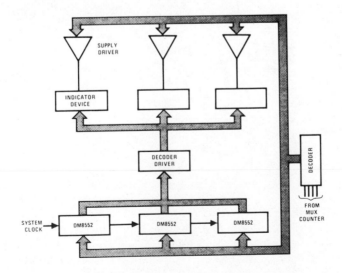

Figure 5-62 Display system with only one decoder–driver.

159

such complex devices rather than simpler conventional gates for the ordinary functions — as data are bused.

As has been shown, the first three-state devices are mostly MSI designs. The main benefit to the system designer is that TSL greatly simplifies the design and construction of bus-organized systems and gives him many new operating modes that can be employed to enhance the performance and increase the modularity of future systems.

Chapter Six

METAL–OXIDE–SEMICONDUCTOR (MOS) INTEGRATED CIRCUITS

MOS (metal-oxide-semiconductor) technology gets its name from the basic MOS structure of a metal electrode over an oxide dielectric over a semiconductor substrate. The transistors of an MOS IC are field-effect transistors (FETs), which are also used as resistors, because the MOSFET is a high-impedance device. Most MOS monolithic ICs are digital circuits built entirely of MOSFETs.

One of the most exciting features of the MOS transistor is its simplicity. An electric field, applied through the oxide-insulated gate electrode, is used to control the conductance of a channel layer in semiconductor material under the gate. The channel is a lightly doped region between two highly doped areas called the source and the drain. Compared with monolithic bipolar ICs, MOS ICs may be characterized by slower operation and lower output power. As with many different technologies, there are application areas in which the MOS structure clearly excels and others in which is does not compare favorably.

The MOS transistor is an almost ideal switch since, when the gate and source potentials are equal, no current flows between the source and the drain. However, when the gate voltage, with respect to the source, is raised to a critical level (called threshold voltage) the transistor turns ON and current can flow from source to drain. The threshold voltage is negative on the gate of this type of MOS transistor, to establish an electrostatic field that inverts the n material under the gate to a p channel between the source and drain.

Figure 6-1 depicts a typical p-MOS structure. The MOS transistor is a self-isolating device, because all junctions are reverse biased during normal operation.

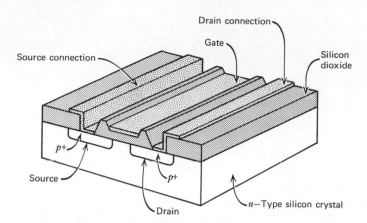

Figure 6-1 The *p*-MOS structure. This device shows the conventional metal–oxide–semiconductor gate sandwich that gives the technology its name. Also apparent are the overlaps between gate and source, and between gate and drain. This overlap increases parasitic capacitances and limits the speed of devices with this structure.

This is extremely desirable when making ICs. Since isolation diffusions are not needed, the packing density of the transistors on a chip is extremely high.

With many MOS transistors on a chip, however, a parasitic problem can arise. The key to avoiding the problem is in the manufacturing process. A parasitic transistor is formed between two adjacent *p*+ regions when a high-voltage metal line crosses them. Unless the "field" oxide under this line is thick enough, the high voltage inverts the surface of the *n*-type substrate and turns on the transistor that is formed. This parasitic transistor is exactly like a conventional transistor, except it has thicker gate oxide and, therefore, a higher threshold voltage.

The manufacturing process must provide a field oxide thick enough to place the thresholds of the parasitic transistors above the maximum voltages in the circuit. Since these voltages are often greater than -30 V, the field oxide must be at least 1.5μ thick. This oxide can be thermally grown, but it takes a long time. More often, the thick oxide is deposited. Methods of deposition include electron-beam evaporation, r-f sputtering, and the thermal oxidation of silane. The last method is the most common because it is the cleanest and most reproducible.

A problem incurred with MOS structures is the limitation in gate voltage due to the relatively thick oxide utilized in these devices. Since the charge necessary for destruction is only 10^9 C, it is necessary to protect each external lead with a breakdown device such as a zener diode. This breakdown device must be included in the mask design for MOS structures.

MOS technology consists of three basic processes: *p*-channel, *n*-channel, and CMOS, as in Figure 6-2. These are now discussed.

p-MOS DEVICES

The most common device, the *p*-type enhancement mode, is built on a substrate of *n*-type silicon into which are diffused two *p* regions: the source and the drain

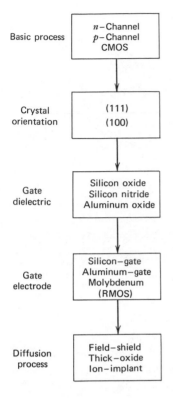

Basic process	n – Channel p – Channel CMOS
Crystal orientation	(111) (100)
Gate dielectric	Silicon oxide Silicon nitride Aluminum oxide
Gate electrode	Silicon–gate Aluminum–gate Molybdenum (RMOS)
Diffusion process	Field–shield Thick–oxide Ion–implant

Figure 6-2   MOS manufacturing processes.

(Figure 6-3). The discussion to follow pertains to the *p*-type enhancement-mode MOS IC. These are normally formed by diffusing two wells of *p*-type impurity (phosphorus) into the substrate, which in operation are connected by an induced *p* region, which is the channel.

The gate or control element covers the region between the source and the drain and is insulated from the semiconductor material by a layer of silicon oxide. The input resistance of the gate is extremely high—on the order of $10^{18}\ \Omega$—and the input impedance at high frequencies is almost purely capacitive. The gate is a layer of metal, usually aluminum, as are the contacts to the source and the drain. Normally the oxide layer under the gate is made much thinner than the protective oxide on the rest of the chip, in order to enhance the effect of the gate field on the conductance of the channel region.

If the gate, the source, and the substrate are grounded and a negative voltage is applied to the drain (Figure 6-3*b*), no current will flow between the source and drain because they are isolated from each other by the reverse-biased drain-to-body *p-n* junction.

If a negative voltage is applied to the gate (Figure 6-3*c*), the surface of the *n*-type silicon inverts, becoming essentially *p* type. The negative gate voltage attracts holes from the *n*-type substrate to the surface. Initially, the channel area very near the surface has an excess of electrons, because the material is *n* type, but the holes drawn into the area by the gate field neutralize these electrons. At some gate voltage the attracted holes just compensate for these electrons, and the channel behaves like the intrinsic semiconductor. At higher gate voltages, the holes pre-

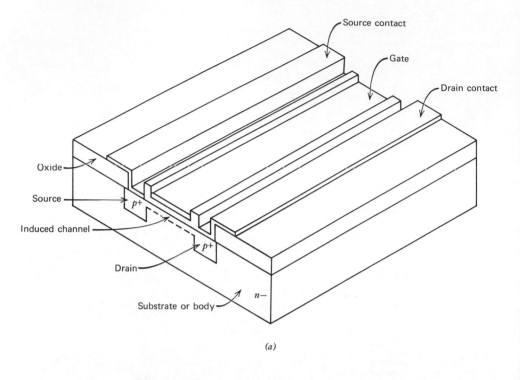

(a)

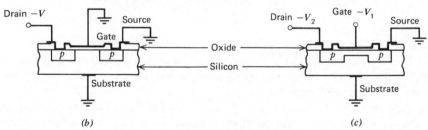

(b)   *(c)*

Figure 6-3   The *p*-MOS device: (*a*) basic structure, (*b*) nonconducting state, (*c*) conducting state.

dominate, and the channel area, less than 1 μ deep, is referred to as "inverted" —
it now behaves like a *p*-type semiconductor, providing a current path from source
to drain.

However, the surface region under the gate does not invert, and no conduction
can occur until the gate voltage is more negative than the threshold voltage V_T,
which is about -5 V for most *p*-channel enhancement-mode devices. This effect
results, in part, from the presence of impurity charge in the silicon, which must be
neutralized before the channel region can invert. In general, the thinner the gate
oxide, the lower the threshold voltage. Figure 6-4 depicts the turn-ON characteristics of a typical MOSFET.

As the gate voltage V_{GS} becomes more negative than the threshold V_T, the conducting channel is formed, and its depth increases with increasingly negative gate
voltage. For low drain current, the channel is an ohmic resistance, and the current
I_D is directly proportional to the drain-to-source voltage V_{DS}. As V_{DS} becomes

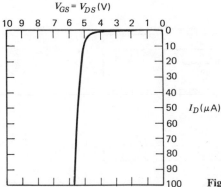

Figure 6-4 Turn-on characteristics of a typical MOSFET.

more negative, however, the channel saturates and the current levels off (Figure 6-5).

The saturation phenomenon is easily understood. Assume that the device is operated with the source grounded and the gate at -12 V. If the drain voltage is zero V, no current flows, even though a channel exists. As the drain voltage is made negative, current flows from the source to the drain through the resistance channel (Figure 6-6). The voltage difference between the gate and the body of the device is -12 V at the left and decreases along the length of the channel (because of the resistive voltage drop) to a minimum of $-12-(-V_{DS})$ V at the drain. This voltage difference determines the extent to which a channel is formed in the substrate material.

If the negative voltage $-V_{DS}$ increases enough, the gate-to-body voltage at the drain $-12-(-V_{DS})$ approaches the threshold voltage V_T, and the voltage near the drain is just sufficient to form a channel at that point. If V_{DS} is made still more

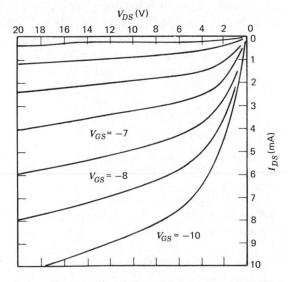

Figure 6-5 Drain characteristics of a typical MOSFET.

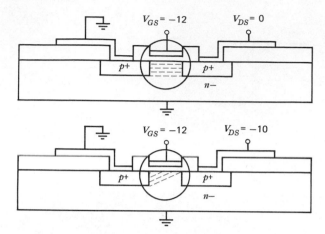

Figure 6-6 The saturation phenomenon: drain-to-source voltage effects are limited by channel pinchoff.

negative, the inversion channel terminates short of the drain; the drain current is limited and becomes independent of further changes in V_{DS}.

If a voltage is applied to the drain when the device is off (i.e., when $|V_{GS}| < |V_T|$), ideally no current flows. There will, however, be the small drain-to-body current, which varies with junction area as well as with the bulk resistivity of the material. For junctions of the size normally encountered in discrete MOSFETs, this current is in the nanoampere range; for the smaller junctions normally used in MOS ICs, it may be as low as a few picoamperes.

p-MOS Load Resistances

Resistance in a double-diffused IC is normally obtained by using the resistivity of doped silicon. A typical value of sheet resistivity is 100 $\Omega/\square$. Thus 200 $\square$ in series is required to obtain a 20-kΩ resistor that would be 2000 μ long and occupy an area of approximately 32 mil². By comparison, the area of a typical MOSFET is less than 1 mil². It is fortunate, therefore, that in an MOS IC the load resistance for an inverter can be supplied by another MOSFET.

In any MOS device, if

$$|V_{GS} - V_T| = |V_{DS}| \qquad (6\text{-}1)$$

a channel exists from source to drain.

If we have

$$|V_{GS} - V_T| \leq |V_{DS}| \qquad (6\text{-}2)$$

the channel terminates short of the drain, and the drain-to-source current I_{DS} is relatively independent of the drain-to-source voltage V_{DS}. The device is then said to be operating in the *saturated* region. If

$$|V_{GS} - V_T| \geq V_{DS} \qquad (6\text{-}3)$$

the device is biased in the *unsaturated* region.

In the case of an MOS device used as a load, as in Figure 6-7, the equations for saturated and unsaturated operation become, respectively:

$$|V_{GG} - V_T| \leq |V_{DD}| \qquad \text{(saturated operation)} \qquad (6\text{-}4)$$

$$|V_{GG} - V_T| \geq |V_{DD}| \qquad \text{(unsaturated operation)} \qquad (6\text{-}5)$$

p-MOS Fabrication

It is worthwhile to discuss the steps used in fabricating a p-MOS device. Since, typically, the processing steps are essentially the same for all MOS technologies, the detailed processing steps are presented for the p-MOS structure only.

The starting material is single-crystal silicon that is doped n type with phosphorus or antimony. The doping level is usually on the order of 10^{15} atoms/cm³, which is several orders of magnitude below the density of the bulk silicon atoms. The first step is to grow a relatively thick oxide layer, maybe $1.5\ \mu$ (15,000 Å). Then holes are etched for the source-to-drain diffusion.

Both the source and drain regions are typically boron doped and are from 2 to 4 μ deep. Diffusion and oxidation are performed at the same time because the diffusion occurs in an oxidizing atmosphere. The next step is to form the gate oxide, the one that is going to serve as the dielectric used for turning on and off the MOS device. This oxide layer is grown on the silicon, essentially forming a capacitor. The dielectric of the capacitor must be of uniform and closely defined thickness and must also possess good electrical characteristics. This is accomplished by etching all the oxide off the gate region and regrowing oxide to the thickness that is desired.

The gate oxide is the most critical part of the MOS IC. Its thickness must be very close to the design value; it can have no "pinholes," or the device will be shorted; and it must be pure, or the device will be unstable electrically. Also, this oxide has to be thick in order to reduce capacitive coupling. The thickness of the oxide also is the major factor determining the breakdown voltage, since there is an inverse relationship between them.

Next, the entire circuit is metallized and etched so that there is metal over the gate, the drain, and the source. The metal referred to in MOS is almost always aluminum, usually between 1 and 2 μ thick, and deposited using an electron-beam evaporator (as was shown in Figure 2-28). Figure 6-8 is a typical process flow

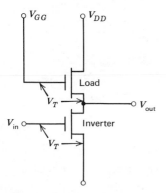

Figure 6-7 Example showing an MOS device used as a load.

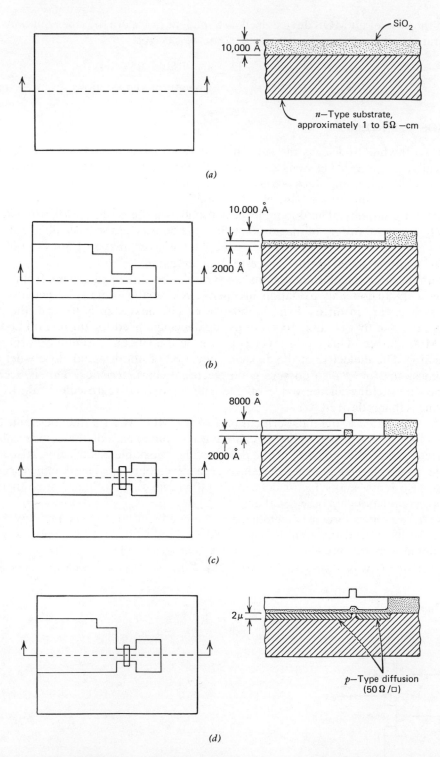

Figure 6-8 Typical MOS process sequence: (*a*) initial oxidation, (*b*) partial etch, (*c*) source-and-drain etch, (*d*) source-and-drain diffusion and oxidation.

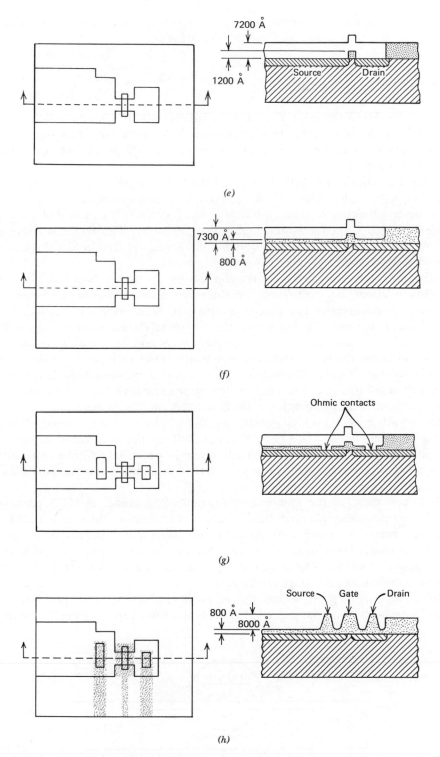

Figure 6-8 (continued)   (e) gate etch, (f) gate oxidation, (g) preohmic etch, (h) deposition of metallization and etch.

169

chart with major steps required to process MOS circuits for p-channel circuits. n-Channel devices would follow a similar sequence.

THE n-MOS STRUCTURE

A typical n-MOS structure is shown in Figure 6-9. ICs made with n-channel transistors seem to offer advantages over the customary p-channel devices. The most obvious advantage stems from the higher mobility of the charge carriers in an n-channel device.

Since the majority of MOS devices available today are p channel, the obvious question is Why, if n channel offers size and speed advantage? p-Channel circuits are much easier to make. p-Channel MOS is a highly controllable, low-cost process; it has good yields and high noise immunity. The wafer surface of n-channel MOS is very sensitive to contaminants, and the threshold voltages are in an undesirable range.

p-Channel transistors use holes for conduction. At normal field intensities, hole mobility is about $200 \text{ cm}^2/\text{V-sec}$. On the other hand, n-channel transistors use electrons to accomplish the charge conduction. Since electron mobility is $400 \text{ cm}^2/\text{V-sec}$, twice that of hole mobility, an n-channel device will have one-half the ON resistance or impedance of an equivalent p-channel device with the same geometry and under the same operating conditions. Thus n-channel transistors need be only half the size of p-channel devices to achieve the same impedance. Therefore, n-channel ICs can be smaller for the same complexity or, even more important, they can be more complex with no increase in silicon area.

Along with greater packing density, n-channel circuits offer a speed advantage over p-channel circuits. This is a direct result of smaller junction areas, since the operating speed of an MOS IC is largely limited by internal RC time constants. A diode's capacitance is directly proportional to its size, and n-channel junctions can be smaller than p-channel junctions.

However, most of the mobile contaminants (the dread of MOS processing people) are positively charged. Since n-channel transistors operate with the gate positively biased with respect to the substrate, these ions collect along the oxide–silicon interface. The charge (Q_{po}) from this layer of ions causes a shift in threshold voltage which tends to make the transistor normally ON (current flows from source to drain with no voltage on the gate).

Beside the mobile positive ions, a fixed positive charge also exists at the oxide–silicon interface. This charge, called Q_{ss}, results from various steps in the manu-

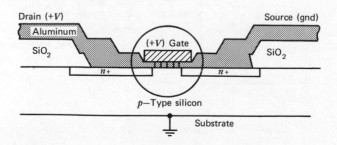

Figure 6-9   A typical NMOS structure.

facturing process and also tends to make the device normally ON. These two charges (Q_{po} and Q_{ss}) exist in p-channel devices too, but the positive ions are pulled to the aluminum–oxide interface by the negative gate bias. There, they cannot affect the device threshold. In addition, the fixed charge in a p-channel device gives higher thresholds. Although this is undesirable, it is not as severe a condition as having the device permanently biased ON.

For an n-channel device, the oxidation of the silicon takes place at the silicon–silicon dioxide interface. No real abrupt change occurs between silicon and silicon dioxide; rather, there is a transition zone. This transition zone contains positively charged silicon atoms which increase the absolute magnitude of the threshold voltage for a p-channel device and decrease the absolute magnitude of the threshold voltage for an n-channel device. The result is that it is difficult to make an n-channel device that is OFF at zero gate voltage. This is why it is more difficult to make an n-channel device than a p-channel device.

Most n-channel devices have moderate doping levels and are, in fact, normally ON devices. When used, the circuit designer provides a negative bias to the source to turn the device OFF. The p-MOS ICs are not TTL compatible; the n-MOS ICs are TTL compatible. All in all, this presents a fairly complicated picture. It is difficult to design circuits using n-channel (an extra bias supply is needed, for one thing), and it is difficult to control the process required to manufacture them. Today however, n-channel devices are becoming more economically competitive with p-channel devices.

THRESHOLD VOLTAGE

The threshold voltage of an MOS transistor is the most important process-dependent device parameter. It is desirable in most cases to have a process that produces low threshold voltages. An IC with low-threshold transistors will operate at lower power-supply voltages than a high-threshold circuit (-5 and -15 V for the low-voltage circuits versus -13 and -27 V for the high-voltage circuits). This means lower-cost power supplies, and lower systems cost. An even more desirable feature is that the low-voltage circuit is directly compatible with bipolar ICs; that is, it requires and produces the same input and output signal swings. This compatibility gives a system designer more flexibility. He can design using both MOS and bipolar circuits without worrying about signal level shifting or interfacing.

In addition, low signal voltages also imply higher operating frequencies. If a voltage only has to swing between 0 and -5 V, it can change state faster than a voltage swinging between 0 and -10 V can.

The advantages of low-voltage MOS are apparent; unfortunately, the best method of making it is not. There are several ways of modifying processes and device structures to achieve lowered threshold voltages.

LOW– VERSUS HIGH–THRESHOLD TECHNOLOGY

Low-threshold MOS circuits have three major advantages over the high-threshold circuits. First, they simplify the task of establishing an interface between the MOS circuit and a bipolar circuit. Second, they substantially improve the MOS circuit's speed–power product. Finally, they facilitate generating bias and clock levels. To appreciate the actual importance of these three advantages, the high-threshold circuit's shortcomings must be fully understood.

Because the low-state transistor turns ON with only 2 V on its gate – in some designs the voltage is even less – a simple interface between a driving bipolar circuit and a driven MOS circuit is possible. (The interface between a driving MOS circuit and a driven bipolar circuit does not involve the MOS threshold and is equally simple with both kinds of MOS circuits.) This simplicity is attributable to the capacity of the MOS chip's input signal to have positive and negative levels quite close together – as close as 0 and 4 V – and still have enough over-drive to nearly saturate the input MOS transistor. Normally, the substrate of either bipolar or MOS circuits is connected to system ground; the main supply voltage for DTL and TTL circuits is +5 V; and the signal swing for the most common variety of p-channel MOS is between ground and a negative voltage. Therefore, by biasing the MOS substrate to +5 V, which is the bipolar supply voltage, the MOS input signal swing goes from 0 to −4 V and from +5 to +1 V. This transition makes the MOS input easy to generate with bipolar circuits.

With the low-threshold circuits, the supply voltage can be as low as 5 V for slow circuits that switch in a few microseconds, and 10 V for more typical speeds of 1 μsec or less. The lower voltages result in lower power dissipation and a speed–power product lower than that obtainable with high-threshold circuits. (In speed–power products, little numbers are better than big numbers, because speed in this context actually is the switching time in seconds, rather than frequency in reciprocal seconds, the usual dimension of speed.)

In addition to the interface and speed–power advantage, the low-threshold circuits require smaller bias (half the power supply requirements of high-threshold process) and clock levels. Low-threshold MOS circuits, however, have some disadvantages, including a lower field inversion threshold, slow speeds, and low noise immunity.

There are three ways to build low-threshold circuits: by properly orienting the crystal structure in the semiconductor layer, by using silicon nitride as the insulating material instead of silicon oxide, and by using doped silicon instead of metal in the top layer.

Originally the low-threshold process was developed around the use of silicon crystal cut along the (100) plane, rather than the (111) plane traditionally used for MOS transistors (see Figure 6-10). The threshold is reduced because the silicon's surface-state charge is less along the (100) plane than along the (111) plane. This surface-state charge is established, in part, by uncommitted bonds between atoms in the crystal; the (111) plane has more of them than the (100) plane. The (100) orientation is required for beam-lead products so that the antisotropic etch can be performed to separate the chips.

The low threshold voltage obtained with the (100) crystals is also present in parasitic transistors that invariably show up between adjacent MOS transistors on a single chip. These parasitic transistors operate because the clock voltage is considerably higher than the supply voltage, relative to system ground; thus it can establish a field-effect channel in an undesirable area. The tendency of these undesired channels to form can be offset, but every compensation has its trade-offs. For example, the insulating oxide layer can be made thicker in the places where

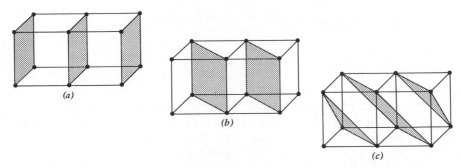

Figure 6-10 The simplest cubic lattice contains three basic crystal planes: (*a*) (100), (*b*) (110), (*c*) (111).

parasitic transistors are likely to form; this increases the separation between the metal layer and the substrate, thus decreasing the electric field intensity in the substrate and retarding the tendency of spurious channels to form. But increasing oxide thickness results in a nonuniform and "rough"-appearing metallization, which possesses the possibility of developing "microcracks." Or the substrate material can be chosen with a different resistivity before any diffusions or depositions are made; this can alter the characteristics of any spurious channels that do form so that they cause less trouble. But it also adversely affects the characteristics of the other transistors.

Another method of compensating for parasitic transistors is to diffuse a barrier of *n*+ material between the channel areas. Since these barriers have a higher threshold voltage than the *n* material of the substrate, spurious channels are less likely to form across them. In addition, by taking advantage of the processing step required to add the *n*+ material, the designer gains the possibility of making *npn* bipolar transistors on the same chip with the MOS devices. Table 6-1 compares the salient features of the (100) process with those of the (111) process.

A second way to obtain a low threshold voltage is to use silicon nitride in place of silicon oxide as the insulating layer between the gate and the channel. Threshold voltage is inversely proportional to gate capacitance, and nitride increases capacitance because its dielectric constant is twice that of oxide. Furthermore, nitride can be used on silicon with a (111) cut; here, parasitic transistors are a less severe problem because there are fewer uncommitted bonds in the (111) plane. But silicon nitride has its disadvantages, too. Among these is a tendency for the threshold to shift when the transistor is strongly biased.

A third way to obtain a low threshold voltage is to fabricate the gate structure from heavily doped silicon instead of metal, again with the substrate material cut along the (111) plane (Figure 6-13). Silicon gates are made of *p*-type polycrystalline silicon, whose work function is less than that of the aluminum used in ordinary MOS circuits. The difference between the work functions of the gate and the semiconductor therefore is less, and this influences the threshold voltage both directly and through a reduced surface-state charge.

In addition, silicon gates offer several advantages in fabrication: stable process, automatic gate alignment, the possibility of mixing both bipolar and MOS circuits on the same substrate with the *n*+ barrier diffusion, higher packing densities, and a

Table 6-1   Comparison of the (100) and (111) MOS Processes

	(100) (Low Threshold)	(111) (High Threshold)
Power supplies	17 V (+5, −12)	27 V (−27, −13, or +10, GND, −17)
Leakage current	ca. 10 times lower than (111)[a]	
Input Protection	Improved by crystal orientation[a]	
Max operating temperature	−55°C to +125°C	−55°C to +85°C
Min operating frequency (guaranteed)	600 Hz at 25°C 2.5 kHz at 70°C	10 kHz
Breakdown voltage	22 V[b]	30 V
Threshold voltages	−1.8 to −3.0 V[c]	−3.0 to −6.0 V
Bipolar compatibility	No interface required	Level shift logic required
Number of process steps	Less	More
Speed	Slower	Improved − depending on product type
Cost	Less	More
Yield	Higher	Less

[a]The (100) crystal has fewer surface recombination states than the (111) crystal, which provides lower inherent leakage current. It also provides a much sharper diode knee, which improves the input protection of the device. Input breakdown is virtually eliminated with the (100) process.

[b]The breakdown voltage is often considered a disadvantage of the (100) process.

[c]The lower threshold voltage is the parameter that allows bipolar compatibility with the (100) process. It also explains why many companies have difficulty using the process. Threshold-voltage variations caused by impurity ions in the silicon–silicon dioxide interface become more pronounced with the low-threshold process than with the high-threshold process.

50% increase in speed over nitride. The gate is self-aligning because it acts as a mask during the diffusion of p dopant into the source and drain areas. Thus the overlap of the gate electrode and the diffused source and drain areas can be small and accurately controlled.

As with the other low-threshold processes, the silicon-gate process has its drawbacks. For example, the silicon deposition is an extra step; metal deposition is still required because external connections cannot be made directly to the silicon. But it is possible to deposit a metal that can join such external connections to the silicon. On the other hand, with both the silicon layer and the metal layer available, an extra level of interconnection can be made directly on the chip, reducing the number of outside connections and the area of circuits.

COMPLEMENTARY–SYMMETRY MOS CIRCUITS

Conventional MOS circuits are constructed with single-polarity MOS devices — in most cases p channel, enhancement-mode transistors. As discussed previously, these devices consist of a single crystal of n-type silicon which has p-type drain and source regions diffused into it. Complementary MOS (CMOS), on the other

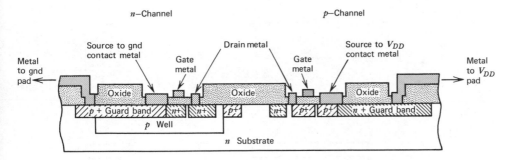

Figure 6-11   Cross-sectional view of a complementary pair.

hand, combines devices of both polarities on the same chip. In this case, the basic building block is not one, but two MOS transistors — one n-channel, enhancement-mode transistor and one p-channel, enhancement-mode transistor (Figure 6-11).

CMOS ICs are normally fabricated on an n-type substrate which serves as the substrate material for all p-channel MOS devices as well as $p+$ tunnels, diodes, and resistors. A p-type substrate is provided for the complementary n-channel MOS devices, $n+$ tunnels, diodes, and resistors by diffusing a lightly doped p-well region into the original n-type substrate. The n-channel units exhibit the higher carrier mobility associated with electrons, and they have approximately twice the transconductance of p-channel units with identical geometry. Therefore, the matching of a p-channel with an n-channel unit requires that a p-channel unit with a given channel length L have approximately twice the channel width W of the n-channel unit with which it is to be matched.

Protective guard bands surround separate MOS devices, tunnels, wells, and diodes or combinations of MOS devices that are interconnected through common diffused regions for the purpose of preventing leakage between the entities named. All p-channel devices, tunnels, and diodes must be surrounded by a continuous $n+$ guard band, which also serves as a tunnel to help conduct current from the external supply voltage V_{DD} across the n-type substrate to every p-channel device tied to the external supply. Similar heavily doped $p+$ guard bands surround all n-channel devices, tunnels, and diodes to help conduct current from the external ground supply V_{SS} across the p-well to every n-channel device tied to ground. Contact to the n-type substrate may be made through the $n+$ guard band and returned to the V_{DD} pad; contact to the p-well substrate may be made through the $p+$ guard band and returned to the ground pad. Figure 6-12 depicts this fabrication cycle. Variations of the CMOS process include the use of silicon gates and ion implantation to attain low threshold voltages, lower power consumption and higher packing densities.

Complementary MOS offers some unique advantages over single polarity devices. Probably the most notable of these is in the area of power dissipation. Because they have two transistors in series, complementary circuits consume very little quiescent power. In either logic state, one transistor or the other is OFF. This means that there is no dc path to ground in the circuit and the only power dissipated is due to the leakage through the OFF transistor. This feature makes CMOS particularly well suited for aerospace applications and in battery-operated equip-

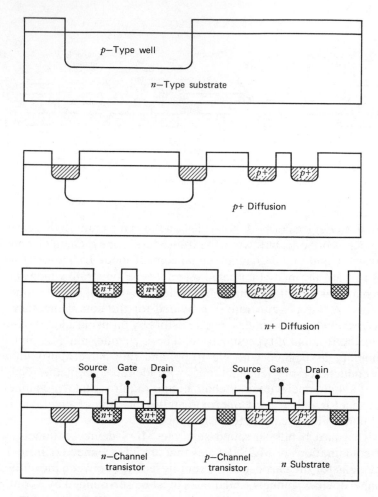

Figure 6-12   Typical CMOS device structure.

ment. Additional advantages include low standby power, high noise immunity TTL compatibility, a single power supply, wide operating voltage ranges, insensitivity to temperature variation, and ease of use. The advantages of CMOS come with a price tag attached. In this case, the penalties are larger chip size and thus lower packing density and higher cost.

The problem of packing density is fairly obvious. CMOS uses two transistors to do the job that one does in many p-channel circuits. For a relatively complex function, a CMOS chip is on the order of 25 to 30% larger than an equivalent p-channel chip. This reduced packing density also contributes to increased cost, since the price of any IC is directly related to the chip size. As chips get bigger, yields go down and costs go up.

The other factor that contributes to higher costs is processing complexity. A CMOS circuit goes through more steps because of the additional diffusion steps required. Each step in the manufacturing process adds its own increment to the total cost.

THE SILICON–GATE MOS STRUCTURE

The combination of low threshold voltage with good speed capability thrust considerable attention on the silicon-gate MOS IC. The low threshold voltage was particularly appealing because it made compatible operation with bipolar ICs possible.

As mentioned previously, the key physical difference between a conventional MOS IC and a silicon-gate IC is the latter's use of polycrystalline silicon instead of aluminum for the gate electrode. The cross section of a silicon-gate MOS device appears in Figure 6-13.

The work function of the silicon is lower than the work function of the aluminum commonly used in MOS ICs, and this results in substantially reduced threshold voltage. The silicon-gate MOS ICs that have become available since the introduction of the first silicon-gate standard IC in 1969 have used (111) crystal orientation, but other orientations have been investigated.

Fabrication begins with placement of an oxide (silicon dioxide) layer on the n-type silicon wafer. After the oxide layer is grown, windows are etched for the MOS transistors and the layer of polycrystalline silicon is deposited over the wafer. Next the polycrystalline silicon layer is etched to define the gate structure and open windows above the source and drain locations. This is followed by diffusion of a p-type dopant (such as boron) into the source and drain regions, and formation of the source and drain junctions. The dopant also diffuses into the polycrystalline silicon gate, turning it into heavily p-doped silicon.

Then another oxide layer is put down and windows are opened for the source and drain contacts. Now the IC is ready for deposition of aluminum metallization over the entire surface. To ensure that adequate gate metallization has been deposited, the gate aluminum layer of the conventional MOS transistor is permitted to overlap nearby portions of the diffused source and drain regions. This causes overlap capacitances and reduces switching-speed capability. Figure 6-14 depicts the processing steps for the silicon-gate structure. A typical silicon-gate process begins with the standard wafer of n-type silicon. The first step grows a layer of silicon dioxide (SiO_2) over the entire surface. An oxide etch removes the oxide in what will be the source, drain, and gate areas. Then the wafer goes back into the oxidizing furnace and a thin layer of oxide is grown in the thick oxide opening. At this point the device resembles Figure 6-14a.

The next steps are back-to-back depositions. The first leaves a thin layer of silicon nitride (Si_3N_4) over the entire surface. This is followed by a layer of amorphous silicon on top of the nitride (see Figure 6-14b).

Photomasking techniques remove the silicon and nitride from the entire surface

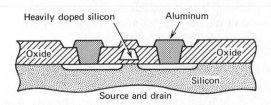

Figure 6-13   A silicon-gate MOS structure.

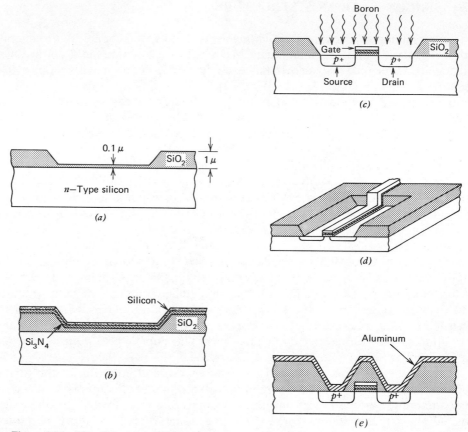

Figure 6-14   The silicon-gate MOS transistor processing sequence.

except for the gate area and where the silicon is to be used as an interconnect. Now another etch removes the thin oxide except under the silicon gate. The process then diffuses boron, a p-type impurity, onto the surface of the wafer, as in Figure 6-14c.

Boron diffuses rapidly into the silicon, but slowly in the oxide and nitride. Since the gate itself is the diffusion mask for the source and drain areas, close alignments of gate to source and gate to drain are maintained. The only overlap is due to the sideways diffusion of the boron in the silicon. A perspective view of the device after it leaves the diffusion is presented in Figure 6-14d.

Another layer of oxide over the entire surface follows. Openings for the connections to the source and drain are etched, and aluminum is evaporated into these openings. Completing the interconnection patterns finishes the device; its final cross section appears in Figure 6-14e.

The polycrystalline silicon that remains in the gate structure after etching sits on top of the oxide region created during the first oxide deposition. The polycrystalline silicon thus protects this oxide region from subsequent etching. This type of gate structure is called a self-aligning gate. The conventional type of MOS IC gate structure must include allowance for misalignment of the three masks used for source-and-drain diffusion, gate-oxide growth, and gate-metal deposition. The conventional device, therefore, requires a considerably larger gate and more gate-metal deposition than is needed for the silicon-gate transistor.

Silicon-gate MOS has three distinct advantages. First, the silicon-gate technology produces high yields. There are more good ICs on each wafer when the circuits are made of silicon-gate transistors than there would be if a more conventional technology were used. This is because the gate oxide, the most critical portion of an MOS transistor, is covered up immediately after it is grown. The deposited layer of silicon protects this oxide from etchants, contaminants, and mechanical damage.

Second, silicon-gate transistors can operate at higher speeds than conventional MOS transistors or nitride MOS (see next section) transistors. This is not because the gate is silicon, but because the gate is smaller than aluminum gates. There is no overlap of the gate conductor over the source and drain regions because the gate itself is the diffusion mask that defines the active edges of these $p+$ regions. Conventional MOS transistors must have an alignment tolerance built into the mask set. This usually produces a gate overlap of from 3 to 5 μ, which increases the feedback capacitance between the gate and the drain. Increased feedback capacitance reduces the usable gain of the transistor.

A third big advantage of the silicon-gate IC is its compatibility with other technologies. Since the gate oxide is covered as soon as it is grown, the wafer in process can go through additional high-temperature oxidations and diffusions after the MOS transistors are formed. Therefore, MOS and bipolar transistors can be designed and manufactured on the same piece of silicon. This presents many options to IC designers; it lets them use the right transistor for each job, rather than compromising performance or size because the right device could not be made.

Many manufacturers deposit another layer of oxide over the completed wafers to protect the aluminum interconnections from mechanical damage during assembly. This passivation also increases the reliability of the circuit, because contamination, which might be present in the package, cannot migrate to the active portions of the circuit.

THE SILICON–NITRIDE MOS STRUCTURE (MNOS)

As mentioned previously, the low threshold voltage necessary for compatibility with bipolar circuits can also be achieved if a layer of silicon nitride (Si_3N_4) is grown above the channel of the MOS transistor. The nitride approximately doubles the dielectric constant of silicon dioxide, increasing the capacitance between the channel and the gate and thus cutting the threshold voltage in half (from 3 to 6 V nominal to 1.5 to 3.0 V). The total mechanical thickness of the dielectric remains approximately the same. Since the threshold voltage is reduced, power-supply requirements are also reduced for MNOS devices; they require $+5$ and -12 V in normal operation.

The nitride passivation also doubles the gain factor of the device. (Gain factor, defined as the amount of output current relative to the voltage on the device, is directly proportional to the dielectric constant of the gate insulator and inversely

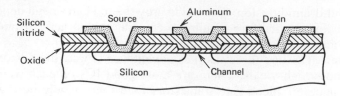

Figure 6-15   The basic MNOS structure.

proportional to the insulator thickness.) The increased gain factor produces lower ON resistance devices and reduces the size of the output device required to supply the 1.6 mA of a normal TTL load.

In an MNOS device, the gate insulator is a lamination of silicon dioxide and silicon nitride, as in Figure 6-15. The silicon dioxide–silicon interface is well understood and easily controlled. The silicon dioxide, about 400 Å thick, effectively separates the nitride from the silicon, and the resultant gate insulator is slightly thinner than the pure silicon dioxide. These devices, called MNOS transistors, have threshold voltages about 2 V less than conventional MOS devices.

ION–IMPLANTED MOS INTEGRATED CIRCUITS

Ion implantation of MOS transistors improves operating speed by reducing the internal capacitance between gate and source and between gate and drain. In the conventional diffused device (see Figure 6-16a), the gate must overlap the diffused regions to allow for photomask registration tolerances and for the lateral spread of the diffused ions.

With ion implantation (Figure 6-16b), only a portion of the source and drain regions is diffused, and the gate is fabricated to the desired width (it can be about three times smaller than in the diffused transistor). A beam of boron ions, accelerated to high velocity, impinges on the entire wafer. Most ions striking the oxide regions penetrate the oxide layer and are deposited in the bulk silicon, doping these areas and forming the rest of the source and drain. Ions striking other

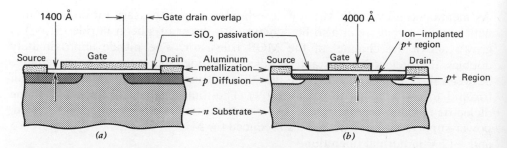

Figure 6-16   (a) A conventional diffused passivated MOSFET; (b) a diffused, passivated ion-implanted MOSFET. Note the gate overlap in (a) and the perfect alignment in (b), where ion implantation forms a critical portion of the source and drain.

regions are collected harmlessly by metallization layers. Alignment of the gate is automatic, since the gate serves as the mask for the critical portions of the source and drain. This alignment produces two to four times lower input capacitance and about 40 times less Miller capacitance, resulting in a device that is about five times faster than an equivalent diffused transistor.

Another approach to ion implantation is one in which ions are implanted only in the channel region (Figure 6-17). This results in reduced charge per unit area and reduced threshold voltage. This technique can lower the threshold voltage sufficiently that the MOS transistor becomes compatible with bipolar circuits.

In either approach, the ion implantation is carried out for areas covered with thin oxide. Areas reject implantation if they have metallization, thick oxide, or photoresist. Ion implantation in the source and drain regions can be accomplished in conjunction with a self-aligning gate and can be applied after the gate metallization has been put down.

MOS devices have achieved significant benefits from the ion-implantation process. Indeed, 20-MHz registers have been fabricated by eliminating the parasitic capacitance between the gate–source and gate–drain junctions (see Figure 6-16). Unlike the diffusion process, ion implantation does not cause the dopant to spread laterally within the wafer. This makes the ion-implanted device potentially faster than the conventional diffused MOS transistor. In diffusion, the lateral spread causes the source and drain regions to underlap the gate, resulting in undesirable capacitance. By using the metallized gate as a mask during implantation, the gate–source and gate–drain junctions are perfectly aligned without any special processing. The ions are implanted directly through the thin silicon dioxide.

Because of the high degree of process control, ion implantation has made possible the fabrication of MOS depletion-mode loads on the same chip as enhancement-mode transistors. Depletion-mode loads coupled with enhancement-mode drivers switch at twice the speed of conventional enhancement-mode circuitry. The dual combination is made possible because dopants can be controlled over a greater range in ion implantation as opposed to the diffusion technique. Hence it is possible to implant a higher dosage of ions (by way of a second, highly controlled light doping step) in the channels of the load transistors so that they operate in the depletion mode rather than the conventional enhancement mode. (Note: Enhancement-mode transistors exhibit no channel conductance at zero gate bias. Depletion-mode transistors conduct at zero gate bias.)

In a similar operation, ion implantation is used to lower the threshold voltage of MOS transistors by implanting ions into the channel region. Threshold voltages as low as 1.0 to 2.0 V have been achieved.

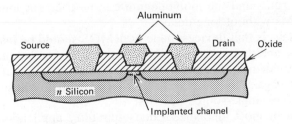

Figure 6-17   An ion implanted-channel MOS transistor.

Ion implantation also adds a third dimension to the basic resistor technologies used in ICs, the others being diffused resistors and thin-film resistors. The process provides a greater range of sheet resistivities together with a low absolute value tolerance (2%) and matching tolerance (0.1 to 0.2%). An application for ion-implanted resistors is in precision ladder networks used in digital-to-analog converters. Other applications include micropower amplifiers and active filters.

As in any process, there are both advantages and disadvantages. Of the former, the ion implantation process offers the following:

- Better control of doping because the dose rate can be electrically monitored.
- Allows making shallow surface junctions and also buried layers by use of the channeling principle.
- Junctions can be formed under thin oxide layers, eliminating oxide etch.
- A wide choice of dopants opens the door for continuing research; material that does not diffuse is no longer a limitation.
- No lateral diffusion. Implant pattern is sharply defined.
- Lower processing temperature as compared to diffusion (500°C, as opposed to 800 to 1000°C).
- The technique lends itself to automated control. For example, beam cutoff is automatically performed by a mechanical shutter electrically activated when the beam-current integrator records a specified value.

The disadvantages of ion implantation are the greater equipment costs as compared with diffusion and the limited information about reliability in a high-production environment.

REFRACTORY MOS (RMOS) CIRCUITS

From a process standpoint, RMOS is virtually the same as silicon gate and can even use the same masks. It has the same advantages as silicon gate with regard to the following features: the self-aligned gate gives low threshold voltage and high speeds, there is an extra layer of interconnect, there is less chance of gate contamination because no photoresist is used on the gate oxide, and the layout rules are similar. But here the resemblances end. Molybdenum is used in place of polycrystalline silicon. The low gate–conductor sheet resistivity provides a number of benefits, including shorter delays because of the lower voltage drops across the gate electrode. Also, the molybdenum can serve as a power conductor without excessive voltage drops, clock signals can be distributed by either aluminum metal or molybdenum runs, and the molybdenum can be used for long buried runs in RAMs and ROMs without long delays.

What all this says is that devices made by RMOS should be faster than equivalent silicon gates, and that designers may find that they can lay out circuits differently and more efficiently. However, the process is so new that it is simply too early to tell. Once again, the disadvantages that crop up are those associated with any new, unproven, single-source process.

RMOS should fit most high-speed, dense-circuitry applications quite well. It avoids part of the controversy raised by silicon gates over frequent oxide cuts

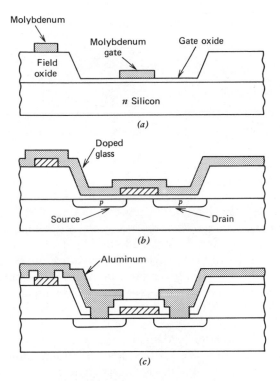

Figure 6-18   Three basic steps in the fabrication of a typical RMOS device.

needed in some circuits, and the operating voltages should be in the $+5$ and -12 V range.

There are three basic steps to the RMOS process. The RMOS process starts with an oxide layer on which molybdenum is deposited (Figure 6-18a); boron is diffused into the p regions of the source and drain by way of layers of boron-doped glass (Figure 6-18b), which also serve as an insulator. Opening contact holes in the oxide and depositing the aluminum electrode pattern complete the process (Figure 6.18c).

THE FIELD–SHIELD MOS STRUCTURE

Field-shield MOS is simply an n-channel process with a silicon guard or "field shield" added that extends over the entire chip, except at gate and contact regions (see Figure 6-19). The shield prevents low-field inversion, which is so low with high-resistivity substrates that all devices tend to short together and be depletion devices. Use of a low-resistivity substrate partially solves this, but at the expense of speed. The combination of a thin-oxide–nitride gate dielectric with the field shield allows the use of high-resistivity substrates, yet ends up with field inversion voltages essentially at infinity and with very high gain factors. The former is possible because the field shield isolates the silicon substrate from the metal and oxide above, and the latter results from the high dielectric constant of the oxide–

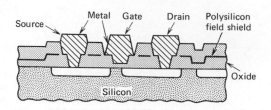

Figure 6-19   Cross-sectional view of a field-shield MOS.

nitride sandwich. The shield is normally biased to the substrate, and the substrate, as in conventional NMOS, has a small negative bias for threshold voltage shifting.

This process has several advantages over conventional n-channel devices. The processing provides self-aligned gates and very small gate areas. This reduces all capacitances and results in high-speed, high-density devices. The low threshold voltage ensures TTL compatibility, and the shield provides such added benefits as reduced noise and cross-talk at high frequencies, a convenient ground bus, and the possibility of making low-capacitance resistors by etching (to isolate a piece of silicon of the desired size).

Among the disadvantages of field-shield MOS are the novelty and complexity of the process, the lack of usage history, and the requirement for a bias supply. The process uses the same number of masks (five) as a conventional n-channel device, and it operates at $+15$, $+5$, and -5 V.

THE SELF–ALIGNED THICK–OXIDE (SATO) MOS STRUCTURE

Self-aligned thick-oxide (SATO) is a process used by Texas Instruments Incorporated that appears somewhat similar to the nitride self-aligned gate process. However, Texas Instruments uses a proprietary dielectric material in place of silicon nitride. This process allows the option of using silicon gates, not as a mask for self-alignment, but to gain the extra level of interconnect. It has a very low threshold voltage [making it TTL-compatible because of the (100) crystal structure], low gate overlap capacitances, and high dielectric constant of the thin dielectric.

Because the process is self-aligning, it is faster and more dense than MNOS even without the silicon-gate feature, and it costs about the same. Its performance is supposed to be equivalent to silicon-gate processes but lower in cost. When silicon gates are added to SATO, improved packing densities, higher speeds, lower threshold voltage, and lower power are claimed as compared with conventional p-channel silicon gates, but at equivalent costs.

SATO (Figure 6-20) without silicon gates is best used for RAMs and ROMs, whereas SATO with silicon gates (Figure 6-21) is suitable for very high-speed RAMS and shift registers.

Operating voltages are $+5$ and -12 V. Disadvantages are the same as for any new process — little history and lack of long-term reliability data, and there is only a single supplier.

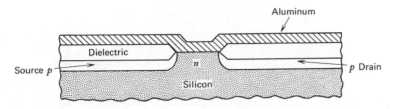

Figure 6-20   SATO without a silicon gate.

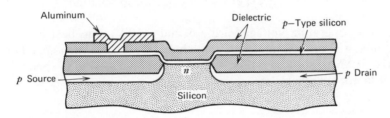

Figure 6-21   SATO with a silicon gate.

DOUBLE–DIFFUSED MOS (DMOS)

DMOS was recently developed by Signetics. It departs from conventional MOS processing in that it is not mask limited to obtain very precise, narrow channels. Rather, channels are defined by two diffusions of n and p dopants through the same mask opening (Figure 6-22). The drift region is always depleted; thus when the narrow channel is turned ON, the device switches at very high speeds. The extremely narrow channels and the self-alignment features are independent of masks, etching, and photolithography, and they give very high speeds because of resultant low feedback capacitances. DMOS is TTL-compatible at input and output; it can also have depletion loads, and it has good threshold voltage control and gives large current drives. Silicon gates or ion implantation can be used for threshold control.

As for disadvantages, DMOS requires one extra diffusion and a precise diffusion source; it has nonsymmetrical drain characteristics and has not been proven

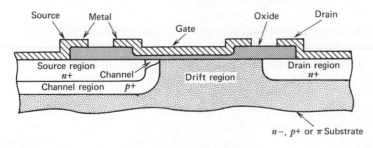

Figure 6-22   A double-diffused MOS (DMOS).

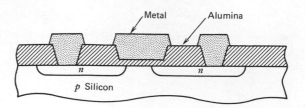

Figure 6-23 A metal–alumina–silicon MOSFET.

in production. The process is expensive; it also has no reliability history, and there is only one supplier.

As of now, operating voltages are $+5$ and -10 V. The best applications of DMOS, once it is proven in production, should be in such high-speed operations as ROMs, RAMs, shift registers, and random logic—where speed is a more important factor than cost.

METAL–ALUMINA–SILICON INTEGRATED CIRCUITS

A metal–alumina–silicon MOSFET (Figure 6-23) is the same as a conventional MOSFET IC except that alumina (Al_2O_3) is used in place of oxide as the gate dielectric material. The characteristics exhibited by the alumina can be varied by changing the voltage applied to the gate. The reaching of a critical applied voltage results in altering of the device's threshold voltage. In a memory application, the shift from one threshold voltage to another can correspond to the storage of a bit of information.

Depending on the polarity of the applied voltage, electrons can be injected into the alumina from either the gate metal or the silicon. Once the electrons are present in the alumina, a certain value of gate voltage must be reached in order for a channel to be created between the source and drain.

The change in threshold voltage is nonvolatile. Volatility refers to the loss of contents by a semiconductor memory when power is lost. Even if both the gate voltage and the power are removed, the threshold voltage remains a fixed characteristic.

SILICON–ON–SAPPHIRE (SOS) MOS INTEGRATED CIRCUITS

The use of sapphire rather than silicon as the underlying structure of an integrated circuit offers one very important advantage: extremely high insulation resistance between components of the IC. The drastic decrease in substrate parasitic capacitance has led researchers to claim that MOS circuits, if built on sapphire, could be as fast or even faster than bipolar ICs.

Parasitic interaction between active elements is also virtually eliminated by SOS construction because of the superior isolation of the elements, which float like islands on a sea of sapphire. Fabrication techniques under study for SOS ICs vary considerably. The main idea, however, is to put individual islands of

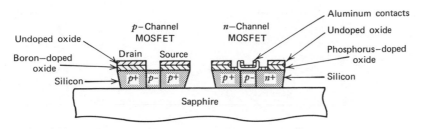

Figure 6-24 A silicon-on-sapphire (SOS) MOSFET.

silicon elements on the sapphire and connect the elements by metallization, thus forming a circuit (Figure 6-24). Note that this is also a means of achieving dielectric isolation and will undoubtedly find application in radiation environments.

SILICON–ON–SPINEL MOS INTEGRATED CIRCUITS

Spinel, a crystalline material in many ways similar to sapphire but more compatible physically with silicon, is being examined as a possible substitute for silicon. Spinel, like silicon, has a cubic crystal structure, whereas sapphire has a rhombohedral crystalline form. Sapphire must undergo more critical surface preparation in order to grow single-crystal epitaxial silicon layers on it. However, both types of substrates, because of their insulating properties, eliminate the requirement for IC transistors to have guard bands to ensure adequate *p-n* junction isolation from transistor to transistor.

The term "silicon-on-spinel" means that the silicon is grown as a chemical extension of the spinel crystal structure. Silicon-bearing gases pass over the spinel substrate in a heated chamber and deposit the silicon atoms as a continuation of the spinel crystal lattice. This process starts with a single crystal of spinel and ends up with a slightly larger single crystal of spinel plus silicon.

The same thing happens when silicon is grown on silicon in normal processing. Silicon atoms from the gas phase skid about on the surface of the growing epitaxial film until they find a correct position at the exposed edge of the lattice, to which they then fasten themselves. That also explains how *p*- or *n*-type silicon expitaxial layers can be formed by including the doping atoms in the gas mixture.

The similarity between silicon-on-spinel and silicon-on-silicon is emphasized by comparing the two, side by side, in Figure 6-25. In order to grow silicon on another material, the second material must have a crystallographic structure quite close to that of silicon. Both spinel and sapphire have sufficiently siliconlike crystallographic structures for this growth to take place. But neither material has the exact structure of silicon. The difference is not enough to prevent spinel or sapphire from "seeding" the add-on growth of silicon, but it is enough to cause the grown silicon to have a less-than-perfect crystallographic structure at the spinel–silicon or sapphire–silicon interface. For this reason, the silicon grown on these substrates has so far been suitable only for simpler MOS device structures where the current flows laterally in the thin epitaxial layer; it is unsatisfactory for the more complex bipolar device structures in which the current flows vertically.

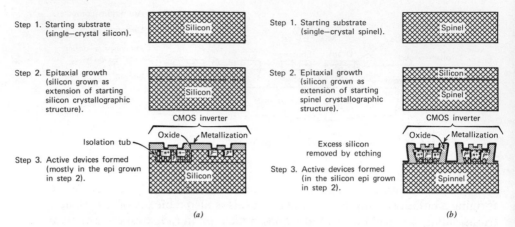

Figure 6-25 Comparison of the SOS and silicon-on-spinel processes. (*a*) Steps in the SOS fabrication process, (*b*) steps in the silicon-on-spinel fabrication process.

Step 3 of Figure 6-25*b* indicates how the silicon epitaxial layer would be processed to form a CMOS inverter (the most elementary form of logic gate), for example. Contrast this to the same step in Figure 6-25*a*, which shows the same inverter being formed conventionally in bulk silicon.

Excess silicon between the active device regions is etched completely away. This is easy because some acids attack silicon but not the very inert spinel. (Processing economies are feasible because the silicon could be "washed off" any defective wafer, a fresh epitaxial layer could be grown, and the substrate could be used again.)

Sapphire substrates boost the speed of CMOS, for example, up to 100 times the speeds of ordinary "bulk-silicon" CMOS. Spinel substrates should be even better for CMOS than sapphire substrates. Spinel is a closer crystallographic match to silicon than sapphire, and it is easier to machine. (Sapphire is one of the hardest materials.) Thus spinel, like sapphire, should make CMOS not only very fast, but also much less expensive to produce. Spinel permits working around defects, obtaining subsystem isolation, and recovering materials from faulty wafers.

Semiconductor material on an insulating substrate is really the best of two worlds. Along with the well-defined electrical performance of the old discrete circuits go the processing economies of modern monolithic circuits. Electrical advantages are:

1. No dc leakage to ground.
2. Much less parasitic isolation junction capacitance.
3. Ability to handle ac voltages referenced to ground.
4. Ability to handle high voltages; no longer limited to the 50 to 100 V limitation of the planar process in bulk silicon.
5. Less lossy silicon material (for microwave circuits).
6. Increased nuclear radiation tolerance.

These advantages have been demonstrated in the so-called dielectric isolation process, too. Here isolation is achieved by etching away unwanted silicon from the back of a bulk-silicon wafer and strengthening the remaining silicon with a glass backing. However, the silicon-on-spinel manufacturing process is much less critical since it does not require careful lapping or grinding. However, the older process may still yield highest-quality bipolar devices with dielectric isolation.

CHARGE–COUPLED DEVICE MOS INTEGRATED CIRCUITS

The charge-coupled device (CCD) is not an MOS process; rather, it uses MOS processing to make potentially very fast (up to 100 MHz) low-power dynamic shift registers. There are no transistors on the silicon substrate, no windows to be cut for diffusing dopant, and no *p-n* junctions in the structure. MOS capacitors are used to process the injected charge. Figure 6-26 is a cross-sectional view of a typical CCD.

Each CCD can produce minority carriers and store them in a special storage region below the interface of the semiconductor and the oxide layer. After being stored, the charges can later be moved along the surface of the semiconductor to other locations, and then their presence or absence can be detected. A biasing voltage is applied to the metal electrode of the MOS capacitor, creating a depletion zone at the semiconductor–oxide interface. The application of a "storage" voltage greater than this bias voltage opens a well at this position and makes possible the acceptance of carriers.

If minority carriers flow into the silicon, they collect at the semiconductor surface in the well. This accumulation of charge may be moved to an adjacent capacitor by placing a larger "transfer" voltage on the metal electrode. The larger voltage deepens the well, and the carriers are attracted to the deeper well and transfer to the new electrode. Once the charge has been transferred, the voltage on the first MOS capacitor is returned to the normal biasing potential and the voltage on the receiving capacitor is lowered to the storage potential. The process can now be repeated by transferring the charge stored by the second capacitor to another adjacent capacitor.

A recent and promising innovation by Bell Telephone Laboratories is the use of buried channels to increase the operating speed of CCDs. Normally, charge is transferred at the oxide–substrate interface of a CCD; consequently, some charge is trapped in "surface states" during each transfer—the condition worsens at higher transfer rates. This means that operating speeds around 20 MHz appeared to be the limit for serial memories. By implanting a lightly doped channel just

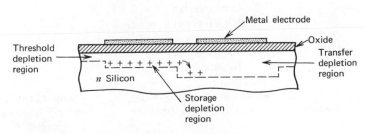

Figure 6-26   A charge-coupled MOS device (CCD).

Table 6-2　Comparison of MOS Technologies

p-MOS	*n*-MOS	Silicon-Gate MOS
• Single-polarity device	• Single-polarity device	• Single-polarity device
• Simplicity	• Smaller circuits	• Smaller geometry transistor than *p*-MOS
• Low cost	• Transistor is half the size of *p*-MOS transistor	• Smaller total die size
• High-voltage transistors	• Difficult process to control	• High-voltage transistor
• Good yield	• Aluminum gate electrode	• High yield
• Highly controllable process	• High drain current and gain	• Polycrystalline gate electrode
• Ease of fabrication	• Device is normally on	• Higher speed than *p*-MOS and *n*-MOS
• Uses five masks	• Uses five masks	• Self-aligning gate
• Aluminum gate electrode	• Requires extra bias supply	• Added processing steps
• Device is normally off	• Wastes less power than *p*-MOS	• Lower gate-to-drain and source-to-drain feed-back capacitance than *p*-MOS or *n*-MOS
• Good noise immunity	• Higher speed than *p*-MOS	• Lower threshold voltage than *p*-MOS or *n*-MOS
• Slow speed	• Threshold voltage is adjustable	• 50% faster than MNOS
• Gate-oxide thickness is critical for:　Breakdown voltage　Threshold voltage　Material resistivity	• TTL compatible	• High packing density
• Requires large gate due to misalignment correction	• Requires large gate owing to misalignment correction	• Stable, reliable, reproducible process
• Long production history	• Easily contaminated	• Layout flexibility due to extra interconnect
• Gate sensitivity to contamination	• Requires tight process control	• Uses five or six masks
• Available in both high-threshold and low-threshold forms	• Requires built-in alignment tolerance causing:　Gate overlap　High feedback capacitance between gate and drain　Reduced transistor gain	• Best suited for large-scale RAMS
• Requires built-in alignment tolerance causing:　Gate overlap	• Requires extra power supply to bias substrate	• Compatible with other technologies and bipolar compatible

Continued on following page

Table 6-2   Continued

p-MOS	n-MOS	Silicon-Gate MOS
High feedback capacitance between gate and drain Reduced transistor gain		
• Highest threshold voltage	• Requires large output buffers to drive TTL	
• Low packing density	• Greater packing density or circuit complexity	
• Bipolar compatible (100); (111) not compatible		

Silicon–Nitride MOS (MNOS)	Ion-Implanted MOS	CMOS
• Single-polarity device	• Single-polarity device	• Devices of both polaries on same chip
• Lower on resistance than p-MOS	• Low-temperature process	• Low quiescent power dissipation (microwatts)
• Highest feedback capacitance	• Wide voltage tolerances	• Low standby power (microwatts)
• 2 V less threshold voltage than p-MOS	• Output buffers half the size of silicon gate	• Insensitive to variations in parameters of individual devices
• Reduced power-supply requirement	• Requires sophisticated and expensive manufacturing equipment	• Good isolation
• Bipolar compatible	• Higher speed than p-MOS	• Complex processing and hard to control process
• Low packing density	• Milliwatt power levels	• TTL compatible
• Low cost	• Automatic and perfect alignment of source and drain under gate	• Large chip size
• Low threshold voltage	• Low gate-to-drain and source-to-drain feedback capacitance	• Ease of user design
• High drain currents	• Low threshold voltage	• Requires seven or eight masks
• Half the power consumption of p-MOS (111)	• Bipolar compatible	• More complex devices
• Higher speed than p-MOS	• Reduces size of MOS devices	• Requires 30% more real estate than single-channel devices
• Prevents sodium ion migration	• Available in low-threshold and self-aligned gate forms	• High speed (20 MHz)

191

Continued on following page

Table 6-2   Continued

Silicon–Nitride MOS (MNOS)	Ion-Implanted MOS	CMOS
• Gate protected against contaminants • Possible threshold-voltage shifts	• Uses six masks • High cost • Complex process • Low throughput	• High cost • Best suited for large arrays like memories • High noise immunity (40% of supply voltage) • Single-phase clocking • Excellent fan-out • Adaptability to wide range of logic levels (swings) • Operation from single power supply • Excellent temperature stability • Limited system interface • Low packing density • Used in ultra-low-power applications

Metal–Alumina–Silicon MOS	Silicon-on-Sapphire MOS	Spinel-on-Silicon MOS	Charge-Coupled Devices
• Single-polarity device • Similar to p-MOS device except that alumina is used in place of gate oxide • Permanent storage (nonvolatile) of threshold voltage	• Single-polarity device • Extremely high insulation resistance between components • Low substrate parasitic capacitance • Ability to handle ac voltages referenced to ground • Ability to handle high voltages • Requires good process control • Requires critical surface preparation • No dc leakages to ground • Very high speed	• Single-polarity device • Similar to p-MOS IC processing • Less expensive than sapphire on silicon • Ability to handle ac voltages referenced to ground • Low parasitic capacitance • Ability to handle high voltages • Requires less process control than sapphire on silicon • No dc leakage to ground • Very high speed; higher than sapphire on silicon	• Single-polarity device • No transistors on silicon substrate • No windows for diffusion • No p-n junction • Less critical process control • Low cost • Use MOS capacitor to process injected charge

    Continued on following page

Table 6-2   Continued

Metal–Alumina–Silicon MOS	Silicon-on-Sapphire MOS	Spinel-on-Silicon MOS	Charge-Coupled Devices
	• Rhombohedral crystal structure	• Poor crystallographic structure at spinel–silicon interface	
	• Eliminates requirement for transistors to have guard bands to ensure adequate *p-n* junction isolation from transistor to transistor	• Presently useful for simple circuits only (i.e., where current flows laterally)	
	• Poor crystallographic structure at sapphire–silicon interface		
	• Presently useful for simple circuits only (i.e., where current flows laterally)		

under the surface of the silicon substrate (ca. 1 μ), CCD designers can confine the charge to the bulk silicon where it cannot be trapped easily. The result is laboratory shift registers that operate at hundreds of megahertz, with the possibility of speed in the gigahertz region. Furthermore, in RAM configurations, buried-channel CCD structures could mean access times of less than a nanosecond for RAM chips containing tens of thousands of bits. A comparison of the salient features of the different MOS technologies discussed in this chapter is presented in Table 6-2.

Chapter Seven

CMOS APPLICATIONS

The increasing popularity of CMOS devices necessitates an in-depth discussion of this basic MOS technology. This chapter[1] presents such a discussion.

During 1972 the 4000A-series CMOS circuits have become widely accepted as a viable production technology rather than a research and development technology for three reasons:

1. Number of functions available.
2. Usage history (reliability data).
3. Development of several second sources.

The total number of all families of 54/74 series TTL circuits (more than 200 different types) and the flexibility and ease of use of these circuits still overwhelms the available number of 4000A-series CMOS circuits, of which there are more than 30 different types. However, the number of functions available using the CMOS technology is rapidly increasing and the reliability of the technology has been proven.

Compared with most bipolar devices, MOS circuits consume less power. The reason for this is the permissible utilization of MOS transistors as load elements. This not only results in space savings but yields higher values of load resistance; hence, lower operating current. In the ON mode, current flow through the load causes dc power dissipation only on the order of 0.5 to 2.0 mW.

[1]Portions of this chapter are excerpted from RCA's *COS/MOS Integrated Circuits Manual*, 1971, through the courtesy of RCA's Solid State Division. Somerville, N.J.

Even this small amount of dc power dissipation can be reduced significantly through CMOS circuitry. Indeed, by combining p-MOS and n-MOS on the same chip in a complementary-symmetry configuration, dc power dissipation can be reduced to virtually zero. This type of micropower operation is particularly desirable for battery-operated equipment where unattended operation for long periods is required.

Since CMOS will compete for design sockets with low power TTL it is germane to compare the salient features of each. Figures 7-1 and 7-2 compare the speed and load capacitances for 54 low-power gates versus CMOS gates and for 54 low-power flip-flops versus CMOS flip-flops, respectively. The curves indicate that CMOS is slower than the 54L series under the same conditions. However, Figure 7-3 shows that CMOS dissipates less power than the 54L series (lowest power 54 series) for a dual D flip-flop under identical test conditions.

CMOS circuits can be designed for the low-power extreme if frequency is unimportant — for example, 100-kHz operation can be achieved at only a few microwatts of power drain. On the other hand, CMOS circuits can be designed to operate at speeds in excess of 10 MHz, at the expense of greater power dissipation due to the increased rate of signal transitions. In the standby state, power dissipation is virtually zero.

The advantage of CMOS circuits is offset somewhat by their increased processing complexity and low packing density. They cannot hope to compete with single-channel circuits in applications that require the utmost in circuit complexity.

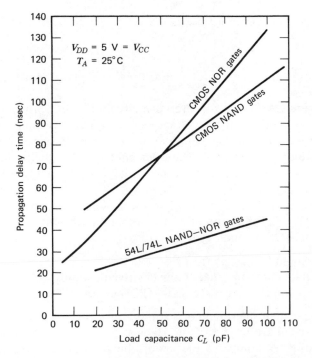

Figure 7-1   Comparison of speed and load capacitance for low power TTL gates versus CMOS gates.

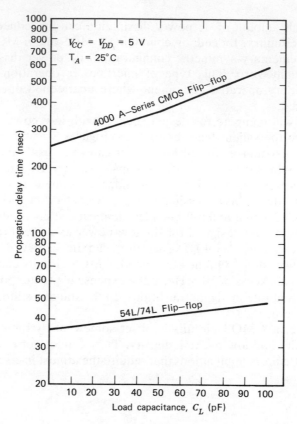

Figure 7-2 Comparison of speed and load capacitance for low power TTL flip-flops versus CMOS flip-flops.

Their high speed and low power drain, however, give them significance in the MOS technology.

The metal-gate CMOS circuits are normally fabricated with a threshold voltage of about 2 V and with a supply voltage between 5 and 16 volts; but silicon-gate complementary circuits can be fabricated by similar processes. The self-alignment feature can be used to enhance operating speed at higher voltages, but the primary emphasis is on very low supply voltage (< 1.5 V) and very-low-power ($< 5\,\mu$W) applications.

CMOS can be combined with other technologies and processes to provide more refined electrical characteristics. Ion implantation can be used in conjunction with the basic CMOS technology to provide:

1. Low threshold voltage shifts.
2. Depletion load pull-ups that are insensitive to supply-voltage variations.
3. High-value implanted resistors 10 k$\Omega/\square$ and greater tolerance resistors.

Additionally, use of the silicon on sapphire process in conjunction with CMOS can increase speed from 45 to 2 to 10 nsec for the 4000 A series as shown in Table 7-1. Table 7-2 compares the major characteristics of the basic MOS processes and the bipolar logic families.

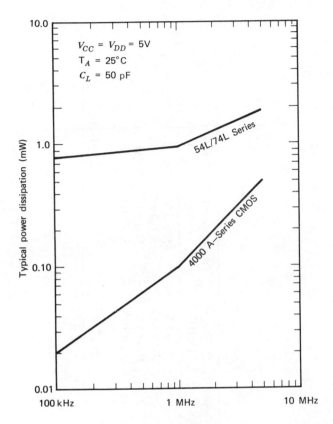

Figure 7-3 Power dissipation versus input frequency for a typical dual − D flip-flop: $V_{CC} = V_{DD} = 5\,V$; $C_L = 50\,pF$, $T_A = 25°C$.

Much of the CMOS logic growth will come at the expense of TTL bipolar circuits with which it is directly competitive. Some of the reasons for this are:

1. The CMOS process is inherently simpler and less critical.

2. CMOS has greater packing density than any (current) bipolar logic form, therefore permitting more circuitry in a given area and reducing the cost per function.

3. CMOS power dissipation is only a small fraction of the lowest power required to drive bipolar logic. This permits the utilization of larger chips (with greater packing densities) without exceeding the thermal limitations of the package, reducing, as well, the need for exotic cooling methods.

CMOS is competitive with TTL in respect to performance, too. Its speed is expected to nearly reach that of the fastest TTL, it generates very little noise, and it operates over a wide supply-voltage range, with logic-swing amplitudes from ground to the full power-supply voltage. But the appeal of CMOS logic is far greater than the mere replacement of bipolar logic or other forms of MOS circuits. Actually, it is in the development of new markets for which other forms of logic are not suitable.

CMOS circuits are projected to penetrate the huge consumer market, where

Table 7-1 Comparison of CMOS Processes

Process Technology	Power Supply (V)	Typical Threshold Voltage (V)	Toggle Frequency (MHz)	Propagation Delay (nsec)	Relative Cost
Standard metal gate (CMOS I)	3–15	±1.8	5, at 5 V 10, at 10 V	45, at 5 V 25, at 10 V	Low
Low-threshold voltage (CMOS II)	1.0–15	±0.7	10, at 5 V 20, at 10 V	25, at 5 V 15, at 10 V	Medium
Silicon on sapphire	1.0–15	±0.7	50–250	2–10	High

Table 7-2 Major Characterisitics of Basic MOS Processes and Bipolar Logic Families

MOS Process	Threshold Voltage (V)	Supply Voltage $V_{DD}(V_{GG})$ (V)	Propagation Delay (nsec/gate)	Frequency (MHz)	Power Dissipation/ Gate (mW)[a]	Noise Margin Logic "1"	Logic "0"
p-channel MOS							
High threshold { medium power	−3.5 to −5	−17 to −27	75	2	1.7	3	1.5
{ low power			300	0.5	0.45		
Low threshold {(100)	−1.5 to −2.5	−12 to −17	70	2	1.0	2	0.7
{ silicon gate			60	5	1.0		
Ion implant, depletion loads	−1.5 to −5	−12 to −27	35	5	1.5	1.5	1
n-Channel MOS							
Metal gate	1–2	5–20		10	1.0	1	1
Silicon gate							
Complementary (CMOS)							
Metal gate	+1.5 to +2.5	3–18	40	20	50 nW	$V_{DD}/2.2$	
Silicon gate	±0.5 to ±2.5	1.2–15	25	25	50 nW		
Bipolar lines							
TTL		5.0 +20% −10%	10	60	15[b]	1.2	1.2
ECL		5.2 +20% −10%	1	400	25–35	0.4	0.4
DTL		5.0 ±10%	30		8		
RTL		3.0 ±10%	24		12		

[a]In milliwatts except as noted.
[b]The 54L/74L series has approximate power dissipation of ∼ 1 mW/gate.

199

logic circuits have no visibility at all at present. This is primarily due to the large anticipated use of CMOS in watches, clocks, and various automotive safety and control devices that are now under active design. Appliance controls and even the toy market are new areas of applications to be invaded by logic circuits. Surprisingly, today's nonexistent consumer market is expected to become the biggest user of CMOS.

It is also projected that CMOS will penetrate the industrial market. The technical features that make CMOS particularly useful for this market are as follows:

1. High noise immunity for operation in noisy industrial environments.
2. Operation from unregulated power supply.
3. Microwatt power dissipation for portable controls, instruments, telecommunications systems where battery supported backup is required, and battery-operated medical electronic devices such as heart pacemakers.
4. Complex functions that are not available in other high-noise-immunity families.
5. Low noise generation, which appeals particularly to manufacturers of medical electronics equipment where noise-free operation is highly important.

Additionally, the economical large array potential of CMOS has opened markets for battery equipment that were not feasible two or three years ago. For example, presently available pocket-playing receivers now decode one unique pattern out of more than 6600 using a CMOS chip with complex logic that draws so little power that it does not affect the life of the receiver battery. Another example: meter readers (water, gas, or electric) that are CMOS logic arrays draw so little power that the shelf life of the battery is not affected.

INVERTER CHARACTERISTICS

The most basic of all the CMOS circuits, the inverter circuit, consists of one p-channel and one n-channel enhancement-type MOS transistor, as schematized in Figure 7-4. In this figure the operation of a simple p-MOS inverter is compared with an equivalent CMOS circuit. In the circuit in Figure 7-4a, transistor Q_1 is switched while Q_2 serves as a fixed-load resistor. When the input signal turns Q_1 ON, current flow from the power supply is instituted. The output voltage is determined by the ON-resistance value of Q_1 compared with the fixed-resistance value of Q_1 and can never go to zero.

In the complementary circuit (Figure 7-4b) when the input signal is low (ground), the n-channel transistor Q_1, is OFF and the p-channel device is ON. The output, therefore, is "shorted" to the positive-supply voltage. If the load resistance is an MOS gate, which has a very high input resistance, virtually no current is drawn from the supply and the output voltage approaches $+V_{DD}$. When the input signal goes high, Q_1 is turned ON and Q_2 is turned OFF. Again, no dc current can flow from the supply, but the output is drawn to ground through the low ON-resistance of Q_1. The output voltage varies from almost $+V_{DD}$ to zero.

Here it is evident that the single-channel circuit draws current from the power supply during the entire ON portion of the input signal. The CMOS circuit, on the other hand, draws no dc current at all. The only power dissipation occurs

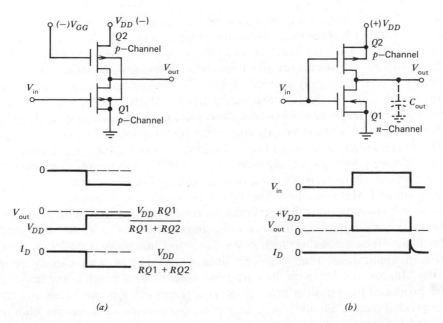

Figure 7-4 Comparison of inverter operations of (*a*) Single-Channel MOS, and (*b*) CMOS. (Courtesy of RCA Corporation.)

during the transitions of the input signal. In either logic state, one MOS transistor is ON while the other is OFF. Because one transistor is always turned OFF, the quiescent power consumption of the CMOS device is extremely low; more precisely, it is equal to the product of the supply voltage and the leakage current.

Because of the complementary nature of the interconnections of the series *p*- and *n*-type devices in the basic inverter, the transfer characteristic of a CMOS logic gate is as shown in Figure 7-5. The high input impedance of the gate causes the input and output signal to swing completely from zero volts (logic "0") to V_{DD} (logic "1") when sufficient settling time is allowed. The switching point is shown to

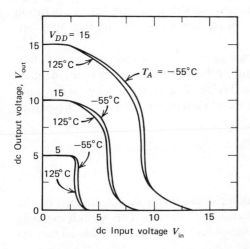

Figure 7-5 Transfer characteristic of a CMOS logic gate. (Courtesy of RCA Corporation.)

be typically 45 to 50% of the magnitude of the power-supply voltage, and it varies directly with that voltage over the entire range from 6 to 15 V for high-threshold devices and 3 to 15 V for low-threshold devices. The CMOS transfer characteristic of Figure 7-5 illustrates the high noise immunity of MOS devices (i.e., typically 45% of the supply voltage). Figure 7-5 also reveals the negligible change in operating point as temperature varies from -55 to $+125°C$. Because of the ideal nature of these switching characteristics, CMOS devices operate reliably over a much wider range of voltage than other forms of logic circuits.

The exceptionally high ac and dc noise immunity of CMOS ICs is attributable to the relatively low output impedances (ca. $600\,\Omega$), moderate-speed operation, steep transfer characteristics, and output voltages within $10\,mV$ of V_{DD} when driving other CMOS circuits. The high values of noise immunity are achieved with power consumptions in the microwatt range, as compared with much higher (milliwatt) consumptions and lower noise immunities (ca. 1 V) for saturated bipolar logic. Because of their high cross-talk noise immunity, CMOS circuits are useful in applications requiring long lines or in circuits with closely coupled wiring. Moreover, CMOS circuits are not susceptible to ground–line noise or to noise produced by improper ground returns; therefore, it is not necessary to use large ground planes. Similarly, the high power-supply noise immunity eliminates the need for complex filtering circuits.

An analysis of noise immunity involves consideration of immunity to both ac and dc noise. Whereas ac noise is usually considered to be made up of those noise spikes with pulse widths shorter than the propagation delay of a logic gate, dc noise spikes are considered to have pulse widths longer than the propagation delay of one gate. Since ac noise immunity, which varies in direct proportion to dc noise immunity, is largely a function of the propagation delays and output transition times of logic gates, it is a function of the input and output capacitances.

Because CMOS logic gates typically change state near 50% of the supply voltage, and because of the steep transfer characteristics exhibited during transitions, exceptionally high input voltages are required to significantly change or falsely switch the output logic state. Typically, then, the input to a CMOS logic gate operating at a dc supply voltage of 10 V may change by 4.5 V before the output begins to change state. In addition ac noise immunity is extremely high because of the high static or dc noise immunity and the moderate-speed operation of standard CMOS circuits. The high ac noise immunity also implies high immunity to cross-talk noise.

Figure 7-6 and 7-7 illustrate CMOS logic-gate noise immunity. These definitions assure that the logic level at the output of the driving device is recognized as the same logic level at the input to the load device. The dc noise level at the junction is equal to or less than the difference in maximum magnitudes of the output and input logic levels.

TRANSMISSION GATES

The CMOS transmission gate is a single-pole, single-throw switch formed by the parallel connection of a p-type device and an n-type device. This switch expands the versatility of CMOS circuits in both digital and linear applications. The per-

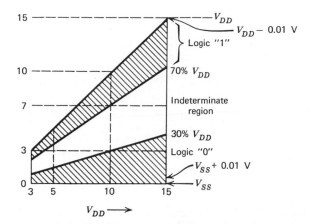

Figure 7-6   Guaranteed noise-immunity values. (Courtesy of RCA Corporation.)

fect transmission gate or switch may be characterized as having zero forward and reverse resistance when closed and infinite resistance when open (i.e., it has an infinite ON/OFF impedance ratio). The CMOS transmission gate approaches these ideal conditions.

The advantages of a CMOS transmission gate can be better understood by consideration first of the single n-channel MOS-transistor transmission gate driving a capacitive load from a positive voltage source, as in Figure 7-8. With zero volts applied to the gate of the n-channel device, no current can flow, and the load capacitance C_L remains uncharged. As the gate voltage to the transmission gate is made positive enough to turn the transmission gate ON the load capacitance begins to charge. However, the load capacitance can only charge to a level

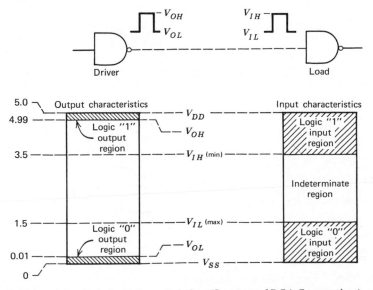

Figure 7-7   Output-to-input logic-level characteristics. (Courtesy of RCA Corporation.)

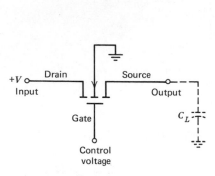

Figure 7-8 Single n-channel MOS transistor transmission gate. (Courtesy of RCA Corporation.)

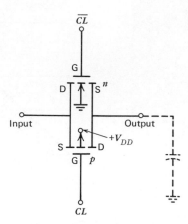

Figure 7-9 Typical CMOS transmission gate. (Courtesy of RCA Corporation.)

equal to the gate voltage minus the threshold voltage of the n-channel transistor because the single n-channel transmission gate operates as a source-follower circuit in which premature gate cutoff occurs.

Another aspect of MOS transmission gates is that they are bilateral (i.e., drain and source are interchangeable). This type of transmission gate also operates with slow speed in large-signal applications (i.e., as the device begins to turn ON, the RC time constant is large). A CMOS transmission gate that overcomes these disadvantages is made by paralleling n- and p-channel MOS transistors (Figure 7-9). This arrangement overcomes the premature-cutoff problem associated with the single-channel transmission gate because one of the two channels is always being operated as a drain-loaded stage regardless of what the input or output voltage may be. If each MOS channel of the CMOS circuit has a 2-V threshold

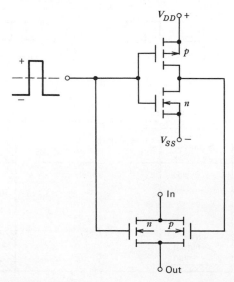

Figure 7-10 Combination of transmission gate and basic inverter to form a switch. (Courtesy of RCA Corporation.)

voltage and if zero volts is applied to the gate of the p-channel unit and 10 V to the gate of the n-channel unit, an increase in input voltage in excess of 8 V (10 V − 2 V) cannot be switched through the n-channel unit. However, proper switching can occur in the p-channel unit because the magnitude of the voltage from gate to source (0 V − 8 V = − 8 V) is greater than the p-channel threshold (− 2 V). As a result, the switch does not turn off prematurely because the gate-to-source voltages of both the n- and p-channel units never equal the threshold voltage of these devices. The full 10-V supply voltage ($V_{DD} − V_{SS} = 10$ V) can, therefore, be switched. The CMOS transmission gate is also considerably faster than the single-channel MOS transmission gate; the RC time constant is always smaller.

The transmission gate can be combined with a basic inverter circuit to form a single switch, as in Figure 7-10. Only one control voltage is required because the inverter provides the control voltage necessary for the complementary unit. The circuit of Figure 7-10 is useful in a variety of analog and digital multiplexing applications.

4000A–SERIES CMOS FUNCTIONS[2]

The 4000A-series CMOS products range from circuits as simple as two-input logic gates through the complexity of a 14-stage binary counter.

NOR Gates

A two-input NOR gate is an inverter with two n-type units in parallel and two p-type units in series (Figure 7-11). Each of the two inputs is connected to the gate of one n- and one p-type transistor. A negative output is obtained when either the A or the B input is positive because the positive input turns OFF the associated

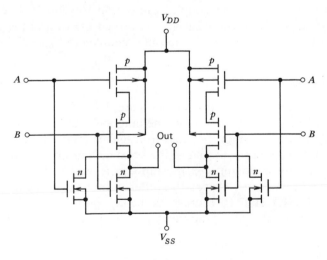

Figure 7-11   A pair of two-input NOR gates. (Courtesy of RCA Corporation.)

[2]*RCA COS/MOS Integrated Circuits Manual*, 1971.

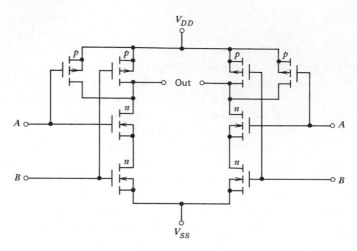

Figure 7-12   A pair of NAND gates. (Courtesy of RCA Corporation.)

p-type transistor (disconnecting the output from V_{DD} supply) and turns ON the associated n-type transistor (connecting it to ground and causing a low output). When both the input signals are at ground potential, both p-type units are ON and both n-type units are OFF. In this case the output is coupled to the V_{DD} terminal and provides a high output.

NAND Gates

A NAND gate is an inverter with two p-type transistors in parallel and two n-type transistors in series (Figure 7-12). The output goes negative only if both inputs are positive, in which case the p-type transistors are turned OFF and the n-type transistors ON. This condition couples the output to ground. If either input is negative, the associated n-type transistor is turned OFF, and the associated p-type transistor is turned ON; thus the output is coupled to V_{DD} and goes high.

Set–Reset Flip-Flops

Two NOR gates, when connected as in Figure 7-13, form a set–reset flip-flop. When the set and reset inputs are low, one amplifier output is low and the other output is high. If the set input is raised to a higher level, the associated n-type unit is turned ON so that the output of the set stage goes high and becomes logic "1" or $\overline{Q}$. Under these conditions, the flip-flop is said to be in the SET state. Raising the reset input level causes the other output to go high, and places the flip-flop in the RESET state (Q represents the low-output state). Thus the circuit of Figure 7-13 represents a static flip-flop with SET and RESET capability.

D-Type Flip-Flops

A block diagram of a D-type flip-flop appears in Figure 7-14a; Figure 7-14b is the schematic diagram for the flip-flop. The block diagram shows a master flip-flop

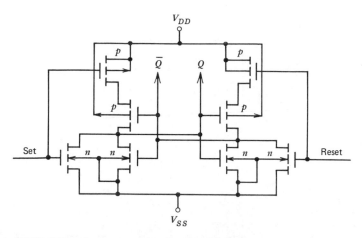

Figure 7-13   A SET/RESET flip-flop in which all p-unit substrates are connected to V_{DD} and all n-unit substrates are connected to V_{SS}. (Courtesy of RCA Corporation.)

formed from two inverters and two transmission gates (show as switches); the master feeds a slave flip-flop having a similar configuration. When the input signal is at a low level, the TG_1 transmission gates are closed and the TG_2 gates open. This configuration allows the master flip-flop to sample incoming data and the slave to hold the data from the previous input and feed them to the output. When the control is high, the TG_1 transmission gates open and the TG_2 transmission gates close, so that the master holds the data entered and feeds them to the slave. The D flip-flop is static and holds its state indefinitely if no pulses are applied (i.e., it stores the state of the input prior to the last clocked input pulse). Both the "clock" CL and "inverted clock" $\overline{CL}$ (Figure 7-14c) are required; clock inversion is accomplished by an inverter internal to each D flip-flop.

J-K Flip-Flops

The circuit diagram for a J-K flip-flop appears in Figure 7-15; the truth table for the flip-flop is shown below. The J-K flip-flop is similar in some aspects to the D flip-flop, but it has some additional circuitry to accommodate the J and K inputs. The J and K inputs provide separate clocked SET and RESET inputs, and allow the flip-flop to change state on successive clock pulses. In the chart below, t_n-1 and t_n refer to the time intervals before and after the positive clock-pulse transition, respectively; CL indicates level change, $\times$ indicates "don't care," and asterisks represent an invalid condition. The J-K flip-flop circuit also has SET and RESET capability; the inverters in the master and slave flip-flops each have an added OR input for direct (unclocked) setting and resetting of the flip-flop.

Memory Cells

The basic storage element common to CMOS memories consists of two CMOS inverters cross-coupled to form a flip-flop as in Figure 7-16. Single-transistor transmission gates are employed as a simple and efficient means of performing the

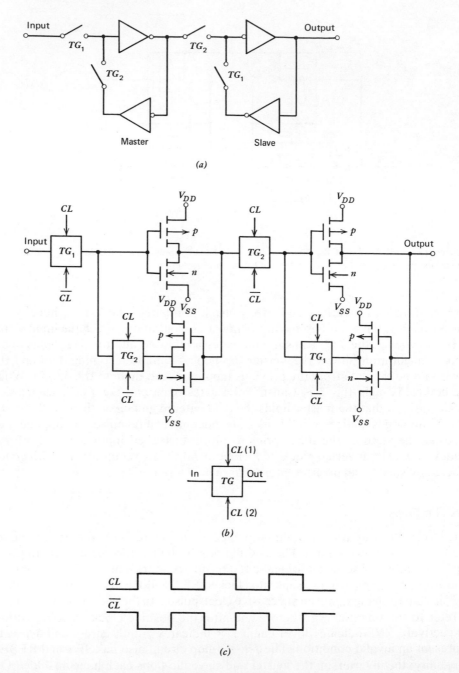

Figure 7-14 Diagrams for a D-type flip-flop: (a) block, (b) schematic, (c) clock-pulse. Here TG represents transmission gate, all p-unit substrates are connected to V_{DD}, and all n-unit substrates are connected to V_{SS}. Input-to-output is a bidirectional short circuit when control input 1 is low and control input 2 is high; it is an open circuit when control input 1 is high and control input 2 is low. (Courtesy of RCA Corporation.)

● t_{n-1} INPUTS						+ t_n OUTPUTS	
CL△	J	K	S	R	Q	Q	Q̄
╱	I	X	O	O	O	I	O
╱	X	O	O	O	I	I	O
╱	O	X	O	O	O	O	I
╱	X	I	O	O	I	O	I
╲	X	X	O	O	X		← (NO CHANGE)
X	X	X	I	O	X	I	O
X	X	X	O	I	X	O	I
X	X	X	I	I	X	✱	✱

△ – LEVEL CHANGE
× – DON'T CARE
✱ – INVALID CONDITION
● – t_{n-1} REFERS TO THE TIME INTERVAL
　　PRIOR TO THE POSITIVE CLOCK PULSE
　　TRANSITION.
+ – t_n REFERS TO THE TIME INTERVALS
　　AFTER THE POSITIVE CLOCK PULSE
　　TRANSITION.

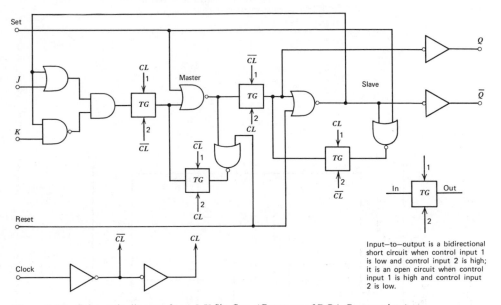

Input–to–output is a bidirectional
short circuit when control input 1
is low and control input 2 is high;
it is an open circuit when control
input 1 is high and control input
2 is low.

Figure 7-15　Schematic diagram for a *J-K* flip-flop. (Courtesy of RCA Corporation.)

logic functions associated with storage-cell selection (i.e., the sensing and storing operations). The resulting word-organized storage cell (Figure 7-17) is composed of six transistors; one word line W, and two digit-sense lines D_1 and D_2. Addressing is accomplished by energizing a word line; this action turns on the transmission gates on both sides of the selected flip-flop. Because the cell in Figure 7-17 has p-channel transmission gates, a ground-level voltage is required for selection. Figure 7-18 shows an eight-transistor bit-organized memory cell employing X-Y selection. A modification of this circuit in which the Y-select transistors are common for each column of storage elements is used in large memory arrays.

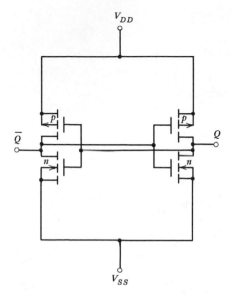

Figure 7-16 The basic storage element common to CMOS memories. (Courtesy of RCA Corporation.)

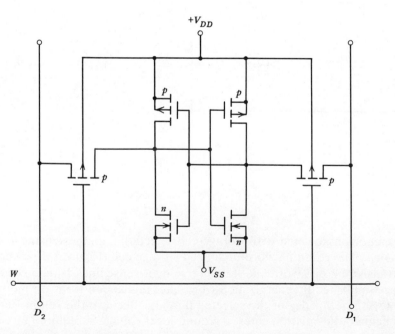

Figure 7-17 A word-organized storage cell: w is word line, D_1 and D_2 are data lines. (Courtesy of RCA Corporation.)

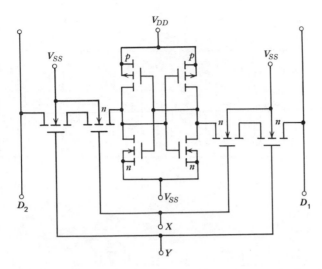

Figure 7-18 An eight-transistor bit-organized memory cell with $X–Y$ selection. (Courtesy of RCA Corporation.)

Dynamic Shift Registers

Figure 7-19a diagrams a two-state shift register; each stage consists of two inverters and two transmission gates. Each transmission gate is driven by two out-of-phase clock signals arranged as in Figure 7-19b, so that when alternate transmission gates are turned ON, the others are turned OFF. When the first transmission gate in each stage is turned ON, it couples the signal from the previous stage to the inverter and causes the signal to be stored on the input capacitance to the inverter. The shift register utilizes the input of the inverter for temporary storage. When the transmission gate is turned OFF on the next half cycle of the clock, the signal is stored on this input capacitance, and the signal remains at the output of the inverter where it is available to the next transmission gate, which is now turned ON. Again, this signal is applied to the input of the next inverter where it is stored on the input capacitance of the inverter, making the signal available at the output of the stage. Thus a signal progresses to the right by one half-stage on each half-cycle of the clock, or by one stage per clock cycle.

Because the shift register is dependent on stored charge, which is subject to slow decay, there is a minimum frequency at which it will operate; reliable operation can be expected at frequencies as low as 5 kHz.

CMOS dynamic shift registers have all the advantages of other CMOS devices including low power dissipation, high noise immunity, and wide operating voltage range; in addition, they are superior in two important ways to the single-channel (p-MOS and n-MOS) dynamic shift registers. First, the CMOS device easily generates the two-phase clock signals required, internal to itself, with just one supply voltage. Second, TTL and DTL logic compatibility is maintained on all inputs and outputs with one supply voltage.

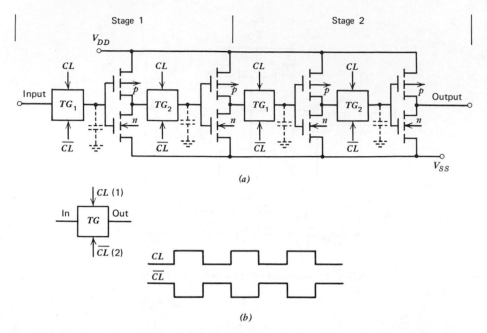

(a)

(b)

Figure 7-19   (*a*) Two-stage shift register, (*b*) clock-pulse diagram. All *p*-unit substrates are connected to V_{DD}; all *n*-unit substrates are connected to V_{SS}. Input-to-output is a bidirectional short circuit when control input 1 is low and control input 2 is high; it is an open circuit when control input 1 is high and control input 2 is low. (Courtesy of RCA Corporation.)

INTERFACING CMOS WITH OTHER LOGIC FORMS

Optimum efficiency in system design is achieved by directly but alternately interfacing CMOS circuits and DTL or TTL circuits (i.e., taking full advantages of the superior CMOS logic characteristics at lower logic speeds and the high-speed capability of saturated bipolar logic when required). Because the recommended operating-voltage range of CMOS circuits in the 4000A series is from 3 to 15 V and the operating voltage range of TTL and DTL circuits is from 4.5 to 5.5 V, all 4000A devices may drive and be driven by TTL and DTL logic devices.

Bipolar Driving CMOS

Figure 7-20 depicts a TTL or DTL gate directly driving one device input. Limit values of bipolar drive parameters are shown, as are the limit parameters of the driven CMOS device. Examination of Figure 7-20 indicates that for the bipolar-output logic "0" case, the CMOS device is safely driven with a safety noise margin of 1.1 V (1.5 to 0.4 V). The net effect is a reduction of noise margin from 30 to 22% of V_{DD} when compared with CMOS driving CMOS. For the logic "1" case of TTL driving CMOS, a borderline situation exists. The minimum no-load logic "1" TTL output is 3.6 V, but the minimum logic "1" input voltage required is 3.5 V. This interface is valid, but it has essentially no noise margin (0.1 V). Therefore, it is recommended that an external pull-up resistor R_p be used to guarantee

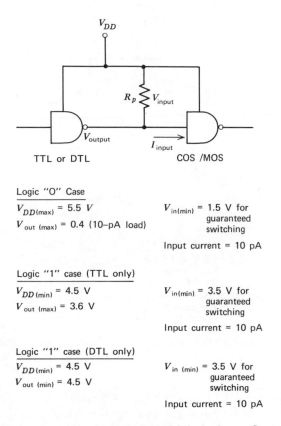

Logic "0" Case

$V_{DD(max)} = 5.5\ V$

$V_{out\ (max)} = 0.4\ (10\text{–pA load})$

$V_{in(min)} = 1.5\ V$ for guaranteed switching

Input current = 10 pA

Logic "1" case (TTL only)

$V_{DD\ (min)} = 4.5\ V$

$V_{out\ (max)} = 3.6\ V$

$V_{in(min)} = 3.5\ V$ for guaranteed switching

Input current = 10 pA

Logic "1" case (DTL only)

$V_{DD(min)} = 4.5\ V$

$V_{out\ (min)} = 4.5\ V$

$V_{in\ (min)} = 3.5\ V$ for guaranteed switching

Input current = 10 pA

Figure 7-20 TTL or DTL gate directly driving one CMOS device input. (Courtesy of RCA Corporation.)

the interfacing of TTL with CMOS. Then V_{out} would be equal to V_{DD}. The value of R_p affects the logic "0" interface condition in that $V_{out(max)}$ is increased. Determination of an optimum value for R_p requires consideration of fan-out, maximum TTL device current, $V_{out(max)}$, TTL device leakage in the high state, power consumption, and propagation delay. The logic "1" case of a bipolar DTL circuit driving a CMOS circuit is a highly satisfactory interface, and full CMOS noise margin is unaffected.

CMOS Driving Bipolar

Figure 7-21 represents one CMOS gate driving one TTL or DTL gate input. When the CMOS output is a logic "1," the TTL or DTL load acts as a reverse-biased diode; maximum load current is 40 μA at a minimum CMOS output voltage of 4.4 V. Because the maximum required TTL or DTL input voltage is 2 V, greater than 50% noise immunity exists.

For the logic "0" TTL or DTL input condition, a maximum input voltage of 0.8 V is required at 1.6 mA of sink current. Driving with 0.8 V permits essentially no noise immunity; therefore, a 0.4-V drive is recommended. Table 7-3 lists CMOS output–sink-current capabilities at 0.4 and 0.8 V. Only the 4009A and

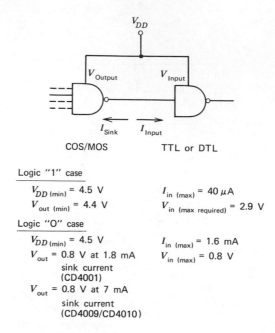

V_{DD}

V_{Output} V_{Input}

I_{Sink} I_{Input}

COS/MOS   TTL or DTL

Logic "1" case

$V_{DD\,\text{(min)}} = 4.5\ \text{V}$ $I_{\text{in (max)}} = 40\ \mu\text{A}$

$V_{\text{out (min)}} = 4.4\ \text{V}$ $V_{\text{in (max required)}} = 2.9\ \text{V}$

Logic "0" case

$V_{DD\,\text{(min)}} = 4.5\ \text{V}$ $I_{\text{in (max)}} = 1.6\ \text{mA}$

$V_{\text{out}} = 0.8\ \text{V at 1.8 mA}$ $V_{\text{in (max)}} = 0.8\ \text{V}$

sink current
(CD4001)

$V_{\text{out}} = 0.8\ \text{V at 7 mA}$

sink current
(CD4009/CD4010)

Figure 7-21 CMOS gate driving TTL or DTL gate input. (Courtesy of RCA Corporation.)

4010A buffers can drive TTL or DTL inputs directly. At an output of 0.4 V, two TTL or DTL gates can be driven; at an output of 0.8 V, four gates can be driven. Low-power TTL logic devices (54L or 74L series) can be driven directly by most CMOS devices. The sink-current requirement of these devices is only 0.18 mA per bipolar input. Table 7-3 indicates that one CD4000A device output will drive three 54L loads at 0.4 V.

Table 7-3   Current-Sinking Limits of CMOS Devices[a]

Type	Description	Sink Current	
		$V_{OL} = 0.4\ \text{V}$	$V_{OL} = 0.8\ \text{V}$
CD4000A	Dual 3-input NOR gate plus inverter	0.40	0.8
CD4001A	Quad 2-input NOR gate	0.40	0.8
CD4002A	Dual 4-input NOR gate	0.40	0.8
CD4007A	Dual complementary pair plus inverter	0.6	1.2
CD4009A	Inverting hex buffer	3.0	6
CD4010A	Noninverting hex buffer	3.0	6
CD4011A	Quad 2-input NAND gate	0.2	0.4
CD4012A	Dual 4-input NAND gate	0.1	0.2

[a]From *RCA COS/MOS Integrated Circuits Manual*, 1971.

4000A–SERIES CMOS USAGE GUIDELINES

Unused Inputs

In a CMOS device, all input pins must be connected to a voltage level between V_{SS} (the most negative potential) and V_{DD} (the most positive potential). In circuit configurations in which input pins connect directly to NAND gates, the unused inputs must be tied to V_{DD} (high-level state) or tied together with another input pin to a signal source. Either connection is necessary because the output of any NAND gate normally remains in the high-level state for as long as any input is in a low-level state; the output only changes to a low-level state when all inputs are high. Conversely, in circuits that connect directly to NOR gates, the unused inputs must be tied to V_{SS} (low-level state, usually ground) or connected together to another input pin driven by a signal source. Either of these connections is required because the output of a NOR gate is in the low-level state if any input is in the high-level state. The output of the NOR gate only goes high when all inputs are low. Floating (unused) inputs guarantee neither a logic "0" nor a logic "1" condition at the output of the device; rather, they cause increased susceptibility to circuit noise and can result in excessive power dissipation. Floating inputs to high-impedance CMOS gates can result in linear-region noise biasing when both the p- and n-type devices are ON.

Paralleling Gate Inputs

The inputs of multi-input NAND and NOR gates are sometimes wired together and connected to a common source. In the case of NAND gates, where as many as four input pins may be wired together (CD4012A), a slight increase in speed occurs when more than one input is tied to the same signal (see Figure 7-22). More importantly, however, the output source current of the device is increased proportionately to the number of inputs wired together.

When the inputs of a NOR gate are tied together to a common input signal, the gate experiences a higher sink current and a slight increase in speed. The speed increase in both NAND and NOR gates results from the lower ON resistance of the paralleled devices. The increase in speed is minimized by a compensating speed decrease caused by the added capacitance of the driving source as well as by capacitance internal to the device itself.

Parallel Inputs and Outputs of Gates and Inverters

Both source and sink-output current are increased by paralleling two or more similar devices on the same chip, as in Figure 7-23. This increased drive capability also increases speed if the growth in capacitive loading is not excessive. When devices are paralleled, power dissipation also increases.

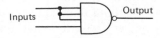

Figure 7-22   Logic diagram of a four-input NAND gate with three inputs interconnected. (Courtesy of RCA Corporation.)

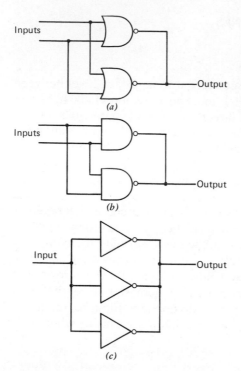

Inputs

(a)

Inputs

(b)

Input

(c)

Figure 7-23 Typical device-paralleling arrangements: (a) NOR gates, (b) NAND gates, (c) inverters. (Courtesy of RCA Corporation.)

Common Busing

Complementary MOS logic function outputs may be "common bused" at transmission-gate outputs. However, because of the complementary nature of the basic CMOS inverter circuit, common busing is not permitted at inverter outputs. Thus the phantom OR function commonly achieved at the inverter outputs of bipolar and p-channel MOS devices is achieved in a similar fashion at CMOS transmission-gate outputs. In Figure 7-24, common bits of four register outputs are common bused. This type of busing uses the quad transmission gate (CD 4016A) approach in which only one output is actively connected to the line at a given instant.

Input Protection

The standard input protection device used in all CMOS ICs appears in Figure 7-25. Protection is required to prevent damage to the MOS input gates that could result from careless handling or testing prior to final installation. Figure 7-25 illustrates the positive, built-in protection afforded by the diode clamps in a CMOS circuit; this approach is in contrast to the widely varying zener-diode breakdown protection used in bipolar circuits.

The breakdown voltage of an MOS gate oxide is on the order of 100 V; the dc resistance is on the order of 10^{12} Ω. In contrast to semiconductor diodes (in which the breakdown limit can be tested any number of times without damaging the device), the MOS gate oxide is shorted as a result of only one voltage excursion to

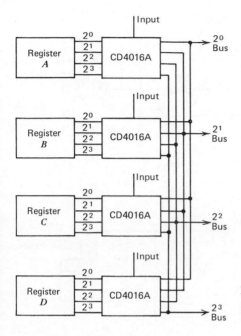

Figure 7-24 Common busing of common bits of four registers using the CD4016A devices. (Courtesy of RCA Corporation.)

the breakdown limit. Because of the extremely high resistance of the gate oxide, even a very-low-energy source (such as a static charge) is capable of developing this breakdown voltage.

The input resistance R (Figure 7-25) is nominally between 1 and 3 kΩ. This value, in conjunction with the capacitances of the gate and the associated protective diodes, integrates and clamps the device voltages to a safe level. Figure 7-26 demonstrates how input circuits can be designed to limit extraneous voltages to safe levels under all operating conditions. Because of the low RC time constants of these circuits, they have no noticeable effect on circuit speed and do not interfere with logic operation.

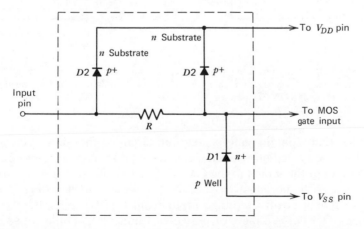

Figure 7-25 CMOS–IC protection circuit showing diode clamps. (Courtesy of RCA Corporation.)

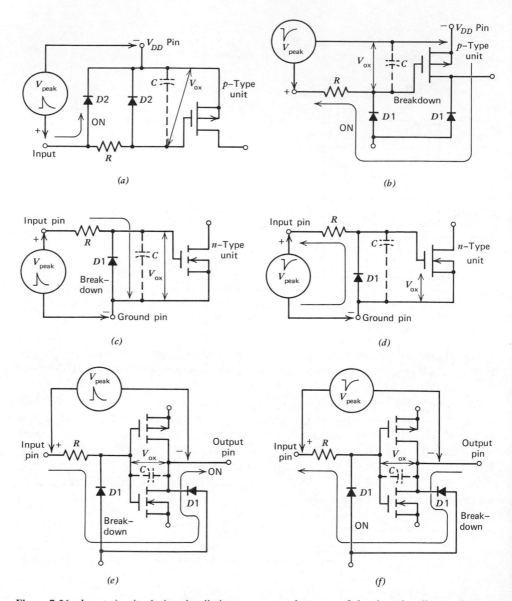

Figure 7-26 Input circuits designed to limit extraneous voltages to safe levels under all operating conditions: (*a*) *p*-type unit, (*b*) *p*-type unit, (*c*) *n*-type unit, (*d*) *n*-type unit, (*e*) *n*- and *p*-type units, (*f*) *n*- and *p*-type units. (Courtesy of RCA Corporation.)

In circuits that contain gate-protection circuits, the power-supply voltage V_{DD} should not be turned off while a signal from a low-impedance pulse generator is applied at any of the inputs to the CMOS IC. The reason for this restriction can be understood with the aid of Figure 7-26*a*, where the V_{DD} line is essentially grounded so that a positive-voltage input from a pulse generator is impressed across diodes D_2. This voltage, of between 3 and 15 V, could cause permanent damage to the diodes or could burn out the V_{DD} metallization. Therefore, if any input excursion in any system design is expected to exceed $+V_{DD}$ or fall below $-V_{SS}$, the current through the input diodes should be limited to 50 mA to assure safe operation.

Power-Source Requirements

Regulation

CMOS devices exhibit reliable switching properties over a wide range of power-supply voltages. Specifically, 4000A-series CMOS devices operate reliably with high noise immunity as long as the power source voltage ($V_{DD} - V_{SS}$) is greater than 3 V and less than 15 V (> 5 V and < 15 V for the 4000 family). This statement implies that an unregulated supply may be used, provided maximum voltage limits are not exceeded and system speed is no greater than the speed that can be supported by the CMOS device operating at the lowest value of V_{DD} expected from the unregulated supply. Maximum system speed, then, is dictated by the minimum-power-source excursion. As an additional system-related consideration, the power-source regulation frequency must be less than the minimum CMOS device frequency at 6 V for the high-threshold types and at 3 V for the low-threshold types.

Battery Operation

Because of their ability to operate over a wide voltage range at a very low level of current drain, CMOS logic circuits can be operated directly from inexpensive batteries having a wide voltage excursion from "beginning" to "end" of life. Additionally, the very high noise immunity of CMOS devices is an advantage in battery operation because the internal impedance of batteries is often greater than the internal impedance of most power supplies.

The low power dissipation of CMOS devices offers another important advantage where battery operation is required. CMOS logic systems will not change state during long periods of time. In essence, then, the battery powering of the devices behaves as a shelf-life battery for quiescent system operation; holding voltages can be as low as 2 V.

Automotive batteries of a 12-V nominal rating can directly drive CMOS logic circuits. Care must be taken, as with most other electronic devices, to protect CMOS devices from high-voltage transients resulting from open-circuited battery terminals; transients as high as 100 V can arise from the high electric field in an automobile generator. The usual means of avoiding such mishaps is the zener-diode protection circuit.

Zener-Diode Reference Operation

Operation from zener-diode references not only protects CMOS devices from momentary line transients but also can provide an effective power-supply source from dc sources exceeding 15 V (e.g., 28-V aircraft supply). The zener-diode source is also effective when accurate digital-to-analog converters are designed using CMOS devices.

Proper design of zener regulators for CMOS logic operations must take into account the transient nature of most CMOS current-drain requirements. Current drains in excess of the microampere quiescent currents are an integral sum of the

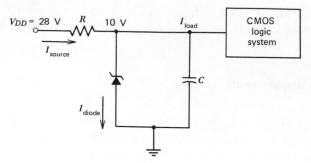

Figure 7-27   Zener-regulating circuit. (Courtesy of RCA Corporation.)

switching-transient current of the CMOS devices. Knowledge of peak current I_P, switching frequency, and current rise and fall times must be judiciously used in calculating the resistance and the capacitance of the zener regulating circuit diagrammed in Figure 7-27.

54C/74C Series CMOS Functions

In 1972 National Semiconductor Corporation introduced the 54C/74C Series of CMOS device. The 54C/74C family provides a CMOS pin-for-pin replacement for the popular 54/74 series TTL. This new concept in CMOS will aid the engineer in design because he is already familiar with the 54/74 family function, features, and pinouts. He does not have to learn how to use a new family of devices.

The 54C/74C family combines the low power of CMOS (10 nW for gates and 10 μW for MSI circuits) with the speed of TTL. 54C/74C is the second generation of standard CMOS circuits. It is superior to the 4000A series in the following ways.

- Typically 50% faster.
- Will sink typically 50% more current.
- Electrical parameters are guaranteed over the entire temperature range.

Chapter Eight

APPLICATIONS OF MOS
INTEGRATED CIRCUITS

The application of MOS technology to systems is fundamentally a problem of large-scale integration. The characteristics of MOS devices suggest their use in arrays of circuit elements. The system then must be arranged so that it can be fabricated from a number of similar functional blocks (arrays).

The motivation for using MOS arrays is reliability no less than economy. Reliability is improved by the smaller number of connections and parts that must be handled. For a given system function, the number of die bonds, wire bonds, and packages is greatly reduced by fabrication on a single die. Since these areas contributed heavily to failures, their reduction leads to increased reliability. The economic aspects also stem from the reduction in packaging effort. The IC package itself as well as the effort of installing the IC represent a significant cost item. The cost per circuit can be reduced if the packaging costs remain nearly constant and the number of circuits is increased. The actual cost of processing MOS devices promises to be lower than that of bipolar devices. Fewer processing steps are required, and these should eventually be simpler than those used in bipolar devices. Fewer steps could lead to higher yield as well as reduced processing cost.

LARGE–SCALE INTEGRATION (LSI)

Existing magnetic storage systems are the basis for the popular organizations of semiconductor memory systems such as random-access, read-only, and serial-storage (delay lines, drums, and disc). However, determining the means of achieving these organizations economically is the major source of controversy among the manufacturers of semiconductor memories.

One key to the success of a semiconductor memory is LSI, because even a small memory contains many circuits. Without LSI, the role of semiconductor memories would be limited to flip-flop configurations and small blocks of logic circuits. The attractiveness of semiconductor memories lies in speed, size, and compatibility with logic circuits.

The two basic types of LSI memories are bipolar and MOS. In bipolar memories the basic flip-flop with multiple-emitter transistors is the simplest storage cell. For increased speed, emitter-coupled logic (ECL) is used. Either type is well suited for LSI fabrication. However, the multiple-emitter configuration is the most commonly used because it lends itself to higher densities per chip. Furthermore, this type of memory cell is compatible with both DTL and ECL.

In MOS memories, dynamic storage (charged capacitor) cells and dc flip-flops serve as MOS storage elements. The flip-flop configuration requires six FET geometries per storage cell—two for gating, two for flip-flop circuits and two representing load resistances. This technique requires a large number of lead connections per chip. Obviously, flip-flop arrays are not the ultimate in low cost.

Dynamic cells composed of only three minimum-geometry FETs are simple and inexpensive. However, an updating scheme is required to retain information in the storage element (charged capacitor). Refreshing memory techniques have been used for years in other memory systems. In fact, ferrite-core systems use a "postwrite" operation for data regeneration because of the core's destructive-readout properties.

Many memory organizations now under development use a 256-bit array—256 words of 1 bit for main frame and 64 words of 4 bits or 128 words of 2 bits for scratchpad application. Typical power dissipation for these chips is 2 mW/bit, and the cycle time is approximately 200 nsec. Also included on these monolithic chips is some form of address decoding, which accounts for most of the cycle time.

Certain applications require sequential addressing. For example, a 16,834-word memory (each word, 32 bits) constructed from 256-bit chips, could transfer in and out of memory at the rate of 2048 bits/memory at 1 bit/chip/cycle. This system can be loaded in 64 μsec, assuming a 250-nsec cycle. If 16 words are accessed in parallel, the access time is reduced. By arranging the memory so that each set of 16 words has consecutively numbered addresses, the average access time is reduced significantly to 4 μsec. Such an organization may provide higher data rates with access times much shorter than now are available with serial memories. The only feasible means of handling such a memory is through LSI.

STATIC AND DYNAMIC LOGIC

Digital logic functions can be implemented by two general types of MOS circuits: (1) conventional dc or static logic that uses techniques similar to those employed in bipolar ICs, and (2) dynamic or ac logic that uses temporary memory and clocked load resistors.

The big difference between conventional static logic and dynamic logic is that the latter is a sampling type logic. That is, a dynamic circuit samples the state of the signal at its input periodically by means of a clock pulse and provides a corresponding output in the form of a stored charge on a capacitor, rather than as a continuous dc output voltage. The dynamic circuit, therefore, needs to draw drain current only during the sampling period, and that is how the very significant savings in power is achieved.

Dynamic logic is feasible with MOS because the input to an MOS transistor is almost purely capacitive. Thus a logic circuit working into an MOS load needs only to provide a path for quickly charging or discharging the input capacitance of the load to establish a specific logic level.

dc or Static Logic

In static logic a relatively high-impedance MOS device that needs very little area is substituted for a diffused-load resistor (Figure 8-1), which requires much more area. This MOS load (or pull-up) device is held in conduction at all times. There are two methods of accomplishing this result: one uses a single power supply applied to both the gate and the drain of the load device; the other uses two supplies, with the more negative one tied to the gate. By using a second supply that is at least one gate threshold drop more negative than the drain supply (V_{DD}), the drop across the load device becomes essentially zero when the inverter is OFF.

Figure 8-2 shows the conventional static MOS inverter. In this circuit, transistor Q_2 is biased "ON" and serves as a fixed resistance. Its value is many times that of the ON-resistance of switched transistor Q_1, so that the output voltage can approach zero when Q_1 is turned on by V_{in}. Because the inverter requires a high ratio between Q_2 and Q_1, it is called a ratio inverter.

When V_{in} is at a logic "1," Q_1 is turned on, causing current to flow from V_{DD} to ground. Due to the high resistance value of Q_2 compared with Q_1, in the ON condition, the voltage drop across Q_1 is very small, causing V_{out} to be virtually at ground. When V_{in} goes to zero, Q_1 is cut off, causing C_1 to charge to V_{DD}. Because R_{Q2} is high, the charge time for C_1 is relatively slow. When V_{in} returns to a logic "1," C_1 discharges rather quickly through the low resistance of Q_1. The high

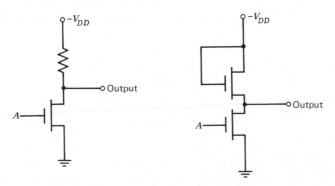

Figure 8-1 An example of static logic: the load resistor is replaced with another MOS device so that the circuit can be made entirely of MOS devices.

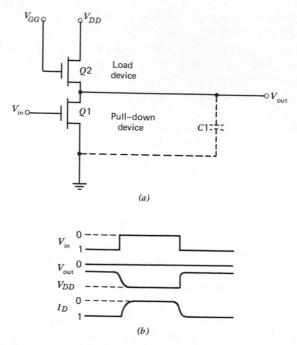

Figure 8-2 (*a*) Typical static MOS inverter circuit, (*b*) waveforms. The following conditions apply: (1) transistors are *p*-channel, (2) V_{GG} and V_{DD} are negative.

resistance of Q_2 reduces current amplitude, but the substantial ON-time results in considerable power dissipation. Moreover, the long charge time for C_1 reduces circuit speed.

The load and pull-down devices in a MOS inverter differ in size. The load device is designed to have an ON impedance about 20 times that of the pull-down device. This ratio has been selected to assure that the output node goes to near ground when the inverter is conducting.

Using the convention of negative true logic, a two-input NOR gate (Figure 8-3) differs from the logic inverter by having two pull-down devices in parallel, whereas the two-input NAND gate (Figure 8-4) differs by having two pull-down devices in series. These two basic logic functions can be expanded and interconnected to perform many more complex logic functions.

For all MOS devices, if some capacitance is required, the residual or parasitic capacitance that exists between the gates and the substrate or between the *p* diffusions and the substrate is all that is needed.

ac or Dynamic Logic

Dynamic logic uses clocked load devices in conjunction with the inherent gate capacitance of the MOS device to give one simpler circuits and greatly reduced power consumption. Since power is consumed only when the clocks are on, dynamic logic lets the designer trade off power consumption for speed of logic operations.

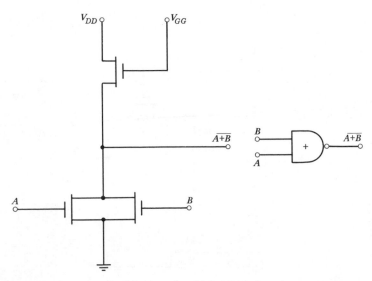

Figure 8-3 The MOS NOR gate (dc) is similar to the inverter except that it has two pull-down devices in parallel.

Dynamic logic has a characteristic that is similar to storing a charge on a capacitor; therefore, it has to run at a certain minimum frequency, because the charge leaks off owing to the junction leakage. Therefore, the dynamic logic concept can be used only for circuits that operate above a minimum frequency of about 10 kHz.

The MOS device is uniquely suited for this application because its gate looks like a perfect capacitor with virtually no leakage. This intrinsic gate capacitance serves as temporary storage for incoming data. Dynamic logic takes advantage of this temporary storage in conjunction with two-phase clocking to establish a built-in delay in any gating structure. Unlike dc logic, in most cases there is no need to add extra devices to the intrinisic gate structure in order to provide delay. The function of the two-phase clocking is to direct the flow of information from one gate to the next. One bit-of-delay (one clock period) occurs in the case where two adjacent logic gates are clocked with the same phase.

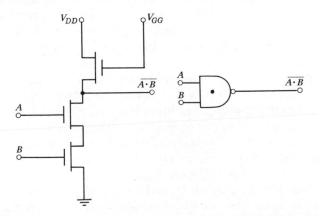

Figure 8-4 The MOS NAND gate (dc) has two pull-down devices connected in series.

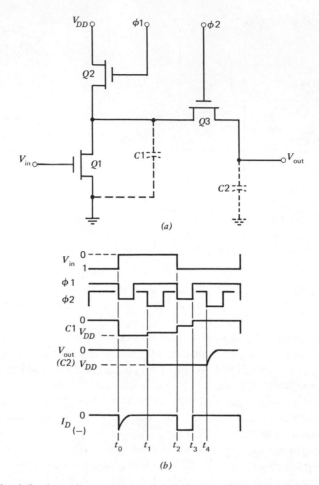

(a)

(b)

Figure 8-5   (a) Circuit for dynamic, two-phase ratioless inverter, (b) waveforms.

Figure 8-5 shows the dynamic inverter. Note that an additional transistor Q_3, is employed. This is a coupling transistor, which gives the impression that dynamic logic is more expensive than static logic simply because it requires more components. This, however, is not the case, since the component count between static and dynamic logic quickly equalizes as circuit complexity increases beyond that of the basic gate.

With the dynamic inverter shown, performance does not necessarily depend on a high resistance ratio between Q_2 and Q_1. From the timing diagrams it is seen that signal transfer from the input to the output is accomplished only during the time that the phase-two clock turns ON transistor Q_3. During phase two, transistors Q_1 and Q_2 cannot be turned ON simultaneously, so that the resistance ratio between the two does not affect the output in any way. The operation of the dynamic inverter of Figure 8-5 is as follows: At t_0, V_{in} goes to zero, turning OFF Q_1. At the same time, ϕ_1 goes low, turning on Q_2 and causing C_1 to charge to V_{DD}. At t_1, ϕ_2 turns ON Q_3, causing the charge on C_1 to be transferred to C_2. Because C_1 is made much larger than C_2, this charge transfer can occur without reducing the output voltage. At t_2, V_{in} turns ON Q_1 and ϕ_1 turns ON Q_2. This causes C_1 to discharge partially, but this discharge cannot be transmitted to C_2 because Q_3 is turned OFF. When ϕ_1 turns OFF Q_2 at t_3, a partial charge still remains on C_1. This charge is discharged through Q_1, which is still turned ON by V_{in}. Thus at t_4, when ϕ_2 turns ON Q_3, C_2 also is permitted to discharge through Q_1 to ground.

During this cycle, drain current flows only at t_1, to charge up C_1, and from t_2 to t_3, the short amount of time during which both Q_1 and Q_2 are turned ON simultaneously.

The ac inverter structure, combined with ac NOR and NAND gates (Figure 8-6), provides the basic building blocks required for logic design. A point to note is that a NOR gate clocked at phase-1 time in series with an inverter clocked at phase-2 time constitutes an OR gate with some delay. On the other hand, a NAND gate followed by an inverter and clocked the same way gives a logic AND gate, including the small delay. Essentially the input is clocked by phase 1 through the first series-coupling device to the gate capacitances of the second inverter pull-down device, where it is stored between clock pulses. Next, phase 2 clocks the data through this inverter and stores them on the next gate until the subsequent phase-1 time.

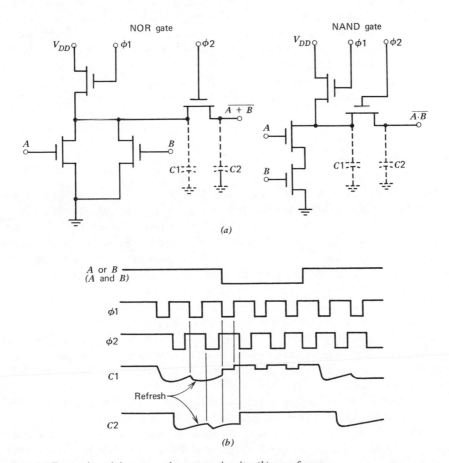

(a)

(b)

Figure 8-6 (a) Dynamic ratioless two-phase gate circuits; (b) waveforms.

If no delay is desired, all the clocks can be returned to a common phase. In this case, all series devices between gates of a common phase can be omitted, since their main function is to isolate data until the alternate clock time. Thus a gating structure with no delay would need no series devices.

To ensure the proper flow of data, the two clocks ϕ_1 and ϕ_2 should never be a logic "1" (V_{DD}) simultaneously. Also, the capacitive memory time constant must be greater than the time period between the trailing edges of ϕ_1 and ϕ_2, or vice versa, whichever is greatest, so that a logic "1" stored on the gate capacitor is not degraded by leakage to a voltage below the gate threshold level. High-speed operation is limited by the ability to fully charge the gate capacitor to a logic "1" level during the clock ON time. This ratioless dynamic circuit draws a very small amount of power compared with a static curcuit. Nevertheless, during every second phase-1 clock pulse, both the transistors cascaded across the power-supply line are turned ON at the same time. This consumes dc power for the duration of the phase-1 pulse.

The power-saving attributes of dynamic logic are quite obvious. As a trade off, however, it demands at least two clock phases, which adds a considerable amount of complexity to circuit design. The clock pulses not only cause a sampling of the input signal, they also provide for "refreshing" the charge on the capacitors in the event that specific voltage levels must be maintained for more than a few milli-seconds. Because of this, the clocks must run continuously to maintain proper operation, and the frequency and spacing of the pulses are critical. In addition to an upper frequency limit, set by the high-frequency capabilities of MOS circuitry, dynamic circuits have a minimum clock-frequency limit, in order to prevent loss of data due to excessive discharge of the capacitors as a result of leakage between clock pulses.

SHIFT REGISTERS

Common to all data-processing equipment is the ability to transfer digital data. The transfer of such data generally involves temporary storage combined with the ability to move the data by a prescribed number of bit positions.

A shift register holds N bits of data that can be entered, stored, and sampled on command. In addition, the register can shift the data from one storage position to an adjacent position. For an N-bit shift register, N clock pulses are necessary for the data appearing at the input to be transferred to the output.

Shift registers can be static or dynamic. Either static or dynamic registers produce output logic levels within the standard MOS range. So the output can be tied to the input, to recirculate data, or shifted directly into another register or other logic circuits. The registers are readily interfaced with bipolar logic circuits, such as TTL. There are, however, several important differences between static and dynamic registers.

A static register is implemented with flip-flops which are bistable devices. Therefore, when a bit is clocked into the register, it remains there until a second clock pulse shifts it into a succeeding flip-flop, or out of the register. Once a static register is loaded, the clock can be stopped and the information will remain in the register, available for repeated retrieval, until it is deliberately obliterated. The price paid for the nondestructive storage feature of static registers is more transistors per storage cell. Static cells are about 1.5 times as large as dynamic cells. Cost per bit and power dissipation are higher.

In a dynamic circuit this is not the case. Since the bit information is stored in a capacitor, the circuit must be continually clocked in order to restore any charge that might leak off between inputs. This continuous succession of clock pulses causes each bit to ripple through the register and finally to disappear when it has reached the last stage of the register chain.

Nevertheless, the dynamic register can be used for data storage simply by feeding the output back to the input.

Dynamic Two-Phase Shift Registers

Dynamic shift registers made with MOS devices are simply inverter structures connected in series by clocked transmission gates. Because the information is held in the register by the charge on capacitors instead of in latched flip-flops, a minimum operating frequency is needed. The discharge rate from the capacitors determines this frequency. If the register is operated at slower speeds, the data stored in the capacitors leak off and are lost.

There are two basic types of dynamic shift register forms: the ratioless and the ratio. The derivation of the ratio name indicates that when the inverter is ON, the load device and the pull-down device form a voltage divider between the supply voltage and ground. Therefore, the voltage that appears at the output of the inverter is determined by the ratio of the resistance of the pull-down device and the load device.

A circuit that has no dc current path regardless of the data stored or the state of the clock offers a significant savings in power dissipation, even over that of the clocked inverter. Such a circuit has a still more important advantage; namely, a reduction in chip area. Since it is not necessary to maintain a ratio of resistances, all devices can be of minimum geometry. This type of circuit has been implemented, and shift registers that use it are referred to as ratioless-powerless.

The ratioless circuit of Figure 8-7 is based on a capacitor precharge concept. During ϕ_{in} clock time, node B is precharged by transistor Q_3 (i.e., Q_3 is turned ON by ϕ_{in}, creating a low-impedance path from node B to V_{GG}, which charges the node capacitor C_2 to a negative voltage). Data are coupled at the same time through transfer transistor Q_1 to node A, the gate of Q_2. If the incoming data are a positive or logic "0" level, Q_2 will be in a high-impedance OFF state, and node B will charge to a negative-voltage logic "1" threshold more positive than the ϕ_{in} clock amplitude.

When ϕ_{in} returns to a positive level, Q_3 is shut OFF, isolating the precharged voltage of node B. The stored charge of node B, coupled with an additional increment contributed by C_4, redistributes between nodes B and C when the ϕ_{out} clock turns ON transistor Q_4. The redistributed charge develops a negative-voltage logic "1" level across C_3 which becomes isolated when ϕ_{out} returns to a logic "0" level. The logic "1" level turns Q_5 ON, resulting in a low-impedance path between the output of the cell and V_{SS}, establishing a logic "0" level at the output.

In the ratioless cell, two nodes (nodes B and C) become isolated from any charge-replenishing source during normal operation of the circuit. These are the

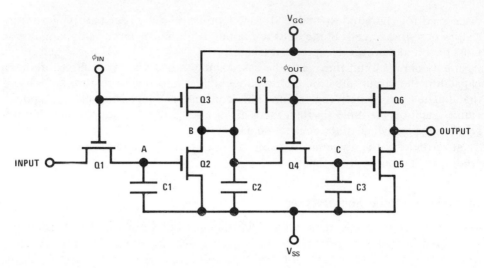

Figure 8-7   Ratioless dynamic shift register cell.

nodes that establish the low-frequency limitations of the cell. In most designs node C, the gate of the logic transistor Q_5, is the limiting node because total capacitance is less. If the initial data coupled by Q_1 during ϕ_{in} had been a logic "1" level, then node A would have been the limiting node of the cell.

The ratio dynamic shift register cell of Figure 8-8 has only one isolated node which limits minimum-frequency operation. It, like the ratioless cell, is the gate node of the logic transistor. The ratio cell does not rely on stored precharge to establish a logic "1" level on a succeeding logic gate node. If a logic "0" level had been transferred to node A of the ratio cell by Q_1 during ϕ_{in} time, Q_2 would be OFF. A ϕ_{out} logic "1" level would turn ON Q_3 and Q_4, creating a charging path between node C and V_{DD}, resulting in a logic "1" level at node C. The node

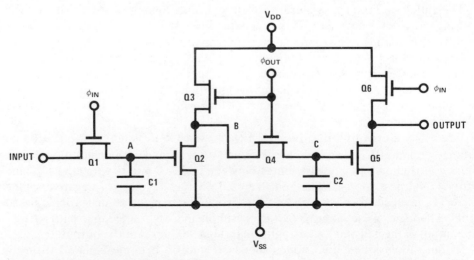

Figure 8-8   Ratio-type dynamic shift register cell.

would be isolated by Q_4, just as in the ratioless cell, when ϕ_{out} returns to a logic "0" level. If the data coupled by Q_1 had been a logic "1," both Q_2 and Q_3 would be ON during ϕ_{out} time. To establish a logic "0" at node B in that case, an electrical ratio between the ON impedance of Q_2 and Q_3 would have to be considered by the cell designer.

In many dynamic shift register applications, it is advantageous to operate the circuit at low clock frequencies or in clock burst modes where high-frequency clock rate periods are followed by long intervals in which the clocks are absent. To ensure that his system will operate correctly under these conditions, the designer should be aware of the limitations of the type of shift register he is using.

Charge must be stored at the logic transistor gate node of the ratioless cell for the period of time between leading edges of the two phase clocks. This is because no charge enters the node B and C network after the leading edge of the transfer clock (ϕ_{out}), and there is no way for charge that leaks off the nodes to be replaced. This portion of the clock period is defined as a partial bit time. The partial bit time between the leading edge of ϕ_{in} and leading edge of ϕ_{out} is the t_{in} period, and the time between the leading edge of ϕ_{out} and the leading edge of ϕ_{in} is t_{out} (Figure 8-9). The period of the minimum operating frequency is[1] the sum of the two, or

$$\phi_{f\min} = \frac{1}{t_{in} + t_{out}} \tag{8-1}$$

Obviously, the lowest operating frequency can be attained when both t_{in} and t_{out} are at their respective maximum limits and therefore equal. This says that for minimum frequency, 50% clock phasing should be used (i.e., the clocks should be equally spaced within the bit time).

The ratio cell has a similar storage requirement, but with one difference. During the time the transfer clock (ϕ_{out} in Figure 8-8) is ON, a source of charge is available to node C through the ON transistors Q_3 and Q_4, assuming that Q_2 is OFF.

Therefore, charge must be stored on the critical capacitor C_2 only after the transfer clock has returned to a logic "0" level and has isolated the node. This required storage time is usually referred to as clock phase delay time ϕ_d. The phase delay time between the trailing edge of ϕ_{in} and the leading edge of ϕ_{out} is ϕ_d; the time between the trailing edge of ϕ_{out} and the leading edge of ϕ_{in} is $\bar{\phi}_d$ (Figure 8-9). Minimum clock operating frequency is[2]

$$\phi_{f\min} = \frac{1}{\phi_{in}PW + \phi_d + \phi_{out}PW + \bar{\phi}_d} \tag{8-2}$$

assuming clock rise and fall time $\ll \phi_{PW}$.

Optimum low-frequency operation can be obtained when the clock pulse widths and phase delays are maximized and made equal. In most cases this would mean 10-μsec clock pulse widths and 50% clock phasing. For power or system application reasons, it is usually not convenient to use such wide pulse widths, and the

[1]Low Frequency Operation with Dynamic Shift Registers, by Bob Johnson, National Semiconductor Corporation, 1971, AN-55.
[2]Ibid.

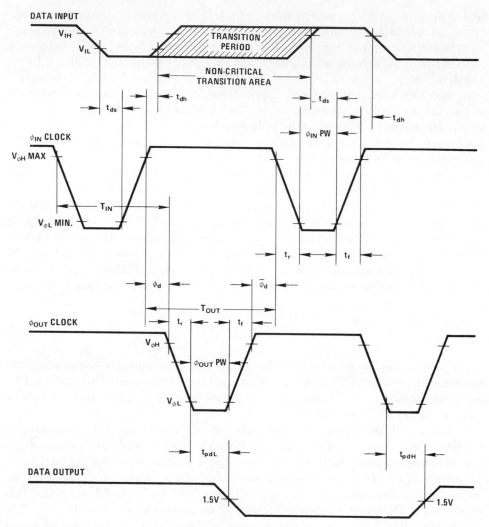

Figure 8-9   Timing diagram for two-phase dynamic shift registers.

minimum clock frequency is simplified to[3]

$$\phi_{f\min} \cong \frac{1}{\phi_d + \overline{\phi}_d} \qquad (8\text{-}3)$$

assuming $\phi_{PW} \ll \phi_d$ or $\overline{\phi}_d$.

The shift-register user can often increase his margin of safety when operating at low frequency, or for long periods of time with the clocks stopped, by designing the system with that operation in mind. The ambient operating temperature of the registers should always be minimized. The cell requires a minimum voltage at the

[3]Low Frequency Operation with Dynamic Shift Registers, by Bob Johnson, National Semiconductor Corporation, 1971, AN-55.

critical node to operate, and the time to discharge the node to that value is dependent on the initial voltage, as well as capacitance and leakage:[4]

$$t_d \approx \frac{C_{node}(V_{initial} - V_{min})}{I_L} \tag{8-4}$$

$$
\begin{aligned}
t_d &= t_{in} \quad \text{or} \quad t_{out} \quad \text{for} \quad \text{ratioless cells} \\
&= \phi_d \quad \text{or} \quad \overline{\phi_d} \quad \text{for} \quad \text{ratio cells}
\end{aligned}
$$

C_{node} = total capacitance at critical node

$V_{initial}$ = voltage at critical node immediately after isolation of that node by transfer clock

V_{min} = minimum voltage required at critical node for operation

I_L = total leakage current at critical node

The initial voltage can be optimized in two ways: by using the highest clock amplitude possible, and by allowing something greater than minimum clock pulse width to ensure that the maximum amount of charge is coupled to the node (and, in the case of the ratioless cell, that the maximum precharge voltage is obtained before transfer). A high value of V_{GG} or V_{DD}, the negative supply voltage, increases on-chip power and therefore junction temperature, as well as increasing the minimum required node voltage. It is a good idea, therefore, to stay away from very high supply voltages. When both the clock-driver reference voltage and V_{GG} or V_{DD} are the same supply, the best trade-off is toward the higher end of the specified range, however.

One other consideration that applies during operation at any frequency, but particularly at low frequency, is excursions of the clock line more positive than V_{SS}. This forward-biases internal junctions, which results in parasitic pnp transistors. If the collector of the parasitic pnp happens to be a critical node, the circuit will fail. Because critical nodes are often closer to the minimum required voltage during low-frequency operation, registers are usually more sensitive to positive clock spikes.

When calculating temperature effects of a system operating in a clock-burst mode, the designer must remember that power dissipation in the shift register is approximately double at 2.5 MHz what it is at 100 kHz. High-frequency bursts will heat the chip, causing high junction temperatures, which reduce the time the clocks can be off.

Dynamic Four-Phase Shift Registers

In the previous section, two-phase logic was discussed. Any function can be implemented with dynamic logic with a resulting saving in power dissipation and chip area. This saving can be further compounded through four-phase rather than two-phase clocking, albeit at the expense of greater design complexity and more complex and critical timing.

Four-phase logic reduces power dissipation to purely reactive power. Moreover, since component geometries need not conform to any minimum dimensions,

[4]Low Frequency Operation with Dynamic Shift Registers, by Bob Johnson, National Semiconductor Corporation, 1971, AN-55.

the chip area they occupy is set by process limitations rather than functional demands. As a result, massive systems can be designed with relatively few packages.

A four-phase circuit has certain distinct advantages over the two-phase logic circuit. These advantages are:

1. Low power. There is never a direct path between the power supply and ground. There is a direct path in the two-phase circuit.

2. MOS devices are essentially all the same size. The two-phase circuit contains different sized MOS devices.

3. Easier to "lay out."

4. More compact circuitry (high packing density).

5. High speed.

Four-phase logic can be implemented in two ways: with nonoverlapping clocks, for maximum versatility, and with overlapping clocks, for greater simplicity and smaller chip-area requirements.

In Figure 8-10, a simple shift register is implemented with overlapping clock four-phase logic. The circuit functions in the following manner. At t_1, ϕ_1 energizes transistor Q_3 and ϕ_2 turns Q_2 ON. If the master capacitor C_1 is discharged, it will

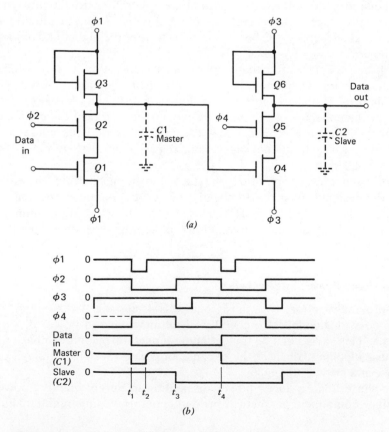

Figure 8-10   (a) Four-phase dynamic shift register circuit, (b) waveforms.

charge to a logic "1" level, regardless of the state of the input data. Should the DATA IN terminal be at a logic "1," the capacitor would charge through the parallel paths of Q_3 and Q_2 in series with Q_1. Should the DATA IN terminal be at a logic "0" level, the capacitor would charge only through Q_3, because Q_1 would be cut OFF.

At time t_2, ϕ_1 turns OFF, and the charge on the capacitor will reflect the logic level at the DATA IN terminal. Should this be at a logic "1" level, as shown, C_1 will discharge through the series-connected transistors Q_2 and Q_1, since both these are turned ON by the logic "1" levels at their gate terminals. Should the DATA IN be at logic "0," there would be no discharge path, and C_1 would remain charged. At the end of the ϕ_2 pulse (at time t_3) capacitor C_1 (master) contains the logic complement of DATA IN.

The second section of the register operates exactly the same way, except that it is made operable by ϕ_3 and ϕ_4. At t_3, the "slave" capacitor C_2 is unconditionally precharged to a logic "1" (regardless of the charge on C_1). During ϕ_4, after ϕ_3 has turned OFF, C_2 either remains charged or discharges, according to the state of C_1. The final state of C_2, after one frame of the clock (t_4), will be the logic complement of C_1, or equal to the DATA IN.

Overlapping clock circuits can be made pin-for-pin compatible with their two-phase equivalents. Whereas nonoverlapping clock systems require four separate clock lines, overlapping clock systems require only two, since ϕ_2 is a derivative of ϕ_1 and ϕ_4 is derived from ϕ_3.

Notice that there is never a dc path across the power supplies in this type of logic; thus no dc power is dissipated, and no resistance ratio is required. The transistors simply act as switches to charge and discharge the capacitors. Therefore, transistors can use minimum geometries.

Static Shift Registers

The basic dynamic shift register can be made into a static shift register by adding a clocked feedback loop around the inverters. This type of circuit must use the static inverters discussed previously.

In a typical static register, each storage cell consists of nine MOSFETs: couplers Q_1, Q_6, and Q_7 in Figure 8-11, inverters Q_2, Q_4, and Q_8 and load resistances Q_3, Q_5, and Q_9. The loads are continually biased ON by the input gate voltage $-V_{GG}$, while the complementary clocks ϕ and $\overline{\phi}$ (see Figures 8-11 and 8-12) switch the other MOSFETs.

Each data bit is maintained within a stage by the feedback and latching action of couplers Q_6 and Q_7, which can be considered to be the cross-coupling resistors of an Eccles-Jordan flip-flop. If the external clock ϕ is stopped (i.e., reduced in frequency to dc), Q_6 and Q_7 will remain ON or OFF, depending on the state of ϕ. Aside from making the state of the stored bit independent of leakage time, this arrangement simplifies external clocking and driving circuitry.

When the phase clock goes negative, it switches ON Q_1, Q_{10}, and Q_{19}. Transistor Q_{19} and the 15-kΩ resistor generate the positive pulse of clock ϕ, which turns OFF Q_6 and Q_7. Node E is now at the level appropriate to change the state of the delayed data bit. This bit can now move into the output state, and the storage cell is set up to receive a new input data bit.

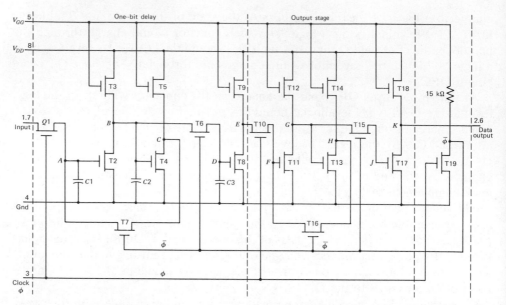

Figure 8-11 Static register contains an Eccles-Jordan type of feedback latch in each stage ($Q6$ and $Q7$ act as cross-coupling resistors) and generates the complementary clock internally.

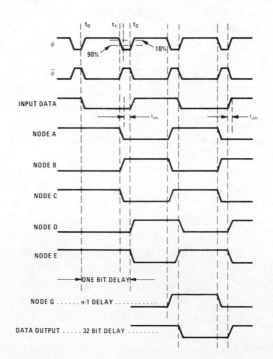

Figure 8-12 Timing diagram of a static register. When the input clock goes negative, the complementary clock goes positive, and vice versa.

Frequency and Power Characteristics

Specially designed and processed MOS registers can operate at a frequency of 5 MHz. As an approximation, we can say that the typical off-the-shelf registers being discussed here run at up to 1 MHz with relatively low signal, supply, and clock voltages (e.g., -10 V rather than the maximum of -25 V). They achieve 2 MHz with reasonable amplitudes and tolerances, $-V_{DD} = 10.5$ V $\pm 5\%$ and $V_\phi = 17 \pm 1$ V, for example.

The minimum frequency, or maximum clock period, of a static register can be whatever the designer desires. The minimum frequency of a dynamic register depends largely on operating temperature, as in Figure 8-13. The lower the operating temperature, the less the leakage current and the longer the charge-storage time available for data retention.

Power dissipation in both types varies with applied voltages. However, a dynamic register dissipates power only when a load resistor is clock on—that is, only when a current path exists between $-V_{DD}$ and ground. Therefore, its power dissipation can also vary with frequency, as in Figure 8-14. This characteristic offers the dynamic register user numerous options for reducing power dissipation. For instance, he can minimize the ON time of the loads by making the clock pulses narrow and lowering the duty cycle (Figure 8-15). After the storage capacity and input–output rates needed by the external system are satisfied, power consumption can be reduced further by slowing down the internal operation of the register.

Power dissipation in a static register (Figure 8-16) is higher per cell than in a dynamic register and varies little with frequency. A static register's load resistors are always ON. Also, loads Q_3 and Q_9 (see Figure 8-11) must be large compared with the ON resistance of inverters Q_2 and Q_8 to assure that nodes B and E will go to the $-V_{DD}$ level. Power can be conserved by minimizing applied voltages and, if data are to be stored for extended periods, by stopping the clock and maintaining the stored data with minimum dc supply voltages.

Lower power dissipation generally implies lower operating frequency. In the dynamic case, lowering the minimum frequency thereby lowers dissipation and die

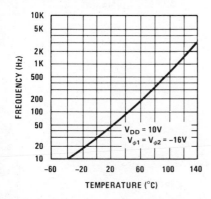

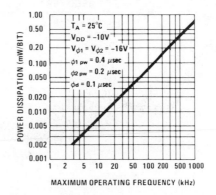

Figure 8-13 Minimum clock period of dynamic register depends on operating temperature, since temperature determines leakage of stored charges.

Figure 8-14 Power dissipation in a dynamic register increases as operating frequency rises.

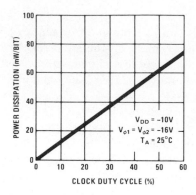

Figure 8-15 Narrow clock pulses conserve power in dynamic MOS registers. This is a typical plot of duty cycle versus power.

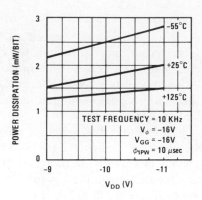

Figure 8-16 Typical static register performance curves. The power-per-bit versus gate-voltage curves are similar. Dissipation is essentially constant with frequency, since load resistors are always on.

temperature and can give the designer more freedom in his clocking techniques. However, higher temperature at a given frequency does not mean higher power dissipation. The temperature coefficients of MOS devices produce a net decrease in power dissipation as temperature rises.

Table 8-1 compares the basic characteristics of static and dynamic shift registers, and Figure 8-17 is a photomicrograph of a typical 512-bit dynamic shift register.

Input Structures and Output Buffers

At some point it is necessary to interface with the outside world. If no data selection or recirculation is required, the input can simply be expanded versions of the

Table 8-1   Characteristics of Typical Static and Dynamic Registers

TYPE OF REGISTER	STATIC	DYNAMIC
Minimum Operating Frequency		
at 25°C	DC	10 Hz
at 125°C	DC	10 kHz
Maximum Operating Frequency	2 MHz	4 MHz
Line Delays per Bit		
Typical maximum	No max	25 msec
Typical minimum	0.5 μsec*	0.5 μsec
Normal Clock Configuration	Single phase	Two-phase**
Signal Input Voltage Levels		
Logic "1"	−4.2V	−4.2V
Logic "0"	−2.5V	−2.5V
Power Dissipation		
at 1 KHz	1.5 mW/bit	0.8 μW/bit
at 1 MHz	1.5 mW/bit	0.8 mW/bit

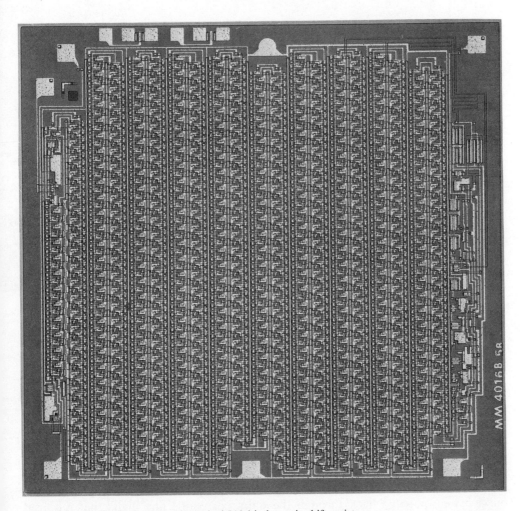

Figure 8-17 Photomicrograph of a typical 512-bit dynamic shift register.

register's cells themselves. If data selection is needed, it can be accomplished with an AND/OR structure.

At the output of the register, it is necessary to transfer data from sources with very little drive capability to the high-capacitive or low-resistive loads of the external environment. To accomplish this, an output buffer is necessary. Figures 8-18 and 8-19 depict two types of buffers. Most of the early dynamic devices used the precharge output in one of various forms. The primary advantage of this approach is the increase in operating speed (usually limited by the output buffer) of 5 to 10 MHz. These precharge output concepts were later moved to the internal circuit areas, where they became the foundation of precharge or dynamic logic.

Although the precharge output is very fast and draws little power, working with it is difficult. Test lights or dc voltmeters are useless, since the presence of a zero output means that the output will be an ac square wave following the clock input. Oscilloscopes are not much better, since examination of complex signal patterns on a scope is difficult at best. As a result, many recent dynamic circuits have incorporated limited-speed, dc-stable output buffers (which operate at 1 or

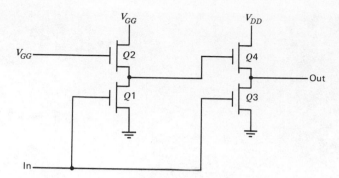

Figure 8-18 A push–pull static buffer having both source and sink capability. This buffer inverts data, thereby correcting for the inversion of data select structures. A noninverting buffer can be implemented by reversing the gate drive of Q_3 and Q_4.

2 MHz) to make complex dynamic devices appear static at the inputs and outputs, as in Figure 8-20. This is known as quasi-static four-phase logic. If speed is not critical, static outputs can make testing, servicing, and maintenance much easier.

Application of MOS Shift Registers[5]

With the exception of a few machines that used core storage, most of the early desk calculators used delay-line storage. Characteristically, although a delay line is economical in terms of cost per bit, its ability to extract the data is very limited and expensive.

Because of this, machines using delay lines for storage tended to use both a bit- and digit-serial structure for the data. In many cases the data had to be interfaced on a bit or digit basis. This forced the system designed to store the bit or digit of one argument until the other was present. During arithmetic operations, two options were possible: to correct the digit and not deposit it in the register until the next cycle, or to go back through the data, correcting any digits that required correction.

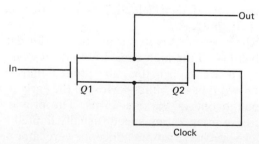

Figure 8-19 A precharge dynamic buffer, where the output load capacity is precharged by the clock going to a logic "1". The output can be sampled after the clock returns to logic "0" and before the input data change. Other dynamic buffers can be constructed using variations of the inverters presented earlier.

[5]MOS Shift Registers in Arithmetic Operations, by Jack Irwin, National Semiconductor Corporation, April 1970.

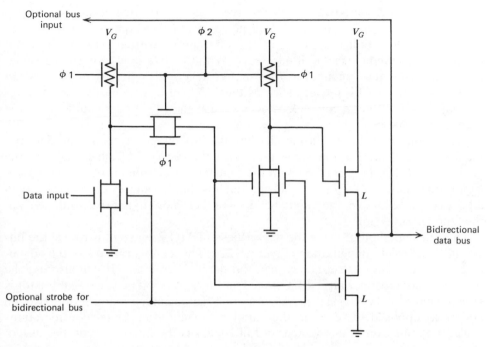

Figure 8-20   A dc stable output does not precharge the output on odd clock cycles. Note the bidirectional data bus strode option used to reduce the LSI pin count.

With the first use of MOS shift registers, the penalty for better accessibility to the data was drastically reduced. Since the introduction of bipolar compatible MOS devices, the configuration of the register area and the adder–subtractor circuit has been optimized to take advantage of this easy access.

This freedom of access may lead in two directions. First, the data could be stored digit-serial, bit-parallel, as in Figure 8-21. Because the bits are in parallel,

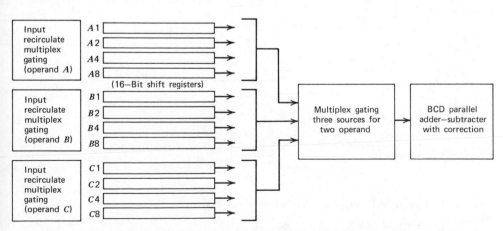

Figure 8-21   A digit–serial, bit–parallel configuration of three arguments, each having 16 digits. This approach offers high-speed operation, but the cost of increased hardwave must be added.

this approach has the advantage of relatively high-speed operation. For the 16-digit example, it requires only 16 clock pulses (cycles) to complete an operation.

In this parallel configuration the adder–subtractor must be fast enough to find the proper sum and correct it, if necessary. With the data-flow gating and complementing circuit, this amounts to four or five levels of logic. At nominal MOS operating speeds, TTL devices such as the 7400 series gate easily meet this requirement. Use of more complex devices such as the 7483 adder could reduce the number of packages.

Performing the addition in parallel has the advantage of making all data immediately available for the correcton. With the corrected value available, the proper result may be placed back in the register within one bit time. The significant disadvantage of the bit-parallel approach is that it requires more hardware in all areas. For example, the parallel adder is much more expensive to implement than the serial version.

The other method of using the data access of MOS registers is to use the bit-serial, digit-serial configuration (Figure 8-22). This configuration has the advantage of fewer data control nodes than the parallel structure. It can also use the more efficient (cost per bit) 64-bit or dual 64-bit registers. The adder–subtractor is affected most by this change in configuration. Significantly, it requires only two full adders as opposed to seven in the parallel configuration. Other considerations aside, the reduction in the number of full adders is the major benefit, because of the cost factor.

Dynamic Shift Registers with Tri-State Outputs[6]

TRI-STATE logic TTL and MOS circuits are similar in principle. The outputs sink and source current in their active logic states. However, in their high-impedance third state, they can neither source nor sink current, and present only

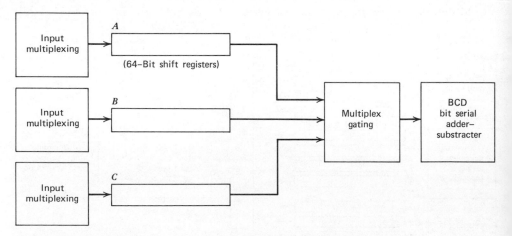

Figure 8-22   The bit–serial, digit–serial arrangement reduces hardware count.

[6]*Design Considerations of MOS Dynamic Shift Registers with TRI-STATE Outputs*, by Bob Johnson National Semiconductor Corporation, July 1972, AN-65.

a small leakage current as dc load to the bus line. Thus outputs switched into the third state do not appreciably load any active output driving the common bus line.

The technique allows numerous push-pull outputs to be bus-connected and operated at high speed compared with conventional "bus-OR'able" passive-pullup or open-collector/drain outputs. TRI-STATE MOS outputs can drive TTL, DTL, or other MOS circuits through a data bus without external current-sinking resistors.

Moreover, the third state provides a convenient method of logically disabling outputs to control data transfer modes. For example, the registers shown in Figure 8-23 can load or recirculate with the output either enabled or disabled. Control signals for the registers can come directly from bipolar control logic.

The main limitation on the number of elements that can be connected to a TTL TRI-STATE bus line is the maximum dc leakage current. In MOS, the effect of the bus capacitance on dynamic characteristics is the main concern when TRI-STATE MOS outputs are used; that is, the available output current limits the number of outputs that can be connected to the bus by determining the total amount of capacitance that can be switched in a given time interval.

The basic MOS push-pull output in Figure 8-24 is modified as in Figure 8-25

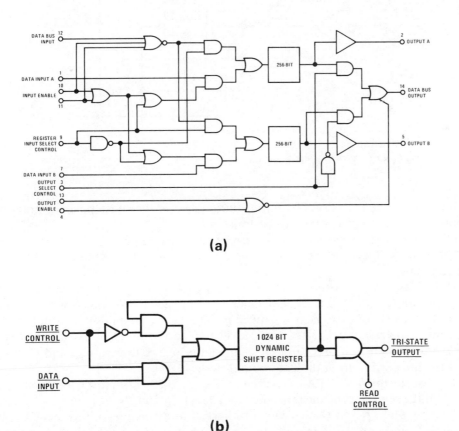

(a)

(b)

Figure 8-23 Two MOS dynamic shift registers which utilize TRI-STATE outputs. (*a*) The MM4012 dual 256-bit register; (*b*) the MM4013 1024-bit register.

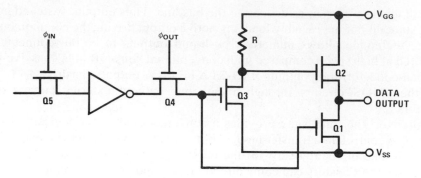

Figure 8-24   Standard push-pull output for a dynamic shift register.

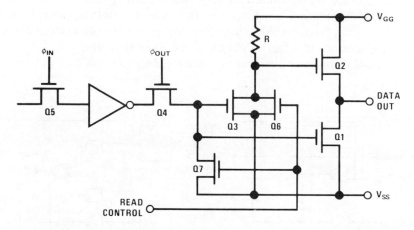

Figure 8-25   TRI-STATE output buffer.

when used as a TRI-STATE output buffer. V_{SS} is the most positive potential applied to the p-channel MOS transistors.

Q_1 of the basic design acts as the current source for a positive level output, and Q_2 provides sink current for a low or negative level output (assuming conventional current flow).

If the logic level transferred to the gates of Q_3 and Q_1 by Q_4 is negative, Q_3 and Q_1 will turn ON. Voltage developed across the resistor R by the drain current of Q_3 pulls the gate of Q_2 to near V_{SS}, turning OFF Q_2. Q_1 supplies the output source current. Conversely, when the logic level transferred by Q_4 is positive, Q_3 and Q_1 are turned OFF. The gate of Q_2 charges through R toward V_{gg}, and Q_2 turns ON to sink current from the data output node.

To disable the TRI-STATE buffer of Figure 8-25, Q_1 and Q_2 both must be OFF. If the read control line is at a negative level, Q_6 and Q_7 turn ON.

The gates of Q_1 and Q_2 are each discharged to a positive voltage near V_{SS} by Q_6 and Q_7, and turn OFF. The output is in a high-impedance condition because the buffer transistors cannot sink or source current. No data is transferred — the only current flow is a leakage of less than 10 μA.

Normal two-state operation resumes whenever the read-control line goes positive, turning OFF Q_6 and Q_7.

When active, the MOS outputs have the voltage-current characteristics shown in Figure 8-26. Considering just the dc aspects, an active output could easily supply the leakage current of more than 100 third-state outputs on the same bus line and still drive a TTL load (as do TRI-STATE TTL outputs).

But each MOS output presents a significant load capacitance that must be charged or discharged each time the data-line changes state. This determines the output transition time and adds to the total output propagation delay, as shown in Figure 8-27. Transition times increase in proportion to the total bus capacitance and therefore the number of connected outputs. The term propagation delay, as used here, includes both actual delay and transition time. The capacitance of each output has a semiconductor junction component that results in the voltage dependency illustrated in Figure 8-28.

Figure 8-29 shows the effect of increased transition time, due to load capacitance, on total propagation delay. It is given as a normalized function and applies to total delays to either a high or low level.

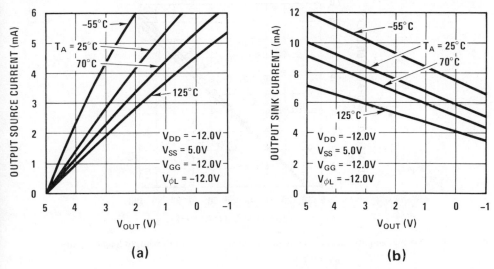

Figure 8-26 (*a*) Typical output source current and (*b*) output sink current characteristics of an MOS push-pull buffer.

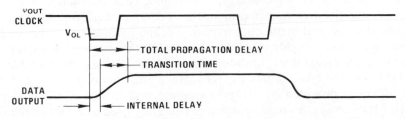

Figure 8-27 Total propagation delay includes transition time, which is a function of the output loading.

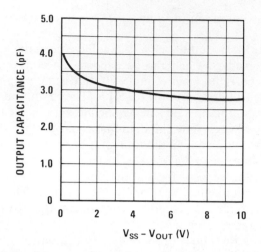

Figure 8-28   Typical TRI-STATE output capacitance as a function of output voltage.

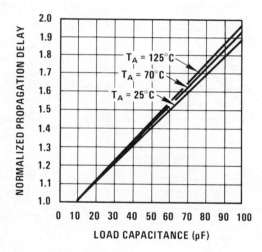

Figure 8-29   Normalized propagation delay factors as a function of the output load capacitance.

READ–ONLY MEMORY (ROM)

One of the fastest-growing markets for semiconductor memories is the ROM. These devices are used for character generation, code translation, and look-up tables. For any given computer design, the ROM has performance characteristics identical to its read–write counterpart. Technologies used for the read–write memories apply to the highly repetitive types of ROMs such as character generators and mathematical tables. Both bipolar and MOS ROMs are available.

A ROM has two basic functions: bit storage and bit isolation. The storage in bipolar arrays is provided by array interconnection in one of the final process steps. In MOS arrays, an oxide pattern is generally used to establish the data. A thin oxide layer over the channel area of an MOS transistor separates the channel from the gate electrode; if the layer is thick, the gate will not work. Thus stored logic "1"s and logic "0"s in the ROM are established by thin and thick regions in the oxide. Oxide thicknesses, in turn, are determined by the mask used at the appropriate process step. Figure 8-30a to f depicts the mask set for a 4096-bit MOS ROM.

Character Generation Using ROMs

TTL-compatible MOS storage circuits solve a dilemma that has plagued display designers: the question of how to generate the display. Eliminating digital-to-analog conversion allows a data system to remain digital right up to the display drivers; but this may exchange one economic headache for another. If the data source generates the digital control signal, its cost rises and communications links also become more expensive. Doing the job in the terminal, on the other hand, has made displays costly in the past.

MOS ROMs reduce to a few relatively inexpensive ICs the hardware required to convert a character communications code to signals that will control a display. Display rates fast enough for most applications can be achieved when the MOS ROMs are controlled by bipolar logic circuits. And when the ROMs and bipolar ICs can be coupled directly, without the use of special voltage translators, the character generator becomes proportionately less expensive.

One method of implementing a complete system is represented in Figure 8-31. All functions are controlled by the system clock, thus assuming proper alignment of the symbols on the display. The dot and space counter provides addressing control to the character generator, the character counter keeps track of the number of symbols displayed on each line in the display, and the line counter monitors the number of lines being displayed.

The MM4240 contains a 2560-bit ROM programmed to generate 64 display symbols when addressed by ASCII code and an 8-bit parallel-in, serial-out shift register. The MM4240 is a $64 \times 7 \times 5$ bit character generator programmed to display 64 characters in a 5×7 dot matrix. Each character is assigned a binary code, thus a 6-bit code describes 64 characters. Figure 8-32 is a photomicrograph of the MM4240 ROM.

The circuit of Figure 8-31 works in the following manner. The ASCII address 100000 for A is applied to the code-input address lines of the MM4240 and the row counter output to character-row input address. Character A requires the sequencing of the row counter through seven states. The (000) state identifies the top row of 5 dots. Generated output 00100 is loaded into the shift register in parallel and unloaded serially to control the Z-axis blanking of the CRT. Dot 3 is intensified on the CRT screen. If the next letter is B, ASCII code 010000 is applied to code-address input, and the row counter is not changed.

Character-generator output 11110 is loaded into the shift register and dumped serially. Dots 1, 2, 3, and 4 are intensified. When the top rows of all characters have been displayed, the CRT beam returns to the left side of the display and the row counter advances one count. The ASCII code for A is applied again, and output 01010 pattern is loaded into the shift register and unloaded serially. This process continues until all seven rows are displayed for the line of data.

The difference between a raster and a vertical-scan character generator is that

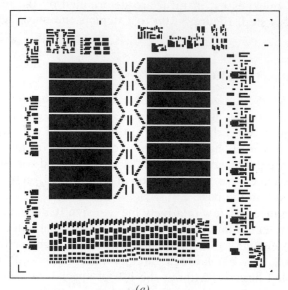

(a)

n+ diffusion mask

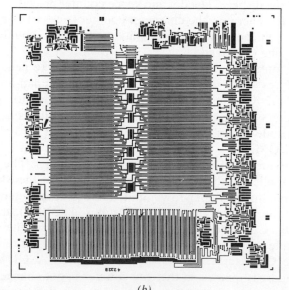

(b)

p-diffusion mask

Figure 8-30 Mask set for a 4096 bit MOS ROM.

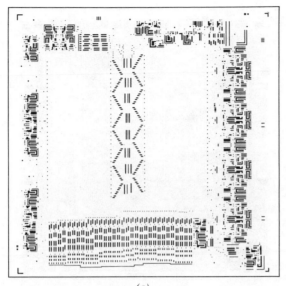

(c)
gate mask

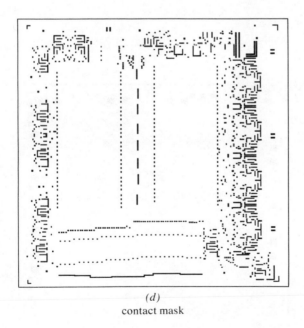

(d)
contact mask

Figure 8-30 contd.   Mask set for a 4096 bit MOS ROM.

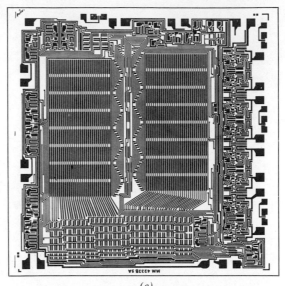

(e)
metal mask

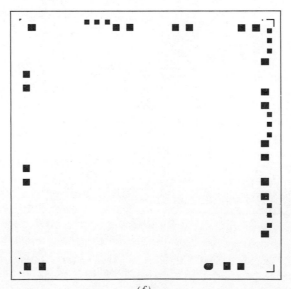

(f)
die passivation mask

Figure 8-30 contd.   Mask set for a 4096 Bit MOS ROM.

(g)

ROM pattern mask

Figure 8-30 contd. Mask set fro a 4096 Bit MOS ROM.

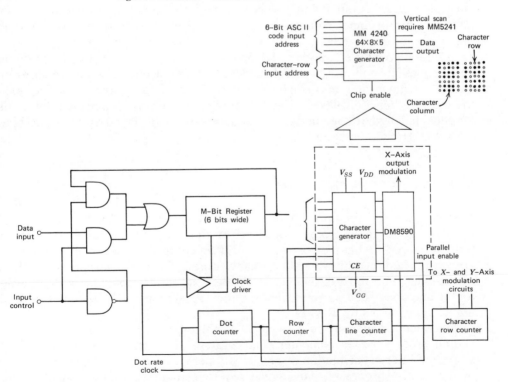

Figure 8-31 Example of complete character generation system using MOS ROMs.

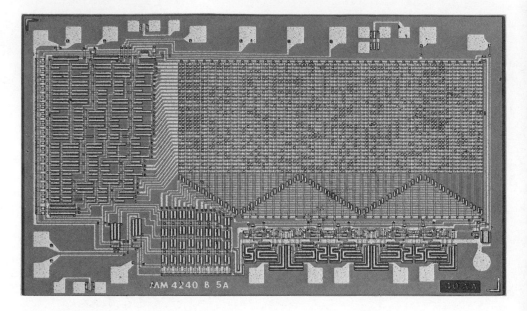

Figure 8-32   Photomicrograph of a typical 2560-bit ROM.

seven rows of five dots per row are used to display a character in raster scan; in vertical scan, five columns of seven dots per column are used for display (row-address input becomes column-address input). Referring to Figure 8-31, the dot counter is the master clock. The line counter (row or column) operates as a five- or seven-bit counter operating at 1/5 or 1/7 the dot rate. The character line counter identifies the end of a line of data. Since phosphor on the screen requires updating at regular intervals (usually at a 50 or 60 cycle rate), some memories store the message being displayed. The "M bit register" is a MOS dynamic shift register equal in length to the number of characters being displayed (one for each bit of 6-bit ASCII code-input address).

RANDOM–ACCESS MEMORIES (RAMs)

Basically, a RAM requires that any location within it be reachable or accessible without regard to any other location. At the selected location, data may be written (stored) in the memory or read (retrieved) from it. Between the time data are written and read, they must be reliably stored. The basic capacity of the RAM to store data and retrieve them at will makes the RAM a popular system design tool. Categories such as scratchpads, buffers, main memories, and mass storage are all applications for RAMs.

A scratchpad memory is a small, fast memory normally associated with the central processor of a computer. The scratchpad, which is used for temporary storage of interim calculation results, must operate at speeds comparable to those of the central processor. This speed requirement means that except for small MOS computers, scratchpad operation is not an MOS strong point.

Buffer memories may be employed between sections of a computer, between a computer main frame and peripheral equipment, or in any digital system where temporary storage is required between operating units. The speed of the buffer must be at least equal to that of the input–output rate of the faster of the operating units, and its information storage capability is related to the data rates of the units. MOS RAMs are a logical choice for many buffer memory applications, especially in computer peripherals.

RAM devices made with MOS technology use two different techniques to store information. Depending on the type of basic memory cell, MOS RAMs can be categorized as being either static or dynamic. Static MOS memories that usually show poorer performance and higher costs are easier to drive than the dynamic memories, which generally require clock signals in addition to power supplies.

Dynamic MOS circuits make use of the very low leakage association with the gate circuits and junctions of well-made MOS devices. These leakage currents are small enough to permit the circuit's parasitic capacitances to exhibit time constants between milliseconds and seconds. These long time constants may be used to provide temporary storage, which may be made permanent by appropriate cycling or "refreshing" operations.

The Static RAM

In the static RAM stage of Figure 8-33a, two static inverters are wired together to make a flip-flop. Devices Q_5 and Q_6 are used as (two-way) transmission gates. When reading, the conducting side of the flip-flop pulls the data line toward ground by way of these gates. Writing is accomplished by forcing the data lines to the value desired in the cell, thereby overriding the contents of the cell. Because of the small current capability of devices Q_3 and Q_4, it is important that neither data line

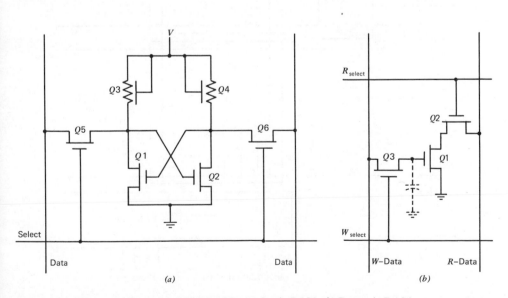

Figure 8-33 Storage cells used in MOS RAMs: (a) Static RAM, (b) Dynamic RAM.

be near ground when the transmission gates are turned ON. With grounded data lines, the charge associated with the capacitance of the data lines may flip the cell.

The Dynamic RAM

The dynamic memory cell of Figure 8-33*b* may also be used as the basis of a MOS RAM. Unlike the memories constructed with the static cell, the data of these dynamic memories must be periodically refreshed in order to maintain its validity. Because the cell is small, many more bits of memory may be produced on a chip of given size than can be made with static cells. The dynamic cell is used as the basis for RAM chips of up to 1024 bits. One possible organization of a 1024-bit chip is suggested in Figure 8-34. With this organization, reading and writing occur for all cells of one row simultaneously. Because only one bit at a time is available for writing, an (internal) read operation must be performed prior to writing. This operation ensures that the refresh amplifiers contain data corresponding to the contents of the row into which writing will take place.

There are three clocklike signals associated with the dynamic RAM: X-enable, Y-enable, and precharge. The X and Y enables both act as a chip select for both reading and writing. Several chips may have the input–output lead OR-tied to realize larger than 1024-bit planes.

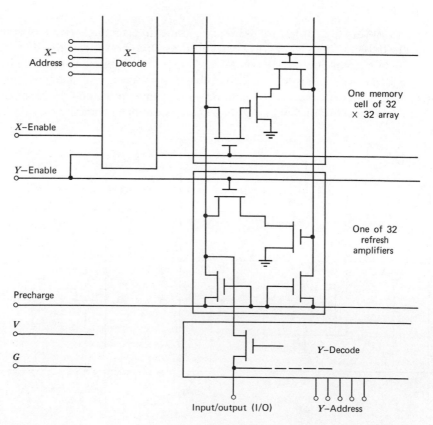

Figure 8-34 A dynamic MOS RAM; the chip contains 1024 bits.

Using silicon-gate MOS technology, the clock signals and addresses are nominally 20 V peak-to-peak. These high voltages are necessary to obtain high-speed performance. The use of TTL-compatible levels would add significantly to the memory cycle and access times. Memory cycle times with the high levels are from 300 to 600 nsec, depending on chip organization and drive signal rise and fall times. Figure 8-35 is a photomicrograph of a typical 2048-bit RAM.

Writing Data In

To change the information stored in the basic cell, the sense-digit lines are appropriately biased and the word-selection line is placed in the active or low state. Assume the sense-digit line condition of Figure 8-36. Sense-digit line A is at a high potential. The complementary signal is present on sense-digit line B, which is connected to a low potential. When the word-select line goes to a low potential, transistors Q_3 and Q_4 turn ON, connecting the sense-digit lines to the cross-coupled transistors Q_1 and Q_2. In the example of Figure 8-35, Q_3 connects node A to the high level on sense-digit line A. This high potential is coupled to the gate of Q_2 and tends to turn Q_2 OFF. At the same time, because Q_4 is conducting, node B is coupled to the low supply. This low voltage is applied to the gate of Q_1 and

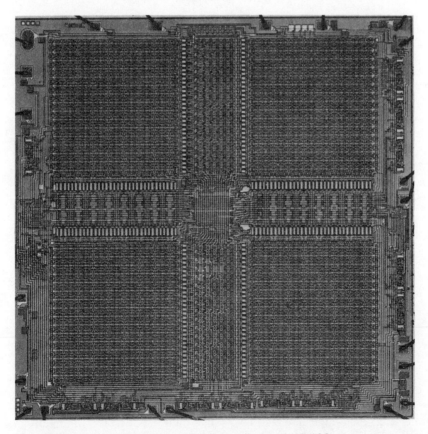

Figure 8-35   A photomicrograph of a 2048-bit fully decoded RAM.

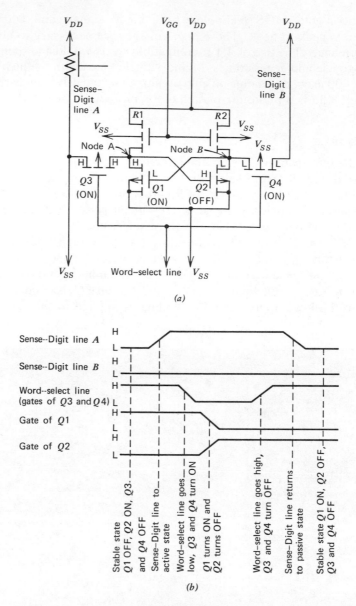

Figure 8-36 (*a*) Writing into the basic storage cell, (*b*) timing diagram for the write operation of (*a*).

tends to turn Q_1 ON. This provides an additional path from V_{SS} to node A, further increasing the potential on the gate Q_2. Therefore, the indicated sense-digit line potentials in Figure 8-35 result in transistor Q_1 turning ON and transistor Q_2 turning OFF. This is the alternate stable state of the storage element. Completing the write operation, raising the word-select line potential, turns OFF Q_3 and Q_4 and isolates the storage cell from the sense-digit lines.

Connecting sense-digit line B to V_{SS}, sense-digit line A to V_{PP}, and activating the word-select line will reverse the state of the flip-flop, turning Q_2 ON and Q_1 OFF.

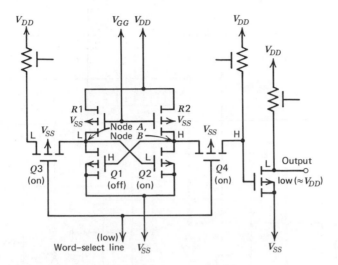

Figure 8-37   Reading out of the basic storage cell.

Applying complementary V_{SS} and V_{PP} signals to the sense-digit lines while the word-select line is activated permits writing into the storage cell. For reading, the word-select line is again activated, but this time both sense-digit lines are terminated with MOS resistors. The read scheme appears in Figure 8-37.

The resistive terminations on the sense-digit lines do not change the state of the storage cell when Q_3 and Q_4 conduct the sense-digit lines to the storage cell. For the node that is in the high state, the sense-digit line resistor appears in parallel with the internal resistance (R_1 or R_2). For the drain node of the OFF storage transistor, the sense-digit line terminating resistance appears as an additional source of V_{PP} potential and, through the cross-coupling of Q_1 and Q_2, tends to keep the conducting transistor turned ON.

In addition to terminating resistors, at least one of the sense-digit lines must contain sensing circuitry to determine the state of the storage cell. At its simplest, this sense circuitry can be an MOS inverter (Figure 8-37).

MOS RAM Main-Memory Application

One fast-growing market for the use of MOS RAMs is the main-frame memory — attributed to the IBM 360/85 computer. Main-frame storage capacities range from 200 kilobits to 1 megabit with access times as high as 300 nsec. Figure 8-38 is a schematic diagram of a main-memory module storing 4096 16-bit words (4K × 16) in four 1K × 16 submodules using the MM5260 MOS RAM. One submodule is selected at a time by two bits of the DM7442 TTL decoder address. The same decoder could access eight submodules with a 3-bit address, and so forth. This is standard decoding practice.

The external data selectors (DM8123) and the read–write bus buffers (DM8093 and DM8094) are three-state TTL devices.[7] These have high-speed, active pull-

[7]*TRI-STATE Logic in Modular System Organizations,* by Don Femling, National Semiconductor Corporation, April 1971, AN-43.

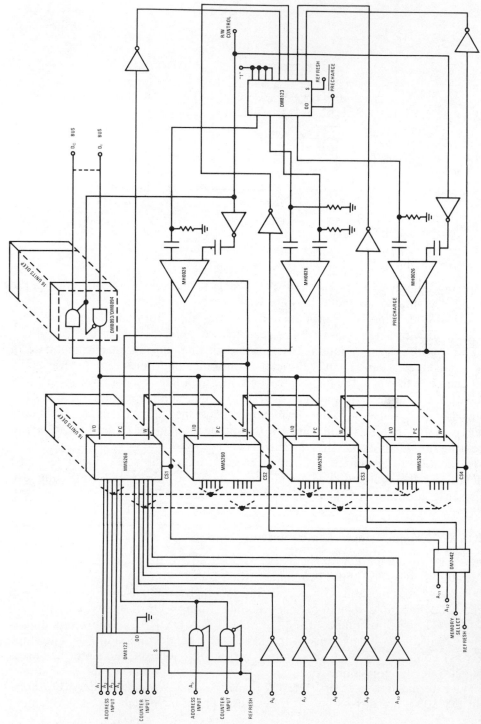

Figure 8-38 Main memory module storing 4096 16-bit words using MM5260.

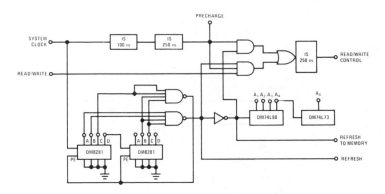

Figure 8-39   System clock timing and control circuit.

up outputs when enabled. The two types of buffers can operate in parallel with the internal MOS input–output mode read–write gates. One control line gates both in complementary fashion because one is enabled by a logic "0" and the other by a logic "1." The data selectors hold off accesses during a submodule's refresh intervals. Figure 8-39 represents the TTL clock-forming and tuning-control circuit for the module.

Precharge Decoding

During standby, each MOS RAM circuit dissipates about 75 mW. To achieve memory access, a pulse called the "precharge" is applied to set up the decoders and other input–output functions. Precharge minimizes system power dissipation by making it unnecessary to energize decoding logic between selects. During a 600-nsec access, MM5260 power dissipation goes up to 400 mW.

It is important to keep as many of the RAM circuits as possible on standby in order to minimize system average power dissipation. Excessive dissipation, without adequate cooling in a high-density system, would overheat the semiconductor junctions. The only ways of preventing overheating, should average dissipation be high, are to reduce speed and lose performance, cut packing density, or increase system cooling hardware. Average power supply and clock-driving requirements are also reduced by the precharge pulse. The added circuitry to effect power minimization comprises only three elements.

Figures 8-38 and 8-39 illustrate the most effective way yet developed to decrease precharge power dissipation. This "precharge decoding" method applies precharge to a submodule only when that module is selected. This is implemented by having the chip-enable of the decoder gate the precharge clock by way of the three-state TTL data selectors in the clock circuitry. Simultaneously, the input–output directions are controlled. Selective precharge decoding can yield an ultimate average power dissipation of 77 mW.

Memory Timing

The MM5260 timing control (Figure 8-40) is quite simple because of precharge decoding and the input–output structure. Maximum cycle time is 600 nsec. A delay of 100 nsec before precharge is allowed for address settling and decoder

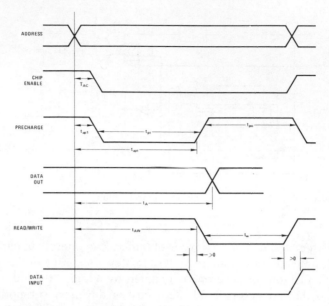

Figure 8-40   MM5260 timing diagram.

operation. This allowable delay will not affect the access time under worst-case conditions. This element design characteristic permits a very straightforward selective precharge decoding technique which does not affect the performance of the memory system. Precharge goes low for 250 nsec to set up the decoders, then returns to conserve power.

If READ is commanded, the READ gates are enabled and the WRITE gates disabled at the outputs. Stored data are available at the output within 350 nsec of the start of the access. WRITE may be commanded just prior to the precharge trailing transition and is completed by the end of the 600-nsec cycle. Address, precharge, and chip-select timing is not critical. A skew of about 50 nsec between address timing by the CPU and leading and trailing edges of chip-select and precharge will not affect access time, cycle time, or overall memory speed.

Computer-Aided Design

There are two problems to consider in the design of a complex digital circuit. One problem is designing the circuit to obtain the required device conductance so that the circuit performs at the required speed and power. The second problem is to lay out the devices and interconnect them into a functioning circuit. This approach is sometimes referred to as random logic design. An alternative approach consists of designing and characterizing a set of standard cells which perform a specific logic function. The design of a circuit using this alternate approach then consists of arranging and interconnecting these standard cells. The task can be accomplished by the use of a computer that decreases the turnaround time and improves the accuracy.

Computer-aided design (CAD) is most useful for custom LSI circuits— primarily MOS arrays. The application of CAD techniques to ICs is a major undertaking. The hardware requirements are impressive, but the major cost is in software, not hardware. In the application of CAD to LSI design, both hardware and software limitations have arisen, indicating that the computer is no panacea. A man must still be plugged into the loop to do layout work that cannot be translated into software.

Unquestionably, CAD has proven its value for fast design turnaround in custom LSI — 10 to 14 weeks, instead of 6 months. Yet, since it is a relatively new tool, CAD has plenty of room for improvement. Manufacturers must still decide on the best method of mask preparation. (See discussion in Chaper 2.) And service charges for CAD work have not changed much — they are still high, partly because the expensive facilities required are a long way from paying for themselves.

Most IC manufacturers realize that the computer cannot do the entire design. The engineer has to interface with the computer to get optimum cell placement and thus more efficient chip layouts. One IC manufacturer asserts that CAD is used for minimum size in a volume-production chip as an aid. However, in the end, the chip size is reduced by 20 to 40% with human layout. The way in which designers or customers interact with the software varies, as does the number of available CAD programs.

One of the most economically pertinent and highly active uses of CAD for ICs is graphic design. To prepare a production mask for an IC, master artwork is generated for a single circuit, reduced in size, and stepped across the actual mask by automatic step-and-repeat equipment. Preparing the mask manually is a long and tedious procedure, particularly prone to human error.

In a more up-to-date method, a computer-controlled drafting machine generates the mask masters directly. Not only is this method more accurate, but it also produces highly satisfactory work at 100 times final size instead of the 500 times required with the manual procedure.

CAD systems can operate either in an on-line or off-line mode. In a typical off-line mode, the designer sketches the circuit and then reduces his design to a dimensional drawing on a worksheet. For all its apparent complexity, it is relatively a straightforward task to describe a complex IC for drafting.

In the on-line system, a computer-controlled CRT display is used as an interactive sketchpad to experiment with various geometric configurations. When an acceptable layout and metal routing has been designed, the computer is instructed to prepare drafting instructions.

The specific descriptions of devices such as diodes and transistors, including their shapes on each mask layer, are stored in the disc library. These are the basic building blocks in both the on- and off-line systems. All shapes, no matter what layer they are on, are defined relative to the same point. When a device is moved, all shapes on all layers move with it.

These devices are often arranged into more complex structures (called cells) such as gates or flip-flops. Those commonly used can also be stored, giving the designer a more powerful set of tools. A cell is considered to be a logic entity. All the constituent parts are defined by one placement and need not be individually handled.

One major limitation of CAD is the lack of more varied standard logic cells. Some manufacturers feel that each application area (e.g., minicomputers, telemetry, and the specifications of telephone companies) needs a cell set that emphasizes speed, low power, or whatever feature the individual user needs most. No matter how many standard cells a company has in its catalog, more are always needed for a custom job.

CAD Design Sequence

The steps of a typical CAD IC design sequence are now presented. The customer normally carries his requirement to the manufacturer by means of a truth table, Boolean expressions, or a timing diagram to describe the logic; or, his requirement may be as detailed as a fully defined logic diagram, utilizing standard cells and submitted with complete test sequences and electrical specifications. No matter what form it takes, the starting point is a functional description of how the circuit must operate.

In the case of a multichip development, the first step in the design cycle is partitioning the system into individual chips. When partitioning has been completed, logic diagrams of each chip are prepared using standard cells (or logic blocks). Once the logic for a chip has been defined, the network is entered into the CAD system for development. Every input and internally generated signal is assigned an alphanumeric name. The coding of the network consists of simply listing each cell name along with the type of cell and the names of the inputs to the cell. The coded network is then assembled in machine language and stored on a disc for use in other phases of the CAD process.

Next the operation of the network is simulated with a logic simulator program. The inputs to this program are the assembled network description and a sequence of design verification tests, either supplied by the customer or developed by the manufacturer. The computer applies this sequence of tests to the input of the assembled network. When simulation is complete, the computer issues a printout of the logic states of inputs, outputs, and selected internal points as a function of time. In essence, the computer is performing a function similar to breadboarding the circuits, but in this case a more concise picture of the circuit response versus time is achieved. A study of the simulation printout is made to verify the correctness of the stored network.

After the stored network has been verified, a propagation-delay program estimates the capacitance associated with each node in the network and calculates the propagation delay through each logic gate. The designer analyzes this information for possible speed problems and determines which interconnection paths must have minimum capacitance. When the chip layout has been completed, the exact node capacitances are automatically computed, based on layout information. The results are then reexamined as a final check for proper ac performance.

Artwork Generation

Once the circuit has been verified to be functionally correct and to meet all performance requirements, the artwork generation can begin (see also Chapter 2). Standard cells, being of uniform height and variable width, can be placed side by side in rows on a rectangular grid. Interconnections run along grid lines in the channels between the cells. Horizontal interconnections are usually made with aluminum, whereas vertical lines are usually $p+$ diffusions so that two layers of interconnection are available. During this phase, any constraints (e.g., pin configurations, critical speed paths) and any peculiarities of the circuit, are fed into the CAD system. Modifications to the automatic placement and wire routing can be manually entered to fulfill any special circuit constraints. These modifications are automatically checked for layout rule violations. Figure 8-41 depicts a typical computer-generated interconnection layout. After all modifications have been made, a computer-driven plotter produces a symbolic drawing of the circuit. The final drawing is reviewed, and artwork for the masks is cut automatically by the same plotter.

Some manufacturers do not cut Rubyliths. Instead they use optical photo-generation to produce 10X phototools. (Optical pattern generators are large cameras that flash rectangles onto plates to turn out mask artwork, usually at 10 times actual size.) With Rubyliths, changes or error corrections are simple. Only the part of the mask artwork that is incorrect need be changed, whereas a complete master reticle would have to be discarded to make changes when an optical pattern generator is used.

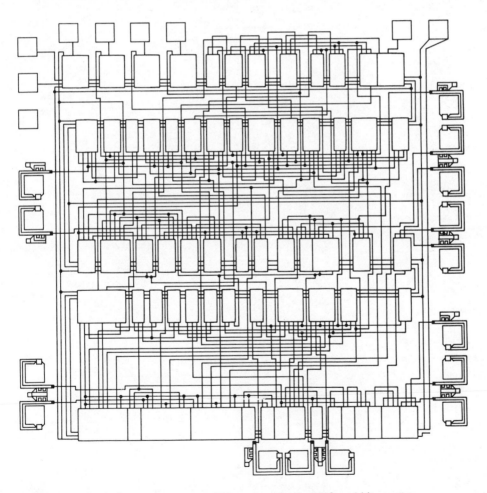

Figure 8-41   Typical computer-generated interconnection layout for a 4-bit counter.

Other manufacturers prefer pattern generators because they eliminate Rubylith stripping (peeling) and checking, as well as the copy-camera step needed to reduce the artwork from 200 or 250 times actual size. Elimination of these steps can save as much as a week of design time.

Peeling of Rubyliths is still a problem area because errors can creep into mask artwork. Although Rubylith cutting is automated on sophisticated flatbed plotters in which cutting tools are substituted for ink pens to generate mask artwork, peeling still must be done manually. This is an area where mistakes can occur.

CAD is not a necessity for LSI, but it is a definite aid. Many complex MOS arrays have been produced using conventional approaches. Indeed, the CAD circuit is very often not the optimum solution to a design problem. However, CAD makes LSI feasible for many applications that would otherwise be prohibitively expensive.

Chapter Nine

MOS/BIPOLAR COMBINATIONS
AND TRADE–OFFS

COMPARISON OF MOS AND BIPOLAR TECHNOLOGIES

For anyone interested in maximum LSI complexity (more than 100 gates) on one chip, MOS is the choice. TTL can go into LSI to some extent, but limits on power dissipation hinder its usage. If the complexity is less than 100 gates but greater than a regular IC, there is MSI, where MOS, TTL, or low-power TTL can be used. If MOS is LSI, then TTL is MSI. Low-power TTL cannot match p-MOS complexity because its resistors become large and start taking up too much room on the chip. MOS begins to lose out at the low end of the complexity spectrum because it does not presently have a sufficient variety of off-the-shelf MSI functions to challenge TTL devices. Furthermore, low-cost, low-complexity MOS may not appear on the scene because it presents bipolar logic with interface problems of both voltage levels and operating speed. Low-level MOS, on the other hand, can be made voltage compatible with bipolar logic. But low-level MOS is more expensive than high-level MOS and is comparable in cost to bipolar logic.

MOS VERSUS BIPOLAR

How is a user to work his way through all the logic families and be reasonably sure that he is coming out right? One position that has been taken is to use MOS for the memory, use bipolar – DTL, TTL, or ECL – for the logic, and use hybrids

265

for clocking circuits and for driving lamps, relays, and readouts. The major controversy in comparing MOS and bipolar circuits is the suitability of asking MOS to do logic. MOS is a memory-oriented technology.

The basic element in bipolar and static MOS memories is the flip-flop. When flip-flops are employed as storage units, a small amount of memory can be combined with drive, sense, and decide circuitry on the same chip. However, for larger memories, the demand on real estate is compounded by the additional requirement for peripheral decode and drive circuitry.

In a dynamic MOS IC, the basic storage cell is a capacitor; its job is to hold or release a stored charge. Because capacitors have leakage, the stored charge must be continually refreshed. A refreshing system, therefore, must be included in dynamic MOS circuitry.

A shift register is really a serial memory. ROMs can be used as a logic device or as a program store in microprogramming. In very small systems, where enough functions can be put together (such as in a MOS arithmetic unit), it is sufficiently complex to be worthwhile in MOS. But can it replace a TTL quad two-input gate with a MOS quad two-input gate? No. Both the economic trade-offs and the performance trade-offs are against it.

Simple processing and low-power dissipation make MOS attractive for main-memory consideration, but speed is bipolar's prime advantage. Where speed is not needed, the user can benefit by choosing MOS. MOS will definitely go into logic, although it is less suitable. Using MOS for logic is not as clean-cut as using MOS for memory. But ROMs will undoubtedly be used for logic and control when the work the logic does can be well enough defined. Microprogramming is an example. Already there is on the market a family of low-cost, off-the-shelf MOS standard logic arrays. Although optimization is needed for individual systems, they allow total design with off-the-shelf MOS.

One of the most important features of MOS is economics. The way MOS makes its low-cost scores is through complexity. The low-cost advantage of MOS is in very complex elements such as long-shift registers (100 bits and longer), where the desired results cannot be achieved on a single chip in bipolar. A quad two-input TTL gate is probably five times cheaper than the same circuit in MOS. But in a long-shift register the opposite is true.

The almost infinite resistance of the channel region means that MOS arrays operate with essentially zero standby power. Consequently, this allows very close element spacing without excessive heating—a decided advantage over bipolar technology.

Circuit and array delays influence cycle time. Circuit delay is the propagation delay of the address decoders, sense amplifiers, and word drivers. To achieve high-speed operation, bipolar technology is used for the word and digit circuits because MOS elements would increase the cycle time by a factor of 5 or more. Line capacitance, readout current, and voltage or current swing used for word selection and digit writing determine the array delays. In order to do this, internal switching and recovery time are added for each storage cell. Schottky-barrier diodes serve to increase recovery time.

Recent advances have made available MOS devices with 1.5 to 2.5 V gate-threshold levels. This means that MOS storage cells now can operate from a 4 to 6 V supply. Consequently, word and digit voltage swings are compatible with popular bipolar circuits.

The use of small geometries with fine-line tolerances (e.g., 0.1 mil) decreases the access delay considerably while increasing the component density. Also, with one power supply for standby current to cells and another for the logic elements, it is possible to achieve higher speed with lower total chip dissipation.

Bipolar devices require isolation of adjacent storage cells, whereas MOS devices are inherently self-isolating. Thus bipolar occupies two times more wafer area for equal mask tolerances. Also, bipolar processing requires more fabricating steps. On investigation, MOS turns out to be a process with a capacity for almost infinite variation. p-Channel enhancement-mode MOS, for example, is the simplest IC process in general use and it requires handling by an operator perhaps 40 times. n-Channel MOS requires 45 to 50 handlings, and the process is more critical than p-MOS. CMOS combines n- and p-MOS on the same chip and thus needs a total of about 100 or more steps. Bipolar circuits are even more complicated, requiring from 130 to 135 process steps.

Today, at least, low-cost MOS means p-MOS. The reputation of p-MOS for low cost was achieved by means of a simple process, high-circuit-density chips, and very-low-power dissipation. The major give-aways in the trade-off are speed and driving power. The particular combination of advantages and disadvantages found in p-MOS are very suitable for memories.

For a small digital system that is repeated many times, MOS is the obvious choice. This is an ideal situation for MOS and can be obtained for the lowest possible cost. What a small digital system repeated many times implies, of course, is a large volume of identical pieces, where the parts count has been brought down to a minimum through MSI and LSI approaches. MOS allows, and probably demands, relatively high circuit complexity before it begins to pay off. MOS is LSI.

MOS devices are inherently high-impedance circuits, and if they are asked to generate current to drive a load, the geometry of the output circuit must be expanded out of all proportion to the results obtained. In terms of silicon real estate, bipolar technology is probably 100 times more efficient in producing current. But even though MOS cannot generate much current, it can still fan out fantastically compared with other MOS devices—a fan-out of 50 is specified in the RCA CMOS line; indeed, the manufacturer claims that fan-outs of 1000 can be handled if desired. This is because the input impedance of MOS devices goes up to 10^{18} Ω, compared with about 10^3 Ω for TTL. Of course, as fan-out is increased, the capacitance, also is increased and thus MOS devices slow down even more. Still, for a fan-out of 1000 and 0.3-pF capacitance, it is possible to operate at up to 100 kHz, worst case. Once above 10 MHz, MOS fades from the picture and the faster families, TTL or ECL become the logical choices.

TTL as a logic family is usable for just about everything except large computer main frames. Some manufacturers are even working on high-speed TTL that will have propagation delays of only 3 to 4 nsec per gate, with clock rates of 75 to 100 MHz.

The main reason for using ECL is speed. Although ECL consumes a large amount of power, it is the fastest logic yet available, it has some advantages other than speed. ECL is a current-mode logic and thus is almost a constant-current form of logic, since it shifts current from one transistor to another, whereas in TTL current is turned ON or OFF. Thus the current in ECL is the same at 1 MHz as at 100 MHz, and once the power supply is designed, the system can be run at any speed. With TTL, a larger power supply is required in order to operate at a higher speed. The uses of ECL at present seem to be mostly in the arithmetic sections of fairly large computers (central processing units). But ECL is expensive compared with DTL and TTL.

The best of all possible worlds, of course, would be MOS and bipolar on the same chip. MOS is a high-resistivity technology. To sink current, low ON-resistance is needed, and for low ON-resistance in MOS, a large-geometry device is necessary. The minute this step is taken, the chip becomes as big as, or bigger than, TTL, and all the advantages have been lost. If MOS and bipolar were combined on the same chip, it would eliminate drive problems. Whereas MOS-plus-bipolar on the same chip has actually been done experimentally for three years or so, its reproducibility in quantity has yet to be proven.

ADVANTAGES AND LIMITATIONS OF MOS DEVICES

The designer must have both advantages and limitations clearly in mind as he applies MOS to his system. First, we consider the advantages.

The MOS Device is Extremely Small

The average transistor in an MOS array occupies as little as 2 mil^2 of chip area. This is a great reduction over bipolar transistors, which average about 40 to 50 mil^2. Most of the size reduction results from the lack of isolation junctions in MOS ICs — isolation of one device from the other is inherent in the MOS structure. In a bipolar IC, as much as 30% of the active area is occupied by isolation junction regions. Roughly speaking, the area taken up by bipolar isolation regions can accommodate 200 MOS devices per chip. A further reduction in size results from the simpler MOS processing. Fewer process steps are necessary, fewer masks are used, and normally only one diffusion is needed. The tolerances that must be allowed in masking for bipolar diffusion steps do not accumulate to the same extent in the MOS process.

Keeping the chip size down is important. The probability of a defect on a chip increases in proportion to its area. As chip area goes up production yield goes down, and the chips become more expensive to produce.

Increased chip complexity has other benefits. It results in fewer chips per system, with fewer interconnections. In one typical MOS, excess-3 code adder–subtractor, the logic is performed on one chip, with a total area of about 3200 mil^2. Going the RTL route, the same logic would require 18 separate chips and a total silicon area of 26,000 mil^2.

The MOS Device has High Input Impedance

With a dc gate impedance of typically 10^{18} Ω, the MOS device behaves as a nearly ideal voltage-controlled resistor. Input–output isolation is excellent. No input current flows in the gate lead, except to charge and discharge the input capacitance, and MOS transistors can be direct coupled with virtually no dc fan-out limitation.

Bilateral Operation is Unique to the MOS Transistor

The device is completely symmetrical, since the source and drain are identical and interchangeable, and current can flow in either direction in the channel. Operating as a switch, with essentially infinite resistance in its OFF state, the transistor is ideal as a coupling device. Its bilateral nature is used to great advantage in MOS multiplexer circuits. This property of bilateral current flow, combined with the extremely high input impedance, makes the MOSFET a nearly ideal analog switching device.

The MOS Device is a Natural Data-Storage Element

The gate-to-source capacitance can be used to store change, since the dc gate impedance is extremely high. The time constant of the gate-to-source capacitance and the gate leakage resistance is on the order of 10 msec. This property makes the operation of low-power dynamic shift registers possible.

The MOS Transistor Makes an Excellent Active Load Resistor

Very high values of resistance — 100 to 400 kΩ — can be achieved in an area as small as 1 mil^2 by using an MOS transistor as an active load resistor. A 100-kΩ resistor built by the standard bipolar diffusion method would be about 0.4 mil wide and 400 mils long, or 160 mil^2. The MOS resistance characteristic is non-linear, since the resistance varies as a function of gate-to-source bias, but this is seldom a disadvantage. The MOS load device can be turned ON and OFF under the control of the gate voltage, and power dissipation becomes a function of the clock duty cycle. An MOS device used as a load resistor is illustrated in Figure 9-1.

The MOS Process is Much Simpler than the Bipolar Process

Only one diffusion step is required in the MOS process. In addition, several expensive and critical high-temperature steps — emitter diffusion, for example —

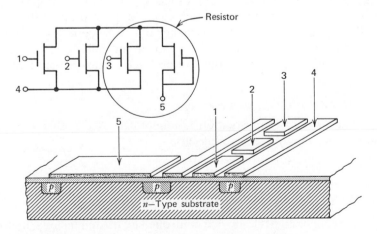

Figure 9-1   MOS device used as a load resistor.

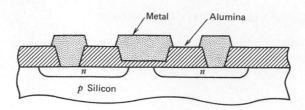

Figure 9-2   A conventional MOS transistor.

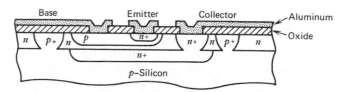

Figure 9-3   A bipolar transistor.

are avoided. And so are the accompanying dangers of crystal dislocations and oxide pitting.

The process does not rely on any critical diffusions as in the case of bipolar circuits (Figures 9-2 and 9-3.) In bipolar ICs we have the base diffusion and the emitter diffusion, and both must be controlled precisely to determine the base width. Thus there are two competing diffusions. In the MOS device there is only a single diffusion, and the characteristics of the diffusion are not extremely critical. Looking at the structure of an MOS device (Figure 9-2), we see that the base material is a wafer of *n*-type silicon. There is a *p*-type source and drain, and the gate oxide, with a metal electrode on it. Diffusion occurs in two directions: down and laterally.

We want to keep the channel length within a certain value because this dimension will become too short if diffusion proceeds for too long a time. That is really the only critical process step. However, since the oxide has to have some special properties, it is necessary to have good process control in growing the gate oxide. Nevertheless, an MOS circuit is a much easier device to make than the bipolar device and this results in our ability to make some very complex circuits, because the yield is higher; and the higher the yield on an individual device, the more complex the circuit can be made and still produce a reasonable yield.

Thus MOS devices have

- No emitter diffusion for gain control
- No isolation diffusion (isolation is inherent in MOS device, since its *p-n* junctions are always reverse biased)
- No high-temperature processing steps

The Gain of an MOS Device is Controlled by its Dimensions

Not only is the gain of an MOS device controlled by its dimensions, it is accurately determined at the layout stage.

The MOS process is normally kept constant, the topology of the device being the only variable. The transconductance is controlled by the width and length of the channel region, and the MOS designer can scale the geometry to achieve exactly the performance he wants in his circuit. This is done in bipolar circuits, too, but it is much more complicated, since bipolar transconductance is controlled by varying the degree of doping in the diffusion steps. Tight control and prediction of the performance in bipolar ICs is much more difficult.

MOS Enhancement-Mode Transistors Have Built-In Noise Immunity

Because of their threshold-voltage effect, MOS enhancement-mode transistors have built-in noise immunity. MOS thresholds vary, depending on the manufacturer's process, in the range of 2 to 6 V. In bipolar devices, the comparable threshold is only about 0.6 V.

MOS Reliability is Good

One of the questions that is frequently asked is, How reliable are MOS circuits? The general consensus of opinion is that, properly made, MOS circuits are as reliable as bipolar ICs. Since much of the interconnection is accomplished on the device, there is a possibility that they will be more reliable. This is because wire bonding represents one of the least reliable fabrication steps of a circuit.

MOS offers Obvious Economics

Since MOS devices are smaller, more can be put on a chip. Chips can be kept small and yields high; thus manufacturing cost is lower. High chip complexity means fewer chips per system, which means that the system interconnection cost is lower, too. However, there are the following limitations.

MOS Devices Have Relatively Low Transconductance and High ON Resistance

A typical MOS driver device has an ON resistance of about $1000\,\Omega$ and must be roughly 300 to 400 mil² in area if it is to achieve a resistance even this low (the ON resistance is inversely proportional to the device area). A bipolar device of this size, on the other hand, would have an ON resistance of 1 or $2\,\Omega$. Furthermore the ON resistance of a bipolar device is inversely proportional to the exponential of the area, and it increases much more quickly, as area is increased, than it does in the MOS device.

MOS Gates are Fundamentally Slower than Bipolar Gates

The MOS circuit is about an order of magnitude slower in switching speed than the bipolar circuit. This is true because the small MOS transistor is operating at a higher impedance level than the bipolar structure, so that the circuit time constants are much larger with the MOS circuit. One way of improving the speed performance of MOS ICs is to employ complementary devices. A circuit combining both a p- and an n-channel device was shown in Chapters 7 and 8 and appears again in

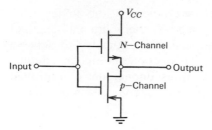

Figure 9-4   A complementary inverting amplifier.

Figure 9-4. When switching either device off or on, the effective load resistance is another MOS transistor, rather than the relatively high-impedance resistor; thus the RC time constants are lower. In addition, this circuit calls for low standby power. Fabrication of both types of structures in the same die, however, necessitates additional processing steps. Moreover, if high speeds are required, the MOS structures must be increased in size to provide the appropriate lower impedance levels, thereby diminishing the area advantage of MOS over bipolar.

MOS Gates Require both V_{GG} and V_{DD} Supplies for Best Operation

Circuits can be easily designed to operate from a single power supply V_{DD}; but such design is not efficient. The circuit dissipates more power for the same logic swing output. Since the power dissipation is proportional to the current drain times the power-supply voltage, power dissipation is higher than it need be to accomplish that logic swing. But if a separate V_{DD} supply is used, it can be typically one-half of V_{GG}, and the logic swing is approximately equal to the difference between V_{DD} and ground. Since power dissipation in this case is proportional to the voltage V_{DD}, the dissipation per volt of logic swing is less. The V_{GG} supply adds only moderately to the system cost because the current drain is very low.

MOS Devices Have Intrinsic Gate and Channel Capacitances

The intrinsic capacitance of MOS devices appears as a shunt to the input signal, and thus the apparent input impedance of the device decreases with the increasing frequency. The capacitance imposes a speed limitation on digital MOS circuits. Since the charge distribution on the gate and in the channel change with the applied voltage, the capacitance also changes with voltage.

Overlap capacitance, between the gate and the source and between the gate and the drain, are also inherent in the MOS device. In all enhancement-mode MOS devices available today, the gate metal must overlap the source and drain regions to allow for alignment and processing variations. The overlap capacitance is roughly 0.01 pF for each $10\,\mu$ of gate width. For a typical gate $40\,\mu$ wide, the capacitance from the gate to each other electrode is 0.04 pF.

Drain-to-body junction capacitances between the source and the substrate and between the drain and the substrate are also intrinsic to the MOS device. This capacitance depends on the amount of reverse bias, decreasing with increasing bias.

Small MOS Devices Pose Process Problems

The alignment of the gate mask to the p-diffusion mask is extremely critical — more so than in the bipolar process because of the smaller dimensions. If the gate is misaligned by more than $2\ \mu$ in an enhancement-mode device, for instance, the inversion layer will not extend completely across the channel, thereby adding series resistance in the channel. And since MOS chips are typically larger than bipolar chips, old-style tooling is not very well suited to lining up big chips to tight tolerances.

Testing Problems are Seldom Anticipated

A complex circuit or array can be put in a 60-pin package, but the testing is very difficult. Testing MOS LSI devices in multilead packages is a serious and costly problem in the IC industry. Figure 9-5 depicts the complexity (and thus capital equipment cost outlay) of a typical LSI testing system.

Packages are a Problem for the More Complex MOS Arrays

The more bits of memory in a package, the more input and output pins are needed. The multilead packages are extremely expensive, and it is possible to lose the cost advantage in the package that is gained by increasing the complexity of the chip.

A comprehensive tabulation of the advantages and limitations of bipolar digital ICs and of MOS ICs is presented in Table 9-1.

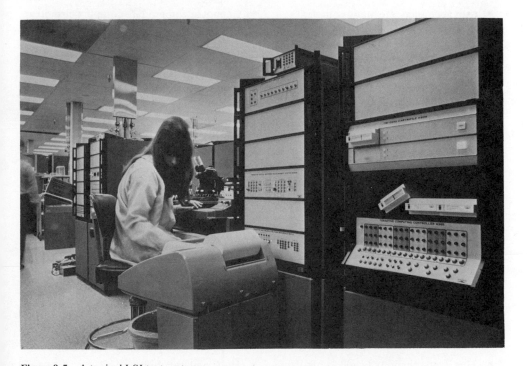

Figure 9-5   A typical LSI test system.

Table 9-1   Comparison of MOS Devices and Bipolar Monolithic Devices

MOS Devices	Bipolar Devices

• Simple fabrication process — fewer fabrication steps (approximately 50 operator handling steps)
    No emitter diffusion
    No isolation diffusion
    Eliminate high-temperature processing steps
	• Complex fabrication process — more fabrication steps (approximately 135 operator handling steps)
	
	• Diffusions are critical
	    Both base and emitter diffusions must be closely controlled to determine base width

• Greatest complexity per chip • Controlling oxide thickness is critical • Low cost in large volume and high complexity	
• High yield process	• Medium to low yield process
• Simplified processing	
• Packaging for complex MOS arrays is a problem	• Packaging is not a problem
• Tight process control required	• Isolation required of adjacent memory storage cells
• Extremely small devices, but larger chips	
• Self-isolating	• Larger devices than MOS, but smaller chips • Occupies greater wafer area for equal mask
• Can accommodate many complex devices on a single chip and still provide good yields; suited to MSI and LSI	• Cannot accommodate complex circuitry
• Very close element spacing	• High power dissipation and packaging difficulty with increasing complexity
• High input impedance	• Lower input impedance
• Lower power dissipation	
• Have bilateral operation	• No bilateral operation
• Have inherent memory	• Does not possess inherent memory
• MOS unable to handle current	• Good high-frequency capability
• MOS transistor makes excellent load resistor	• Isolation must be built (diffused) in
• Inherent isolation, since p-n junction is always reverse biased	
• Gain of MOS device is controlled by its dimensions	• Tight gain control and gain prediction are difficult
• Have high gain at cryogenic temperatures	• Gain decreases as temperature is reduced
• MOS enhancement-mode devices have built-in noise immunity	• Do not have inherent noise immunity
• MOS devices can be characterized by five equations	• There are no equations for predicting bipolar operation accurately (as with MOS)
• Has good reliability (same as bipolar)	• Have high reliability

Table 9-1   Continued

MOS Devices	Bipolar Devices
• MOS devices have low transconductance and high ON resistance	• Have high transconductance and low ON resistance
• MOS devices are about an order of magnitude slower in switching speed than are bipolars	• Fast switching speed
• MOS gates require both V_{DD} and V_{GG} supplies for best operation	• One supply voltage required for best operation
• MOS devices have intrinsic capacitance associated with charges stored on gate and in channel	
• Small MOS devices pose process problems Mask alignment is critical	• Larger devices, process problems are less critical
• MOS devices fail due to static discharge between gate and other elements	
• No standard lines	• Standard lines developed
• High development costs	• High development costs

MOS−BIPOLAR INTERFACE

An IC made with MOS transistors consumes less power and is smaller than the same circuit made with bipolar transistors. On the other hand, the bipolar circuit is faster and can drive more current. Therefore, most designers of logic systems tend to use both types, and somewhere they must interface the two. Interfacing, however, creates problems because MOS and bipolar are two different breeds. A MOS transistor is a voltage-controlled device.

The point of all this is that many MOS ICs made today use higher supply voltages and logic than those used by bipolar ICs. In a conventional MOS logic gate, therefore, it is necessary to change the voltage by a greater amount to distinguish between a logic "1" and a logic "0" than is necessary with a bipolar gate.

The obvious solution is to make MOS devices that operate with the voltages compatible with those of bipolar circuits. Such a device, combined with efforts to improve other MOS parameters and to design "all-MOS" digital equipment, has led to several different constructions. Because of the variety of MOS operating levels, a number of different circuits are needed to interface bipolar and MOS logic.

Chapter Ten

LINEAR INTEGRATED CIRCUITS: THE OPERATIONAL AMPLIFIER

The basic linear IC building block is a broad-band, high-gain, differential-input amplifier, called the operational amplifier. The designation "operational amplifier" was originally adopted for a series of high-performance dc amplifiers used in analog computers. These amplifiers performed mathematical operations applicable to analog computation (summation, scaling, subtraction, integration, etc.). Today the availability of inexpensive IC amplifiers has made the packaged operational amplifier useful as a replacement for any low-frequency amplifier.

Operational amplifiers are often the easiest and best way of performing a wide range of linear functions from simple amplification to complex analog computation. Some of the advantages of monolithic operational amplifiers over discrete amplifiers are: low cost, much higher complexities, improved performance, and temperature stabilization of the amplifier chip.

INTEGRATED CIRCUIT OPERATIONAL AMPLIFIERS

A monolithic "operational amplifier" is a very high-gain dc amplifier, usually with inverting and noninverting inputs and external feedback; it may simply amplify an input signal, or it may operate on a current or voltage in mathematical form. Although it may have one or more inputs, ordinarily there is only a single output, which is internally compensated for both temperature drift and voltage stability. Capacitor phase compensation is usually applied externally (although some op amps have it built in), as are input, feedback, and any offset-voltage resistors used to establish input impedance, to control loop gain, and to minimize offset voltage that occurs between the two operational-amplifier inputs.

The great versatility and many advantages of operational amplifiers stems from the use of negative feedback. Negative feedback tends to improve gain stability, to reduce output impedance, to improve linearity and, in some configurations, to increase input impedance. Another valuable property of negative feedback, which is the basis for all operational-amplifier technology, is that with enough gain, the closed-loop amplifier characteristics become a function of only the feedback components.

IC operational amplifiers generally use several differential stages in cascade to provide both common-mode rejection and high gain. Therefore, they require both positive and negative power supplies. Those ICs which do not require dual power supplies are rarely (if ever) used as operational amplifiers. (Without a differential amplifier, there would be no common-mode rejection.)

Most operational-amplifier ICs require equal power supplies (such as $+15$ and -15 V). However, equal power supplies are not required for all ICs.

Unlike most transistor circuits, where it is usual to label one power-supply lead positive and the other negative without specifying which (if either) is common to ground, all IC power-supply voltages should be referenced to ground.

THE IDEAL OPERATIONAL AMPLIFIER

Figure 10-1 is a block diagram of a typical operational amplifier. The ideal operational amplifier offers characteristics approaching the following limits:

1. Infinite voltage amplification A.
2. Infinite input impedance Z_{in}.
3. Zero output (source) impedance Z_{out}.
4. Infinite bandwidth.
5. Linear output swing above and below zero volts.
6. Symmetrical limiting of output voltage to specified "saturation" levels without damage.
7. $E_{out} = -AE_{in}$.
8. $E_{out} = 0$, when $E_{in} = 0$.
9. Instantaneous recovery from saturation.
10. Differential inputs.

With the ideal operational amplifier, then, the gain A is infinite, the input impedance Z_{in} is infinite, and the output impedance Z_{out} is so low that it approaches zero. This means that almost any signal can turn it on; there is no loading of the preceding stages, and the amplifier itself could drive an infinite number of other devices. In practice, however, Z_{in} is never infinity and the gain is always finite. Thus there exists a given input impedance and a gain, through a feedback loop, which is usually controlled by a simple resistance. Closed-loop gain depends on the ratio of feedback resistance to input resistance; but loop gain — the product of A (gain) and B (feedback attenuation) — is the ratio between open-loop and closed-loop gain. Loop gain decreases with an increase in frequency and with high values

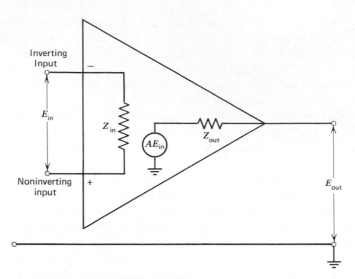

Figure 10-1   An ideal operational amplifier.

of closed-loop gain. Also, the bandwidth of the amplifier decreases as the closed-loop gain is increased.

Figure 10-2 is the schematic diagram of a simple current amplifier. In this circuit the output voltage is equal to the product of the output current and the feedback resistor value, whereas the output current is equal to the ratio of the output voltage to the feedback resistor value. The input and feedback currents are equal.

Since the open-loop input impedance Z_{in} of Figure 10-2 is virtually infinite, all input current I_{in} passes through the feedback resistor R_f. Developing an expression for Z_{in} in terms of R_f, we proceed as follows:

$$Z_{in} = \frac{E_{in}}{I_{in}} \tag{10-1}$$

$$I_{in} = \frac{E_{out} - E_{in}}{R_f} \tag{10-2}$$

$$Z_{in} = \frac{E_{in} R_f}{E_{in} - E_{out}} \tag{10-3}$$

$$Z_{in} = \frac{E_{in} R_f}{E_{in} - (-E_{in}A)} \tag{10-4}$$

$$Z_{in} = \frac{R_f}{1 + A} \tag{10-5}$$

If the absolute magnitude of R_f approaches the absolute value of the gain of the device, we no longer have an operational amplifier, and the derived equations no longer hold. For large values of A, Z_{in} becomes so small that the input is called a virtual ground. This input now becomes a voltage node or null point; therefore, the current from an input source will flow as though the source were returned to ground.

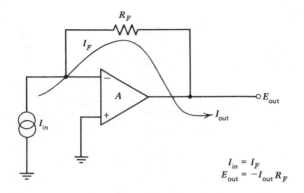

Figure 10-2   Basic current amplifier circuit.

The circuit of Figure 10-3 is similar to that of Figure 10-2 except that an input resistance has been added. This circuit connection facilitates the determination of the gain of the operational amplifier.

Figure 10-4 further develops the circuit of Figure 18-2 by adding a bias resistor R_{bias} to minimize the offset voltage at the amplifier's output resulting from input bias current. The value of the bias resistor is equal to the parallel combination of the feedback resistor and the input resistor. The circuit of Figure 10-4 is the basic operational-amplifier circuit used to calculate closed-loop response.

DESIGN CONSIDERATIONS FOR FREQUENCY RESPONSE AND GAIN

Most of the design problems for IC operational amplifiers are the result of trade-offs between gain and frequency response (or bandwidth). The open-loop (without feedback) gain and frequency response are characteristics of the basic IC package, but these can be modified with external phase compensation networks. The closed-loop (with feedback) gain and frequency response are primarily dependent on external feedback components.

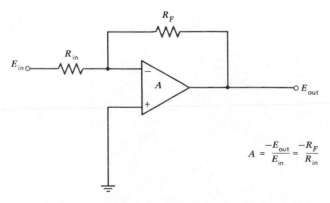

Figure 10-3   Basic amplifier circuit with input resistance added.

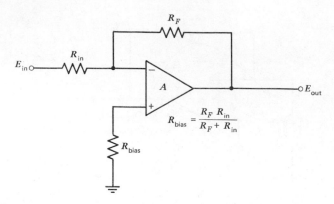

Figure 10-4   Fully developed basic operational-amplifier circuit.

The two basic operational-amplifier configurations, inverting feedback and noninverting feedback, appear in Figures 10-5 and 10-6, respectively. Loop gain in these figures is defined as the ratio of open-loop gain to closed-loop gain, as in Figure 10-7. The relationships in Figure 10-7 are based on a theoretical operational amplifier. That is, the open-loop gain rolls off at 6 dB/octave or 20 dB/decade. (The term 6 dB/octave means that the gain drops by 6 dB each time frequency is doubled. This is the same as a 20 dB drop each time the frequency is increased by a factor of 10.)

If the open-loop gain of an amplifier was as shown in Figure 10-7, any stable closed-loop gain could be produced by the proper selection of feedback components, provided the closed-loop gain was less than the open-loop gain. The only concern would be a trade-off between gain and frequency response. For example, if a gain of 40 dB (10^2) was desired, a feedback resistance 10^2 times higher than the

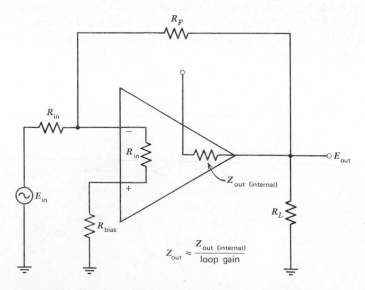

Figure 10-5   Inverting feedback operational amplifier.

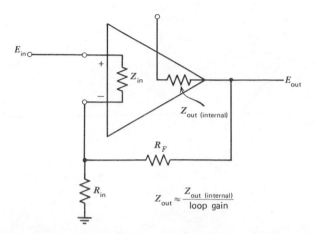

Figure 10-6   Noninverting feedback operational amplifier.

input resistance would be selected. The gain would then be flat to 10^4 Hz and roll-off at 6 dB/octave to unity gain at 10^6 Hz. If 60-dB (10^3) gain was required, the feedback resistance would be raised to 10^3 times the input resistance. This would reduce the frequency response. Gain would be flat to 10^3 Hz, and then roll-off at 6 dB/octave to unity gain.

The open-loop frequency response curve of a practical amplifier more closely resembles that of Figure 10-8. Here gain is flat at 60 dB to about 200 kHz, then rolls off at 6 dB/octave to 2 MHz. Beyond that point, roll-off continues at 12 dB/octave (40 dB/decade) to 20 MHz, then rolls off at 18 dB/octave (60 dB/decade).

In itself, the sharp roll-off at high frequencies would not be a problem in operational-amplifier design (unless the circuit were obliged to operate at a frequency very near the high end). However, the phase response (phase shift between input and output) changes with frequency. The phase response of Figure 10-8 indicates that a negative feedback (at low frequency) can become positive

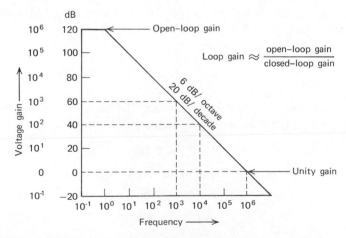

Figure 10-7   Frequency-response curve of a theoretical operational amplifier.

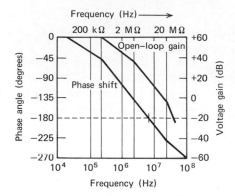

Figure 10-8 Frequency response and phase-shift curve of a practical operational amplifier.

and can cause the amplifier to be unstable at high frequencies (possibly resulting in oscillation). In Figure 10-8, a 180° phase shift occurs at approximately 4 MHz. This is the frequency at which open-loop gain is about 20 dB.

As a rule of thumb, when a selected closed-loop gain is equal to or less than the open-loop gain at the 180° phase-shift point, the circuit will be unstable. For example, if a closed-loop gain of 20 dB or less had been selected, a circuit with the curves of Figure 10-8 would be unstable. Therefore, the closed-loop gain must be more than the open-loop gain at the frequency where 180° phase shift occurs (but less than the flat, low-frequency, open-loop gain). Using Figure 10-8 as an example, the closed-loop gain would have to be greater than 20 dB, but less than 60 dB.

OPERATIONAL–AMPLIFIER TERMINOLOGY

In order to better understand operational amplifiers and to analyze their performance, we must assimilate the associated terminology. The terms presented here are most often used in operational-amplifier specifications as measures of amplifier performance.

Input bias current (I_{bias}) is the average of the currents (I_{B1}, I_{B2}) flowing into the input terminals when the output is at zero voltage. This is represented in Figure 10-9 and in the following equation:

$$I_{\text{bias}} = \frac{I_{B1} + I_{B2}}{2} \tag{10-6}$$

Input offset current (I_{OS}) is the difference in the currents flowing into the input terminals when the output is at zero voltage; this is expressed as

$$I_{OS} = I_{B1} - I_{B2} \tag{10-7}$$

Typical values of the input bias and input offset currents are

$$I_{\text{bias}} = 0.8 \text{ to } 300 \text{ nA}$$

$$I_{OS} = 0.05 \text{ to } 100 \text{ nA}$$

In an IC operational amplifier, the input devices will not be perfectly matched. Therefore, a delta or difference voltage usually exists between the input terminals.

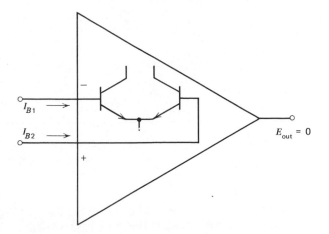

Figure 10-9 An operational-amplifier circuit depicting input bias current.

Input offset voltage (V_{OS}) is the voltage that must be applied between the input terminals through two equal resistances to force the output voltage to zero (Figure 10-10). Typical values of input offset voltage are

$$V_{OS} = 0.3 \text{ to } 7.5 \text{ mV}$$

Operational-amplifier input impedance (Z_{in}) is defined as the ratio of the change in input voltage to the change in input current on either input with the other grounded (Figure 10-11). Typical values of input impedance are

$$Z_{in} = 150 \text{ k}\Omega \text{ to } 30 \text{ M}\Omega$$

Operational-amplifier output impedance (Z_{out}) is the ratio of the change in output voltage to the change in output current with the output voltage near zero (Figure 10-12). Typical values of output impedance are

$$Z_{out} = 0.75 \text{ to } 1000 \text{ }\Omega$$

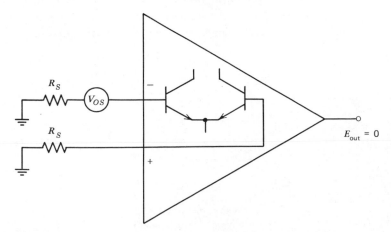

Figure 10-10 In this operational amplifier circuit, showing input offset voltage, $R_S = 10 \text{ K}\Omega$ is an industry-defined source resistance.

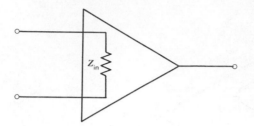

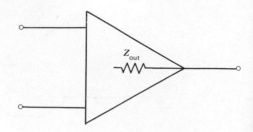

Figure 10-11 Circuit depicting the input impedance of an operational amplifier.

Figure 10-12 Circuit depicting the output impedance of an operational amplifier.

The *common-mode rejection ratio* (CMRR) is defined as the ratio of the input voltage range to the peak-to-peak change in input offset voltage over this range. Typical values of CMRR are 70 to 120 dB.

SPEED AND FREQUENCY RESPONSE

Transition Time

Transition time is defined as the time required for a signal to pass through the 10% and 90% points between two specified levels. Worst-case transition time for an amplifier is the time required for the amplifier output to travel between the 10% and 90% points of the swing between positive and negative saturation limits (Figure 10-13).

Slew Rate

As can be seen from Figures 10-13 and 10-14, transition time is a function of the voltage swing. A better indication of circuit speed is the rate of voltage change. The slope dv/dt is called the *slew rate*. *Slew rate* is defined as the internally limited rate of change in output voltage with a large amplitude step function applied to the input. Typical slew rate values range between 0.1V and 1500 V/μsec.

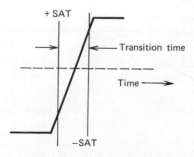

Figure 10-13 Graphical representation of an operational amplifier's transition time.

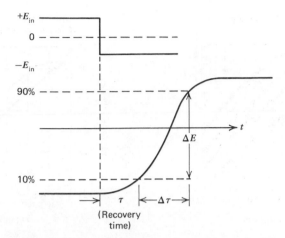

Figure 10-14   A large-amplitude signal derives the operational amplifier from plus to minus saturation.

Frequency Response (cf. Figure 10-15)

The *frequency response* of an operational amplifier is specified by three frequency measurements:

1. $f_{(3\,dB)}$ the frequency at which the voltage amplification is 3 dB below its maximum value.

2. f_t, the frequency at which the amplification factor is unity: $(A_v\,dB = 0)$.

3. $f_{o\,max}$ the maximum frequency of full output swing, is not directly related to the curve of Figure 10-15. Frequently omitted from data, it is nevertheless an important parameter.

$$f_{o\,max} = \frac{\text{slew rate}}{\pi\,(\text{peak-to-peak saturation voltage})} = \text{frequency where we} \qquad (10\text{-}8)$$

expect slew-rate limiting; or, assuming a symmetrical saturation swing:

$$f_{o\,max} = \frac{\text{slew rate}}{2\pi\,(\text{peak saturation voltage})} \qquad (10\text{-}9)$$

This frequency is often less than either of the first two.

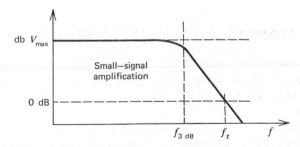

Figure 10-15   General figure illustrating frequency response.

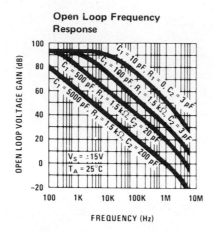

Figure 10-16 Frequency response of a 709 externally compensated operational amplifier.

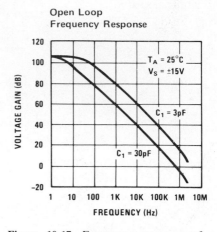

Figure 10-17 Frequency response of a 101/101A externally compensated operational amplifier.

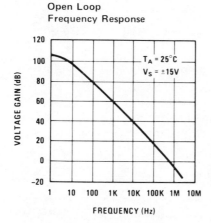

Figure 10-18 Frequency response of a 741 internally compensated operational amplifier.

Typical curves of open-loop frequency response for three popular operational amplifiers are given in Figures 10-16, 10-17, and 10-18. As might be expected, the frequency-compensation components determine the frequency response.

FACTORS TO CONSIDER IN CHOOSING AN OP AMP

In choosing an operational amplifier, finding one that meets the requirements may be only half the job. Often there are many ICs available that could be used. In this case, narrowing down to the one IC best suited for a specific application can be a truly difficult task.

First of all, the specifications on the data sheet do not tell the whole story. Factors often not readily obtainable from the data sheet which can affect one's choice include reliability, acceptability and compatibility of the package and the supply requirements, capability to withstand the application's environmental stresses, and degree to which the device is accident-proof.

Then, of course, there is cost. Device cost is rarely, if ever, cited on data sheets. But if a device has been subjected to a lot of tests that are not necessary for the potential customer's application, he will probably get a better buy if he looks somewhere else. The same rationale applies if the temperature range exceeds requirements.

Do Not Overspecify

Decide which specifications are critical and which are not. For example, in building a voltage regulator with operational-amplifier ICs where there is a large differential voltage during turn-on, this must be considered to be a very critical parameter. For working with a very high-gain, high-source-impedance amplifier, offset current would be more critical than offset voltage. If, in such a situation, both offset current and offset voltage were specifed very tightly, the cost of the amplifier would be needlessly raised.

There are few applications in which dc input impedance is important. Input bias current, which can be easily measured, is the limiting factor in dc applications. Input offset voltage, input offset current, and input bias current measured over the common-mode range of the device will guarantee the integrity of the input stage. Therefore it makes more sense and costs less to test V_{os}, I_{os}, and I_{bias} at common mode extremes than to measure dc input impedance. Where, then, is input impedance important? At frequencies above 1 kHz, the input impedance decreases to the point where it dominates the input bias current. Therefore, if the application is much above 1 kHz, and it is critical in the application, the ac input impedance at that frequency may require testing. In these cases, it is still probably less costly to add a voltage follower such as the LM110 to the amplifier input than to test ac input impedance.

Trade Off Cost Versus Performance

Cost, understandably, keeps entering the picture during the evaluation of an operational amplifier. Throughout the elimination process, cost must be continually weighed against performance. Judging cost is complicated because determining the overall system cost involves guesswork. Nevertheless, it should not be too difficult to decide, for example, if using a dual op-amp IC instead of two separate units or a high-gain amplifier instead of two low-gain amplifiers are economical moves. Although if, as in the case of the dual op-amp IC, the close-temperature tracking of two circuits on one chip is especially desirable, the weighing becomes more complex. Of course, when special testing and screening procedures are imposed on the IC manufacturer, the cost will increase. The user must evaluate whether this increased screening (electrical and/or environmental) is worth the added cost and delivery schedule impact.

There are several ways to select the most-cost-effective op amp in order to save money. The most-cost-effective-means is when the designer specifies standard data sheet limits for his design. An off-the-shelf op amp is best choice if all the user's critical parameters are normally tested by the manufacturer (if they have a min or max limit on the data sheet). This ensures that the same part from any manufacturer will function properly in the user's circuit. Also, since tested parameters are typically much better than the data sheet limits, a "guard band" or derating is built into the device. Therefore, the average circuit is better able to perform within specification during unanticipated worst-case conditions.

Note that it must be determined exactly what electrical tests the IC manufacturer actually performs. Because a minimum or maximum limit appears on the manufacturer's data sheet does not mean that this limit is 100% tested. There are some data sheet parameters that are guaranteed by design and not 100% tested. Also, not all manufacturers test the devices over the guaranteed temperature range. This information is vital to both the designer and the buyer when trying to effectively specify and procure the most-cost-effective op amp.

Another way to select the most-cost-effective op amp is to request the manufacturer to select a part to tightened specifications. However, selection to tighter parameters can result in a yield loss and price increase.

If the designer needs a specially selected and tighter parameter, he must determine if the added cost buys him anything. For example, an op amp with a data sheet offset voltage limit of 2 mV probably can be generally selected for 1 mV or better, and tighter offset voltage drift can generally be tested by the manufacturer. But this costs money.

Evaluate Op-Amp Self-Protection Features

Besides selecting which specifications are needed and which are not, choices regarding safety factors may be required. Problems that will arise under fault conditions should be considered. Will the output become seriously distorted? Will the whole system go up in smoke? For example, op-amp ICs can be bought with or without protection against output short-circuit and input overvoltage. Since the emitter-to-base junctions of input transistors are sensitive to damage by large applied voltages, some form of input protection may be desirable. Ungrounded soldering irons, excessive input signals, and static discharges are all apt to challenge the input of the IC.

Evaluate Means of Frequency Compensation

The frequency compensation that will be required differs widely among commercially available op-amp ICs. At one time, in fact, opponents of these devices cited the need for frequency compensation networks as the prime reason for not using IC operational amplifiers.

The discrete components required for frequency compensation do add to the space requirements and assembly cost. Various arrangements used for commercially available IC op amps can require from zero to seven components depending on the IC and application. Certainly, it is of key importance for the designer to consider frequency compensation. He should look at the IC op amp and attempt to determine how prone it is to oscillation if the supply is not bypassed properly. The IC should be evaluated to see how apt it is to oscillate with varying capacitive loads and to ascertain the probability that stray capacitance around the circuit will send it into oscillation. Too much bandwidth in an amplifier can be detrimental. Extra capacitors may be needed across the feedback resistors to prevent oscillation.

Internal frequency compensation reduces the piece-parts count but also limits the performance of the op amp. For a unity-gain noninverting connection, maximum compensation is needed, and this imposes the limits of a certain slew rate and bandwidth. If the op amp is internally compensated, this limitation remains irrespective of the way in which the amplifier is connected in the circuit. For example, in a gain-of-10 circuit, we would have a 0.5 V/μsec slew rate and a power bandwidth of about 5 kHz (zero dB at 1 MHz) for a 741 internally compensated operational amplifier. If an externally compensated op amp were used, such as the LM101A, we would obtain a higher slew rate and a greater gain–bandwidth product — a factor of 20 better than that obtainable from internally compensated op amps. Furthermore, if used in an inverting mode, the size of the compensation capacitor required for an externally compensated op amp can be cut in half.

Roll-off Curve

The op-amp user should know his amplifier's roll-off curve in order to build a circuit with adequate gain stability over the working frequency range. The manufacturer may be perfectly justified in departing from the conventional 6-dB/octave frequency compensation to achieve such desirable features as fast settling time, high slew rate, fast overload recovery, or increased gain stability over a wide range of frequencies. But to obtain these improved features generally requires fast roll-off characteristics and, therefore, a propensity toward oscillation. The key to preventing instability, of course, is knowing that this kind of amplifier is involved. It is then possible to use one or more well-known circuit techniques to eliminate the oscillations.

CMMR

CMRR is another important parameter. An amplifier's common-mode rejection performance can vary with operation conditions, notably with the value of common-mode input voltage. For example, some well-known FET types boast a common-mode voltage range of ± 10 V. But the common-mode rejection figures are specified for a ± 5 V common-mode range. It is possible for the CMRR to degrade by as much as a factor of 10 when the applied common-mode voltage is raised from 5 to 10 V.

Moreover, since CMRR is a nonlinear function of input common-mode voltage, a single specification number can at best give an average value over the best voltage range. For small-input signal variation about some large common-mode voltage, the specified "average" CMRR gives little indication of the actual errors that can be expected due to the steeper error slope at high voltages.

Slew Rate

Another characteristic that is often overlooked is the slew rate. The slew rate usually cannot be calculated from the small-signal response, or vice versa. The designer should evaluate the devices with the particular frequency-compensating network used to determine the small-signal response and the slew rate. These characteristics may not be important if a low-frequency sine wave is all that has to be amplified; however, if a square-wave amplifier is needed or if a sample-and-hold circuit is required, then these characteristics are very important.

Slew rate for the most part is just another way of looking at rate limiting of the amplifier's circuitry. The slew-rate specification applies to transient response, whereas full-power response applies to steady-state or continuous response. For a step function input, slew rate tells how fast the output voltage can swing from one voltage level to another. Fast amplifiers will slew at up to 1500 V/μsec, but amplifiers designed primarily for dc applications often slew at 0.1 V/μsec or less.

Settling Time

Settling time, a parameter of increasing interest, defines the time required for the output to settle within a given percentage of final value in response to a step function input. Common accuracies of interest are settling times of 0.1 and 0.01%. Heretofore, engineers have been forced to use slew rate and unity-gain bandwidth as rough indicators of relative settling time performance when comparing or choosing amplifiers, since no other data are given. As it turns out, these two parameters have little bearing on settling time, particularly to 0.01%. The problem here is that settling time is really a closed-loop parameter (all other op-amp specifications being open-loop parameters) and therefore depends on the closed-loop configuration and gain.

Chapter Eleven

OPERATIONAL–AMPLIFIER APPLICATIONS

Monolithic bipolar devices used in linear circuits, in particular, operational amplifiers, are required to exhibit characteristics that are demanding by any standard. In order to achieve very low input bias currents, the input stage transistors need betas on the order of several hundred at microampere current levels. Additionally, the differential amplifier transistor pair has to be closely matched to display minimum offset voltage. This close matching is required of several parameters, especially geometry and beta. As an example, a 1% mismatch in emitter area plus a 2% mismatch in beta will lead to a 1-mV offset voltage. Voltage supply and wide common mode requirements dictate that the transistor breakdown voltage exceed 40 V for most applications. Other necessary features are low junction leakages and low noise.

To meet such requirements and still maintain high production yields, heavy demand is placed on process techniques and control. High betas at low current levels, good beta matching, and optimum low noise performance are all strongly influenced by minimizing recombination centers, both at the silicon-silicon dioxide interface and within the bulk silicon.

Intensive development of the monolithic operational amplifier has yielded circuits which are quite good engineering approximations of the ideal for finite cost. The low cost and high quality of these amplifiers allows the implementation of equipment and systems functions impractical with discrete components. An example is the low-frequency function generator, which may use 15 to 20 operational amplifiers in generation, wave shaping, triggering, and phase locking. The availability of the low-cost monolithic operational amplifier makes it mandatory that systems and equipment engineers be familiar with operational amplifier applications.

The monolithic operational amplifiers presented in this chapter do not generally exhibit the frequency-stabilization components.

BASIC OPERATIONAL AMPLIFIER CONNECTIONS

The Inverting Amplifier

The basic inverting operational-amplifier circuit appears in Figure 11-1. This circuit gives a closed-loop gain of R_2/R_1 when this ratio is small compared with the amplifier open-loop gain. The input impedance is equal to R_1. The closed-loop bandwidth is equal to the unity-gain frequency divided by 1 plus the closed-loop gain.

The only precautions to be observed are: (1) that R_3 should be chosen to be equal to the parallel combination of R_1 and R_2 to minimize the offset-voltage error due to bias current, and (2) that there will be an offset voltage at the amplifier output equal to closed-loop gain times the offset voltage at the amplifier input.

Offset error at the input of an operational amplifier is comprised of two components, which are identified in specifying the amplifier as input offset voltage and input bias current. The input offset voltage is fixed for a particular amplifier; however, the contribution due to input bias current is dependent on the circuit configuration used. For minimum offset voltage at the amplifier input without circuit adjustment, the source resistances for both inputs should be equal. In this case, the maximum offset voltage would be the algebraic sum of amplifier offset voltage and the voltage drop across the source resistance due to offset current. Amplifier offset voltage is the predominant error term for low source resistances, and offset current causes the main error for high source resistances.

In high-source-resistance applications, offset voltage at the amplifier output

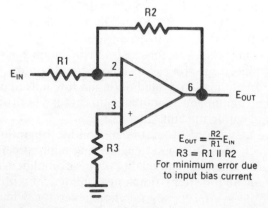

$$E_{OUT} = \frac{R2}{R1} E_{IN}$$
$$R3 = R1 \parallel R2$$
For minimum error due
to input bias current

Figure 11-1   Schematic diagram of an inverting amplifier.

may be adjusted by adjusting the value of R_3 and using the variation in voltage drop across it as an input offset voltage trim.

Offset voltage at the amplifier output is less important in ac-coupled applications. Here the only consideration is that any offset voltage at the output reduces the peak-to-peak linear output swing of the amplifier.

The gain–frequency characteristic of the amplifier and its feedback network must be such that oscillation does not occur. To meet this condition, the phase shift through amplifier and feedback network must never exceed 180° for any frequency in which the gain of the amplifier and its feedback network is greater than unity. In practical applications, the phase shift should not approach 180°, since this is the situation of conditional stability. Obviously, the most critical case occurs when the attenuation of the feedback network is zero.

Amplifiers that are not internally compensated may be used to achieve increased performance in circuits where feedback network attenuation is high. As an example, the LM101A may be operated at unity gain in the inverting amplifier circuit with a 15-pF compensating capacitor, since the feedback network has an attenuation of 6 dB, whereas it requires 30 pF in the noninverting unity-gain connection when the feedback network has zero attenuation. Since amplifier slew rate is dependent on compensation, the LM101A slew rate in the inverting unity-gain connection will be twice that for the noninverting connection, and the inverting gain of 10 connections will yield 11 times the slew rate of the noninverting unity-gain connection. The compensation trade-off for a particular connection is stability versus bandwidth. Larger values of compensation capacitor yield greater stability and lower bandwidth and vice versa.

The Noninverting Amplifier

The high-input-impedance noninverting circuit of Figure 11-2 gives a closed-loop gain equal to the ratio of the sum of R_1 and R_2 to R_1 and a closed-loop 3-dB bandwidth equal to the amplifier unity-gain frequency divided by the closed-loop gain.

The primary differences between this connection and the inverting circuit are that the output is not inverted and that the input impedance is very high and is

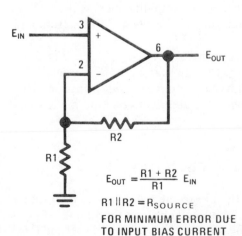

$$E_{OUT} = \frac{R1 + R2}{R1} E_{IN}$$

R1 ‖ R2 = R_{SOURCE}

FOR MINIMUM ERROR DUE
TO INPUT BIAS CURRENT

Figure 11-2 Schematic diagram of a noninverting amplifier.

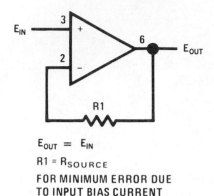

$E_{OUT} = E_{IN}$
$R1 = R_{SOURCE}$
**FOR MINIMUM ERROR DUE
TO INPUT BIAS CURRENT**

Figure 11-3   Schematic diagram of a unity-gain buffer.

equal to the differential-input impedance multiplied by loop gain. (Open-loop gain/closed-loop gain.) In dc-coupled applications, input impedance is less important than input current and its voltage drop across the source resistance.

Applications precautions are the same for this amplifier as for the inverting amplifier, with one exception. The amplifier output will go into saturation if the input is allowed to float. This may be important if the amplifier must be switched from source to source. The compensation trade-off discussed for the inverting amplifier is also valid for this connection.

The Unity-Gain Buffer and Voltage Follower

The unity-gain buffer (Figure 11-3) gives the highest input impedance of any operational-amplifier circuit. Input impedance is equal to the differential-input impedance multiplied by the open-loop gain, in parallel with common-mode input impedance. The gain error of this circuit is equal to the reciprocal of the amplifier open-loop gain or to the common-mode rejection, whichever is less.

Input impedance is a misleading concept in a dc-coupled unity-gain buffer. Bias current for the amplifier will be supplied by the source resistance and will cause an error at the amplifier input because of its voltage drop across the source resistance. Since this is the case, a low-bias current amplifier such as the LM110 voltage follower should be chosen as a unity-gain buffer when working from high source resistances.

Three precautions should be observed in applying the circuit of Figure 11-3: (1) the amplifier must be compensated for unity-gain operation, (2) the output swing of the amplifier may be limited by the amplifier common-mode range, and (3) some amplifiers exhibit a latch-up mode when the amplifier common-mode range is exceeded.

OPERATIONAL AMPLIFIER PROTECTION CIRCUITRY

Maximum voltage ratings must be strictly observed when using operational amplifiers. Input and output voltages must not exceed either of the supply voltages during usage. Transients that occur when the power supply is turned ON and OFF

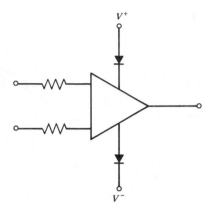

Figure 11-4 Operational amplifier protective circuitry for transients, power-supply reversal, and high common-mode voltages.

often result in a higher input voltage than supply voltage. This condition can cause a catastrophic op-amp failure. The minus supply voltage or ground should always be applied first or simultaneously with the positive supply to avoid "latch-up" or device destruction. Some op amps can be destroyed when an input voltage is applied to the op amp and both supplies are turned OFF. This condition turns ON the collector–base diode of the input transistors. The op amp will be degraded or destroyed because the positive power supply, by being OFF, is essentially grounded.

Device inputs can also be degraded if a dc voltage is placed on the output of an integrator or on the input of a sample-and-hold circuit and the positive supply is again OFF. If a large capacitor is used, the inputs may be destroyed.

What is the solution? A high-voltage diode inserted in each power supply lead will prevent failures caused by transients and power-supply reversals. A resistor inserted in each input signal line will limit the transient inputs caused by high common-mode voltages. These circuit fixes are diagrammed in Figure 11-4.

EFFECTS OF ERROR CURRENT

In an operational amplifier, the input current produces a voltage drop across the source resistance, causing a dc error. This effect can be minimized by operating the amplifier with equal resistances on the two inputs. The error is then proportional to the difference in the two input currents, or the offset current. Since the current gains of monolithic transistors tend to match well, the offset current is typically a factor of 10 less than the input currents.

Bias-Current Compensation

Bias-current compensation reduces offset and drift when the amplifier is operated from high source resistances. With low source resistances, such as a thermocouple, the drift contribution due to bias current can be made quite small; in this case, the offset voltage drift becomes important.

A technique is presented here by which offset voltage drifts better than 0.5 μV/°C can be realized. The compensation technique involves only a single room-

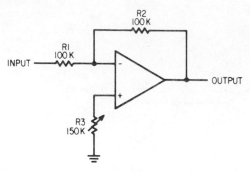

Figure 11-5   Summing amplifier with bias-current compensation for fixed-source resistances.

temperature balance adjustment. Therefore, chopper-stabilized performance can be realized, with low source resistances in a fairly simple amplifier, without tedious cut-and-try compensation methods.

The simplest and most effective way of compensating for bias currents is illustrated in Figure 11-5. Here, the offset produced by the bias current on the inverting input is canceled by the offset voltage produced across the variable resistor R_3. The main advantage of this scheme, besides its simplicity, is that the bias currents of the two input transistors tend to track well over temperature, and low drift is also achieved. The disadvantage of the method is that a given compensation setting works only with fixed feedback resistors, and the compensation must be readjusted if the equivalent parallel resistance of R_1 and R_2 is changed.

Figure 11-6 presents a similar circuit for a noninverting amplifier. The offset voltage produced across the dc resistance of the source due to the input current is canceled by the drop across R_3. For proper adjustment range, R_3 should have a maximum value about three times the source resistance; and the equivalent parallel resistances of R_1 and R_2 should be less than one-third the input source resistance. This circuit has the same advantages as the circuit of Figure 11-5; however, it can only be used when the input source has a fixed dc resistance.

Figure 11-7 is a compensation scheme for the voltage-follower connection. The compensating current is obtained through a resistor connected across a diode that is bootstrapped to the output. The diode acts as a regulator so that the compensating current does not change appreciably with signal level, giving input impedances of about 1000 MΩ. The negative temperature coefficient of the diode voltage also provides some temperature compensation.

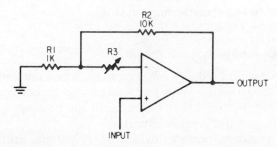

Figure 11-6   Noninverting amplifier with bias-current compensation for fixed-source resistances.

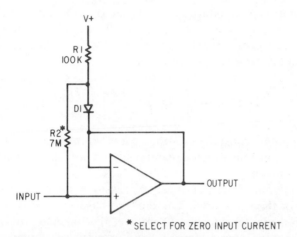

Figure 11-7   Voltage follower with bias-current compensation.

All the circuits discussed thus far have been tailored for particular applications. In the completely general circuit of Figure 11-8, both inputs are current compensated over the full common-mode range as well as against power supply and temperature variations. This circuit is suitable for use either as a summing amplifier or as a noninverting amplifier. It is not required that the dc impedance seen by both inputs be equal, although lower drift can be expected if they are.

As was mentioned earlier, all the bias-compensation circuits require adjustment. With the circuits in Figures 11-5 and 11-6, this is merely a matter of adjusting the potentiometer for zero output with zero input. It is not as simple with the other circuits, however. For one thing, it is difficult to use potentiometers because a very wide range of resistance values is required to accommodate expected unit-to-unit variations. Resistor selection must therefore be used.

Offset-Voltage Compensation

The highly predictable behavior of the emitter–base voltage of transistors has suggested a unique drift-compensation method. The offset voltage drift of a differ-

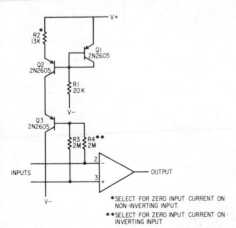

Figure 11-8   Bias-current compensation for differential inputs.

ential transistor pair can be reduced by about an order of magnitude by unbalancing the collector currents such that the initial offset voltage is zero. The basis for this comes from the equation for the emitter–base voltage differential of two transistors operating at the same temperature;

$$\Delta V_{BE} = \frac{kT}{q} \log_e \frac{I_{s_2}}{I_{s_1}} - \frac{kT}{q} \log_e \frac{I_{c_2}}{I_{c_1}} \tag{11-1}$$

where k is Boltzmann's constant, T is the absolute temperature, q is the charge of an electron, I_s is a constant which depends only on how the transistor is made and I_c is the collector current.

It is worthwhile noting here that these expressions make no assumptions about the current gain of the transistors. The emitter–base voltage is a function of collector current, not emitter current. Therefore, the balance will not be upset by base current (except for interaction with the dc-source resistance).

The first term in Equation 11-1 is the offset voltage of the two transistors for equal collector currents. This offset voltage is directly proportional to the absolute temperature. The second term is the change of offset voltage arising from operating the transistors at unequal collector currents. For a fixed ratio of collector current, this is also proportional to absolute temperature. Hence, if the collector currents are unbalanced in a fixed ratio to give a zero emitter–base voltage differential, the temperature drift will also be zero.

For low drift, the transistors must operate from a source resistance that is low enough to render insignificant the voltage drop across the source due to base current (or base current differential, if both bases see the same resistance). Furthermore, the transistors must be operated at a collector current so low that

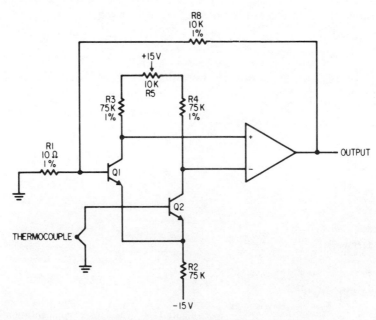

Figure 11-9 Example of a dc amplifier using the drift-compensation techniques.

the emitter-contact and base-spreading resistances are negligible, since Equation 11-1 assumes that they are zero.

In a complete amplifier using this principle (Figure 11-9), a monolithic transistor pair serves as a preamplifier for a conventional operational amplifier. A null potentiometer, which is set for zero output for zero input, unbalances the collector-load resistors of the transistor pair such that the collector currents are unbalanced for zero offset. This gives minimum drift. An interesting feature of the circuit is that the performance is relatively unaffected by supply-voltage variations: a 1-V change in either supply causes an offset voltage change of about 10 μV. This happens because neither term in Equation 11-1 is affected by the magnitude of the collector currents.

In order to get low drift, the gain of the preamplifier must be high enough to prevent the drift of the operational amplifier from degrading performance.

Another important consideration is the matching of the collector-load resistors on the preamplifier stage. A 0.1% imbalance in the load resistors due to thermal mismatches or any other cause will produce a 25-μV shift in offset voltage. This includes the balancing potentiometer, which can introduce an error that will depend on how far it is set off midpoint if it has a temperature coefficient different from that of the resistors.

UNIVERSAL BALANCING TECHNIQUES

IC operational amplifiers are widely accepted as a universal analog component. Although the circuit designs may vary, most devices are functionally interchangeable. However, offset voltage balancing remains a personality trait of the particular amplifier design. The techniques shown here allow offset voltage balancing without regard to the internal circuitry of the amplifier.

The circuit in Figure 11-10 is used to balance out the offset voltage of inverting amplifiers having a source resistance of 10 kΩ or less. A small current is injected into the summing node of the amplifier through R_1. Since R_1 is 2000 times as large as the source resistance, the voltage at the arm of the potentiometer is attenuated by a factor of 2000 at the summing node. With the values given and ± 15 V supplies, the output may be zeroed for offset voltages up to ± 7.5 mV.

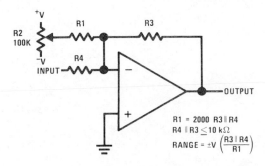

Figure 11-10 Offset-voltage adjustment for inverting amplifiers using 10KΩ source resistance or less.

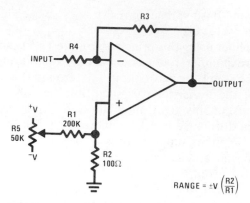

Figure 11-11   Offset-voltage adjustment for inverting amplifiers using any type of feedback element.

If the value of the source resistance is much greater than 10 kΩ, the resistance needed for R_1 becomes too large. In this case it is much easier to balance out the offset by supplying a small voltage at the noninverting input of the amplifier (Figure 11-11). Resistors R_1 and R_2 divide the voltage at the arm of the potentiometer to supply a ± 7.5 mV adjustment range with ± 15 V supplies. This adjustment method is also useful when the feedback element is a capacitor or nonlinear device.

This technique of supplying a small voltage effectively in series with the input is also used for adjusting noninverting amplifiers. As in Figure 11-12, divider R_1, R_2 reduces the voltage at the arm of the potentiometer to ± 7.5 mV for offset adjustment. Since R_2 appears in series with R_4, R_2 should be considered when calculating the gain. If R_4 is greater than 10 kΩ, the error due to R_2 is less than 1%.

A voltage follower may be balanced by the technique of Figure 11-13, where R_1 injects a current that produces a voltage drop across R_3 to cancel the offset voltage. The addition of the adjustment resistors causes a gain error, increasing the gain by 0.05%. This small error usually causes no problem. The adjustment circuit essentially causes the offset voltage to appear at full output, rather than at low output levels, where it is a large percentage error.

Differential amplifiers are somewhat more difficult to balance. The offset adjustment used for a differential amplifier can degrade the CMRR. The adjustment circuit in Figure 11-14 has minimal effect on the common-mode rejection.

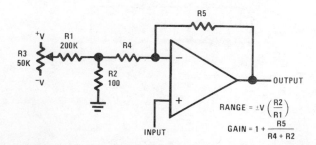

Figure 11-12   Offset-voltage adjustment for noninverting amplifiers.

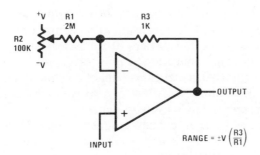

Figure 11-13   Offset-voltage adjustment for voltage followers.

The voltage at the arm of the potentiometer is divided by R_4 and R_5 to supply an offset correction of ± 7.5 mV. Here R_4 and R_5 are chosen such that the CMRR is limited by the amplifier for values of R_3 greater than 1 kΩ. If R_3 is less than 1 kΩ, the shunting of R_4 by R_5 must be considered when choosing the value of R_3.

The techniques described for balancing offset voltage at the input of the amplifier offer two main advantages: First, they are universally applicable to all operational amplifiers and allow device interchangeability with no modifications to the balance circuitry. Second, they permit balancing without interfering with the internal circuitry of the amplifier.

FREQUENCY-COMPENSATION TECHNIQUES

The ease of designing with operational amplifiers sometimes obscures some of the rules that must be followed with any feedback amplifier to keep it from oscillating. In general, these problems stem from stray capacitance, excessive capacitive loading, inadequate supply bypassing, or improper frequency compensation.

In frequency compensating an operational amplifier, it is best to follow the manufacturer's recommendations. However, if the combination of operating speed and frequency response is not a consideration, a greater stability margin can

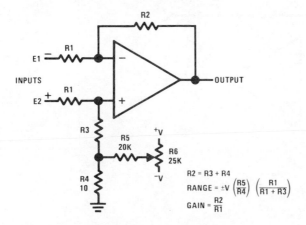

Figure 11-14   Offset-voltage adjustment for differential amplifiers.

usually be obtained by increasing the size of the compensation capacitors. For example, replacing the 30-pF compensation capacitor on the LM101A with a 300-pF capacitor will make the device 10 times less susceptible to oscillation problems in the unity-gain connection. Capacitor values less than those specified by the manufacturer for a particular gain connection should not be used, since they will make the amplifier more sensitive to stray and capacitive loading or, with worst-case units, the circuit could even oscillate.

The basic requirement for frequency compensating a feedback amplifier is to keep the frequency roll-off of the loop gain from exceeding 12 dB/octave when it goes through unity gain. Figure 11-15 illustrates the concept of loop gain. The feedback loop is broken at the output, and the input sources are replaced by their equivalent impedance. Then the response is measured such that the feedback network is included.

Figure 11-15b gives typical responses for both uncompensated and compensated amplifiers. An uncompensated amplifier generally rolls off at 6 dB/octave, then 12 dB/octave, and even 18 dB/octave, as various frequency-limiting effects within the amplifier come into play. If a loop with this kind of response were closed, it would tend to oscillate. Frequency compensation causes the gain to roll off at a uniform 6 dB/octave right down through unity gain. This allows some margin for excess roll-off in the external circuitry.

Some of the external influences that can affect the stability of an operational amplifier are indicated in Figure 11-16. One is the load capacitance, which can come from wiring cables or from an actual capacitor on the output. This capacitance works against the output impedance of the amplifier to attenuate high frequencies. If this added roll-off occurs before the loop gain goes through zero, it can cause instability. It should be remembered that this single roll-off point can give more than 6-dB/octave roll-off, since the output impedance of the amplifier can be increasing with frequency.

A second source of excess roll-off is stray capacitance on the inverting input. This becomes extremely important with large feedback resistors (as might be used

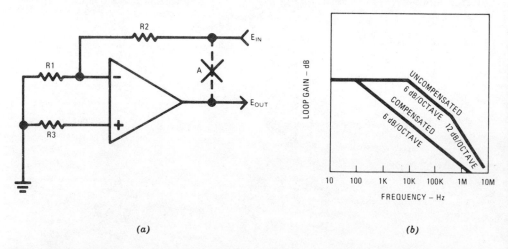

(a)   (b)

Figure 11-15 Illustrations of operational-amplifier gain: (a) measuring loop gain, (b) typical response.

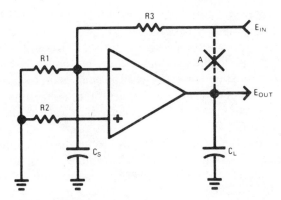

Figure 11-16 External capacitances that affect stability.

with a FET input amplifier). A relatively simple method of compensating for this stray capacitance is diagrammed in Figure 11-17: a lead capacitor C_1 is put across the feedback resistor. Ideally, the ratio of the stray capacitance to the lead capacitor should be equal to the closed-loop gain of the amplifier. However, the lead capacitor can be made larger as long as the amplifier is compensated for unity gain. The only disadvantage of doing this is that it will reduce the bandwidth of the amplifier. Oscillations can also result if there is a large resistance on the non-inverting input of the amplifier. The differential-input impedance of the amplifier falls off at high frequencies (especially with bipolar input transistors); thus this resistor can produce troublesome roll-off if it is much greater than 10 kΩ. With most amplifiers, this situation is easily corrected by bypassing the resistor to ground.

When the capacitive load on an integrated amplifier is much greater than 100 pF, some consideration must be given to its effect on stability. Even though the amplifier does not oscillate readily, there may be a worst-case set of conditions under which it will. However, the amplifier can be stabilized for any value of

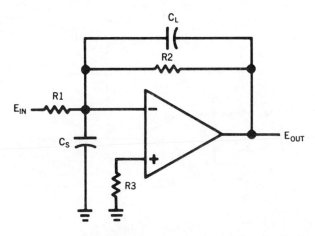

Figure 11-17 Compensating for stray input capacitance.

capacitive loading using the circuit in Figure 11-18. The capacitive load is isolated from the output of the amplifier with R_4.

At high frequencies, since the feedback path is through the lead capacitor C_1, the lag produced by the load capacitance does not cause instability. To use this circuit, the amplifier must be compensated for unity gain, regardless of the closed-loop dc gain. The value of C_1 is not too important; but at a minimum, its capacitive reactance should be one-tenth the resistance of R_2 at the unity-gain crossover frequency of the amplifier.

When an operational amplifier is operated open-loop, it might appear at first glance that it needs no frequency compensation. This is not always the case, however, because the external compensation is sometimes required to stabilize internal feedback loops.

When high-current buffers are used in conjunction with operational amplifiers, supply bypassing and decoupling are even more important, since they can feed a considerable amount of signal back into the supply lines. For reference, bypass capacitors of at least 0.1 μF are required for a 50-mA buffer.

When emitter followers are used to drive long cables, additional precautions are required. An emitter follower by itself—which is not contained in a feedback loop—will frequently oscillate when connected to a long length of cable.

The loop gain of an amplifier is a valuable tool in understanding the influence of various factors on the stability of feedback amplifiers. But it is not very helpful in determining if the amplifier is indeed stable. The reason is that most problems in a well-designed system are caused by secondary effects—which occur only under certain conditions of output voltage, load current, capacitive loading, temperature, and so on. Making frequency phase plots under all these conditions would require unreasonable amounts of time, and invariably the plots are not made.

A better check on stability is the small-signal transient response. It can be shown mathematically that the transient response of a network has a one-for-one correspondence with the frequency-domain response. The advantage of transient response tests is that they are displayed instantaneously on an oscilloscope; thus it is reasonable to test a circuit under a wide range of conditions.

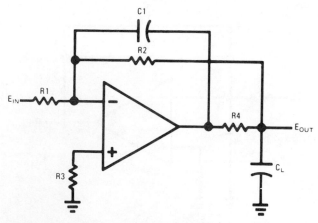

Figure 11-18   Compensating for very large capacitive loads.

FEED–FORWARD COMPENSATION TECHNIQUES

The slew rate of an operational amplifier can be increased by using feed-forward compensation. This means of frequency compensation extends the usefulness of the device to frequencies an order of magnitude higher than the standard compensation network. With this speed improvement, IC operational amplifiers may be used in applications that previously required discrete components. The compensation is relatively simple and does not change the offset voltage or current of the amplifier.

In order to achieve unconditional closed-loop stability for all feedback connections, the gain of an operational amplifier is rolled off at 6 dB/octave, with the accompanying 90° of phase shift, until a gain of unity is reached. The frequency-compensation networks shape the open-loop response to cross unity gain before the amplifier phase shift exceeds 180°. Unity gain for an op amp such as the LM 101A is designed to occur at 1 MHz. This is because the lateral *pnp* transistors used for level shifting have poor high-frequency response and exhibit excess phase shift at about 1 MHz. Therefore, the stable closed-loop bandwidth is limited to approximately 1 MHz.

Usually the LM101A is frequency compensated by a single 30-pF capacitor between pins 1 and 8, as in Figure 11-19. This gives a slew rate of 0.5 V/μsec. Feed-forward compensation is achieved by connecting a 150-pF capacitor between the inverting input, pin 2, and one of the compensation terminals, pin 1, as in Figure 11-20. This eliminates the lateral *pnp*'s from the signal path at high frequencies. Unity-gain bandwidth is 3.5 MHz and the slew rate is 10 V/μsec. Diode D_1 can be added to improve the slew rate with high-speed input pulses.

Figure 11-21 plots the open-loop response using both standard and feed-forward compensation techniques. Higher open-loop gain is realized with the fast compensation, since the gain rolls off at about 10 dB/octave until a gain of unity is reached at about 3.5 MHz.

As with all high-frequency, high-gain amplifiers, certain precautions should be taken to ensure stable operation. The power supplies should be bypassed near the amplifier with 0.01-μF disc capacitors. Stray capacitance, such as large lands on

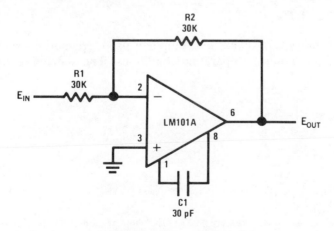

Figure 11-19   Standard frequency compensation.

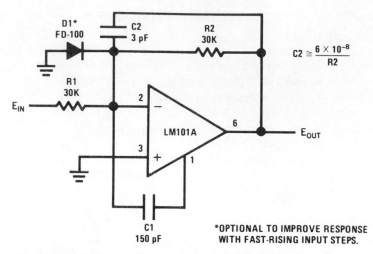

$$C2 \cong \frac{6 \times 10^{-8}}{R2}$$

*OPTIONAL TO IMPROVE RESPONSE
WITH FAST-RISING INPUT STEPS.

Figure 11-20   Feed-forward frequency compensation.

PC boards, should be avoided at the input, balance, and compensation pins. Load capacitance in excess of 75 pF should be decoupled, as in Figure 11-22; however, 500 pF of load capacitance can be tolerated without decoupling at the expense of bandwidth by the addition of 3 pF between pins 1 and 8. A small capacitor C_2 is needed as a lead across the feedback resistor to ensure that the roll-off is less than 12 dB/octave at unity gain. The capacitive reactance of C_2 should equal the feedback resistance between 2 and 3 MHz. For integrator applications, the lead capacitor is isolated from the feedback capacitor by a resistor, as in Figure 11-23.

Feed-forward compensation offers a marked improvement over standard compensation. In addition to having higher bandwidth and faster slew rate, there is vanishingly small gain error from dc to 3 kHz, and less than 1% gain error up to 100 kHz as a unity-gain inverter. The power bandwidth is also extended from 6 to 250 kHz. Some applications for this type of amplifier are: fast summing amplifier, pulse amplifier, digital-to-analog and analog-to-digital systems, and fast integrator.

DESIGN PRECAUTIONS

Regardless of the type of operational amplifier used, precautions should be taken to protect it from abnormal operating conditions. The specific techniques given

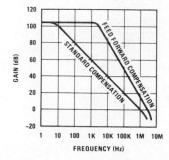

Figure 11-21   Open-loop responses for both frequency-compensation networks.

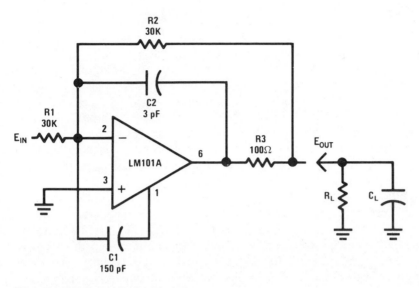

Figure 11-22   Capacitive load isolation.

below apply to the LM101A, but the advice is applicable to almost any IC op amp, even though the exact reasons may vary with different devices.

When driving either input of the LM101A from a low-impedance source, a limiting resistor should be placed in series with the input lead to limit the peak instantaneous output current of the source to something less than 100 mA. This is especially important when the inputs go outside a piece of equipment, where they could accidentally be connected to high-voltage sources.

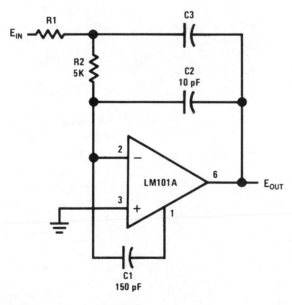

Figure 11-23   Fast integrator.

Large capacitors on the input ($> 0.1\,\mu$F) should be treated as a low-source impedance and isolated with a resistor. Low-impedance sources do not cause a problem unless their output voltage exceeds the supply voltage. However, since the supplies go to zero when they are turned OFF, the isolation is usually needed.

The output circuitry is protected against damage from shorts to ground on either supply. However, if it is shorted to a voltage that exceeds the positive or negative supplies, the unit can be destroyed. When the amplifier output is connected to a test point, it should be isolated by a limiting resistor, because test points frequently become shorted to undesirable places. Furthermore, when the amplifier drives a load external to the equipment, it is advisable to use some sort of limiting resistance to preclude mishaps.

Precautions should be taken to ensure that the power supplies for the IC never become reversed, even under transient conditions. With reverse voltages greater than 1 V, the IC will conduct excessive current, fuzing internal aluminum interconnects. If there is a possibility of this happening, clamp diodes with a high peak-current rating should be installed on the supply lines. Reversal of the voltage between V^+ and V^- will always cause a problem, although in many circuits reversals with respect to ground can also lead to difficulties.

The minimum values given for the frequency-compensation capacitor are stable only for source resistances less than 10 kΩ, stray capacitances on the summing junction less than 5 pF, and capacitive loads smaller than 100 pF. If any of these conditions is not met, it becomes necessary to overcompensate the amplifier with a larger compensation capacitor. Alternately, lead capacitors can be used in the feedback network to negate the effect of stray capacitance and large feedback resistors, or an RC network can be added to isolate capacitive loads.

Although the LM101A is relatively unaffected by supply bypassing, this contingency cannot be ignored altogether. Generally it is necessary to bypass the supplies to ground at least once on every circuit card, and more bypass points may be required if more than five amplifiers are used. When feed-forward compensation is employed, however, it is advisable to bypass the supply leads of each amplifier with low inductance capacitors because of the higher frequencies involved.

TYPICAL OPERATIONAL–AMPLIFIER APPLICATIONS

Operational amplifiers find applications as oscillators, level shifting amplifiers, voltage comparators, high-current output buffers, storage (sample-and-hold) circuits, nonlinear amplifiers, servo amplifiers, computing circuits and root extractors, summing amplifiers, low-pass filters, and differentiators, to name a few. Some of these applications are now discussed in greater detail. Obviously all possible applications cannot be treated.

Voltage Comparators

Comparators are amplifiers operated open-loop. They have a variety of applications, including interface circuits, detectors, and sense amplifiers. Integrated comparators, such as the 710 and the 111, are designed for these applications. General-purpose amplifiers such as the LM101A also make good comparators

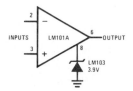

**Voltage Comparator for Driving
DTL or TTL Integrated Circuits**   **Figure 11-24**   Comparator connection.

because they have large differential input voltage ranges and are easily clamped to make their outputs compatible with logic and driver circuits. Internally compensated op amps are generally poor comparators because they switch very slowly.

A comparator circuit using the LM101A appears in Figure 11-24. This clamping scheme makes the output signal directly compatible with DTL or TTL ICs. A breakdown diode D_1 clamps the output at zero or 4 V in the low or high states, respectively. This particular diode was chosen because it has a sharp breakdown and a low equivalent capacitance. When working as a comparator, the amplifier operates open-loop; thus frequency compensation is not normally needed. Nonetheless, the stray capacitance between pins 5 and 6 of the amplifier should be minimized to prevent low-level oscillations when the comparator is in the active region. If this becomes a problem, a 3-pF capacitor on the normal compensation terminals will eliminate it.

ac-Coupled High-Input-Impedance Amplifier

High input impedance and low input capacitance are achieved with a unity-gain buffer amplifier having positive feedback. The circuit in Figure 11-25 has an input impedance of several hundred megohms and an input capacitance that is less than 1.0 pF. The high input impedance is obtained by positive feedback through C_1 to the positive input of the amplifier. The input capacitance plus the capacitance to ground can be canceled by adding the feedback capacitor C_2 and adjusting R_3.

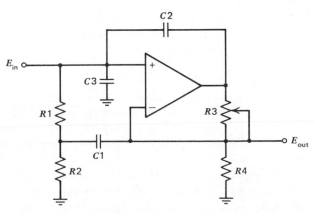

Figure 11-25   A high-input impedance amplifier.

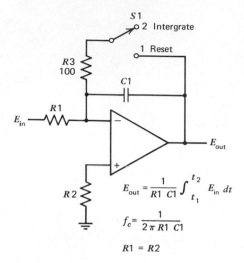

$$E_{out} = \frac{1}{R1\ C1} \int_{t_1}^{t_2} E_{in}\ dt$$

$$f_c = \frac{1}{2\pi\ R1\ C1}$$

R1 = R2   **Figure 11-26**   Integrator schematic.

Since the low-frequency response is determined primarily by C_1, the use of an electrolytic capacitor is advisable. High-frequency response is limited by the operational amplifier. Using an amplifier such as the 709, the circuit can amplify a 5-μsec wide pulse coupled through a 1.0-pF capacitor.

The Integrator

A typical integrator (Figure 11-26) performs the mathematical operation of integration. This circuit is essentially a low-pass filter with a frequency response decreasing at 6 dB/octave. A typical amplitude–frequency plot is presented in Figure 11-27.

The circuit must be provided with an external method of establishing initial conditions. This appears in Figure 11-26 as S_1. When S_1 is in position 1, the amplifier is connected in unity gain and capacitor C_1 is discharged, setting an initial condition of zero volts. When S_1 is in position 2, the amplifier is connected as an integrator and its output will change in accordance with a constant times the time integral of the input voltage.

Two precautions are to be observed with this circuit: the amplifier used should generally be stabilized for unity-gain operation, and R_2 must equal R_1, to keep error due to bias current to a minimum.

Sample-and-Hold Circuit

Although there are many ways to make a sample-and-hold circuit, the circuit illustrated in Figure 11-28 is undoubtedly one of the simplest. When a negative-going sample pulse is applied to the MOS switch, it will turn ON hard and charge the holding capacitor to the instantaneous value of the input voltage. After the switch is turned OFF the capacitor is isolated from any loading by the operational amplifier (LM102), and it will hold the voltage impressed on it.

The maximum input current of the LM102 is 10 nA; with a 10-μF holding capacitor, therefore, the drift rate in hold will be less than 1 mV/sec. If accuracies

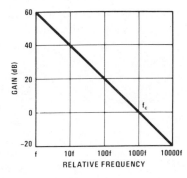

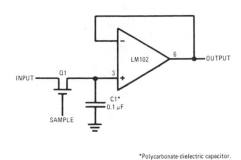

*Polycarbonate-dielectric capacitor.

Figure 11-27   Integrator frequency response.

Figure 11-28   A sample-and-hold circuit.

of about 1% or better are required, it is necessary to use a capacitor with poly-carbonate, polyethylene, or Teflon dielectric. Most other capacitors exhibit a polarization phenomenon which causes the stored voltage to fall off after the sample interval with a time constant of several seconds. For example, if the capacitor is charged from 0 to 5 V during the sample interval, the magnitude of the falloff is about 50 to 100 mV.

Active Filters

Active *RC* filters have been replacing passive *LC* filters at an ever-increasing rate because of the declining price and smaller size of active components. Figure 11-29 is a low-pass filter, which is one of the simplest forms of active filters. The circuit has the filter characteristics of two isolated *RC* filter sections, as well as a buffered, low-impedance output. The attenuation is roughly 12 dB at twice the cutoff frequency and the ultimate attenuation is 40 dB/decade. A third low-pass *RC* section can be added on the output of the amplifier for an ultimate attenuation of 60 dB/decade, although this means that the output is no longer buffered.

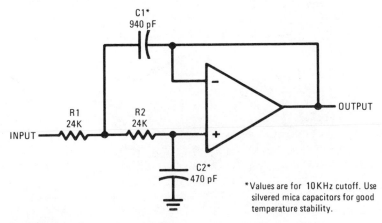

Figure 11-29   Low-pass active filter.

There are two basic designs for this type of filter. One is the Butterworth filter with maximally flat frequency response. For this characteristic, the component values are determined from

$$C_1 = \frac{R_1 + R_2}{\sqrt{2} \, R_1 R_2 \omega_c} \qquad (11\text{-}2)$$

and

$$C_2 = \frac{\sqrt{2}}{(R_1 + R_2) \omega_c} \qquad (11\text{-}3)$$

where ω_c is the cutoff frequency.

The second kind is the linear phase filter with minimum settling time for a pulse input. The design equations for this are

$$C_1 = \frac{R_1 + R_2}{\sqrt{3} \, R_1 R_2 \omega_c} \qquad (11\text{-}4)$$

and

$$C_2 = \frac{\sqrt{3}}{(R_1 + R_2) \omega_c} \qquad (11\text{-}5)$$

Substituting capacitors for resistors and resistors for capacitors in the circuit of Figure 11-29, a similar high-pass filter is obtained (see Figure 11-30).

Self-Tuned Filter

Audio and instrumentation systems frequently employ bandpass filters to improve the signal-to-noise ratio. If the input-signal frequency varies over a wide range — as it does with many types of vibrating transducers — the bandwidth of the filter usually must be large enough to avoid undue attenuation and phase shift at the frequency extremes. But a wide-band filter will be less effective in rejecting noise.

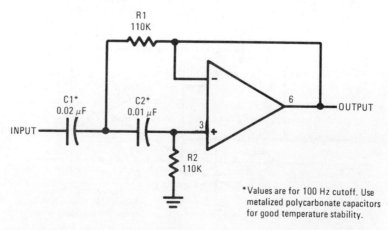

Figure 11-30   High-pass active filter.

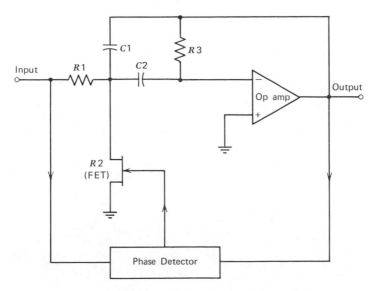

Figure 11-31 Self-tuned filter.

One solution is to use a self-tuning filter that automatically adjusts its center frequency to track the signal frequency. This technique allows the use of a filter that has a bandwidth considerably less than the range of input-signal frequencies. The circuit in Figure 11-31 tunes itself over a frequency range of 2 to 20 kHz. It requires no reference frequency other than the input signal, and there is no internal oscillation or synchronization circuitry. The frequency range can be extended in decade steps by capacitor switching. In this circuit, we have

$$f_{out} = \frac{1}{2\pi C} \left[\frac{1}{R_3} \left(\frac{1}{R_1} + \frac{1}{R_2} \right) \right]^{1/2}$$

Thus f_{out} can be varied merely by changing R_2 and without affecting the gain or bandwidth. If resistance R_2 can be made to vary approximately as the frequency of the input signal varies, the filter can be made to stay tuned to the signal frequency.

If the filter is properly tuned, there will be 180° of phase shift between the input and the output of the filter. If, however, the filter is not tuned to the input frequency, then the phase shift is not 180°, and the phase detector generates an error signal, which is applied to the gate of the FET to control its drain-to-source resistance R_2. The phase detector and the FET form part of a negative-feedback loop around the filter. Because of this configuration, any error in phase resulting from detuning will change the resistance of the FET, thereby retuning the filter. Thus the FET acts as the variable resistance R_2.

One-Shot Multivibrator

The circuit for the operational amplifier used to make a temperature-stable one-shot multivibrator (Figure 11-32) provides a highly repeatable trigger point independent of the input pulse rise time. Also, its output swing can drive FET switches or reed relays directly.

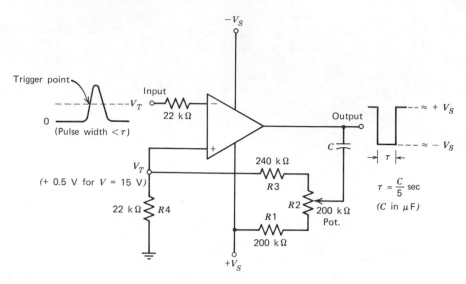

Figure 11-32   One-shot multivibrator.

Voltage divider R_1, R_2, R_3, and R_4 holds the input threshold V_T at a positive value. Therefore, when the input is at ground, the output is saturated positively. When the input level exceeds V_T, R_3 provides regeneration to switch the output to negative saturation. Capacitor C then charges exponentially such that V_T returns toward its normal positive value. When V_T passes through 0 V, regenerative action causes the output to rapidly return to positive saturation.

Because the inverting input must return to ground before timeout, an RC differentiator can be used if the input pulse is too long or if its base level is at some level other than ground. Pulse width, variable from microseconds to seconds, is proportional to the value of C; a value of 0.005 μF gives a pulse width of about 1 msec. Resistor R_2 provides a fine adjustment without affecting V_T.

Sine-Wave Oscillators

Although it is comparatively easy to build an oscillator that approximates a sine wave, making one that delivers a high-purity sinusoid with a stable frequency and amplitude is another story. Most satisfactory designs are relatively complicated and will not work without individual trimming and temperature compensation. In addition, they generally take a long time to stabilize to the final output amplitude.

A unique solution to most of these problems is presented in Figure 11-33, where A_1 is connected as a two-pole low-pass active filter and A_2 is connected as an integrator. Since the ultimate phase lag introduced by the amplifiers is 270°, the circuit can be made to oscillate if the loop gain is high enough at the frequency at which the lag is 180°. To ensure starting, the gain is actually made somewhat higher than is required for oscillation. Therefore, the amplitude builds up until it is limited by some nonlinearity in the system.

Amplitude stabilization is accomplished with zener clamp diodes D_1 and D_2. This does introduce distortion, but the effect is reduced by the subsequent low-

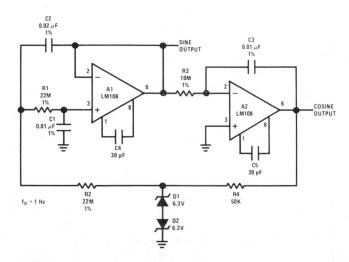

Figure 11-33   Sine-wave oscillator.

pass filters. If D_1 and D_2 have equal breakdown voltages, the resulting symmetrical clipping will virtually eliminate the even-order harmonics. The dominant harmonic is then the third, and this is about 40 dB down at the output of A_1 and about 50 dB down at the output of A_2. This means that the total harmonic distortion on each output is 1 and 0.3%, respectively.

The frequency of oscillation and the oscillation threshold are determined by R_1, R_2, R_3, C_1 and C_3. Therefore, precision components with low temperature coefficients should be used. If R_3 is made lower than the value shown, the circuit will accept looser component tolerances before dropping out of oscillation. The start-up will also be quicker. However, the price paid is an increase in distortion. The value of R_4 is not critical, but it should be made much smaller than R_2 so that the effective resistance at R_2 does not drop when the clamp diodes conduct.

The output amplitude is determined by the breakdown voltages of D_1 and D_2. Therefore, the clamp level should be temperature compensated for stable operation. Diode-connected (collector shorted to base) *npn* transistors with an emitter–base breakdown of about 6.3 V work well, since the positive temperature coefficient of the diode in reverse breakdown nearly cancels the negative temperature coefficient of the forward-biased diode. Added advantages of using transistors are that they have less shunt capacitance and sharper breakdowns than conventional zeners.

Another approach to generating sine waves appears in Figure 11-34. This circuit will provide both sine- and square-wave outputs for frequencies from below 20 Hz to above 20 kHz. The frequency of oscillation is easily tuned by varying a single resistor. This is a considerable advantage over Wein bridge circuits, where two elements must be tuned simultaneously to change frequency. Also, the output amplitude is relatively stable when the frequency is changed.

An operational amplifier is used as a tuned circuit, driven by a square wave from a voltage comparator. Frequency is controlled by R_1, R_2, C_1, C_2, and R_3, with R_3 used for tuning. Since tuning the filter does not affect its gain or band-

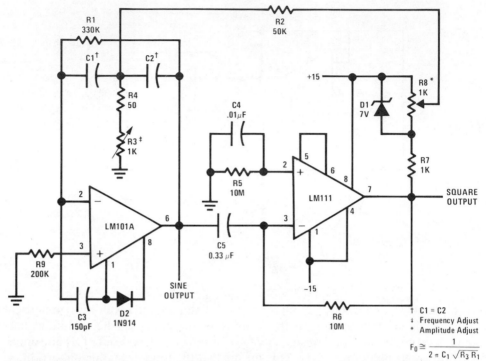

Figure 11-34   Easily tuned sine-wave oscillator.

width, the output amplitude does not change with frequency. A comparator is fed with the sine-wave output to obtain a square wave. The square wave is then fed back to the input of the tuned circuit to cause oscillation. Zener diode D_1 stabilizes the amplitude of the square wave that is fed back to the filter input. Starting is ensured by R_6 and C_5, which provide dc negative feedback around the comparator. This keeps the comparator in the active region.

If a lower-distortion oscillator is needed, the circuit in Figure 11-35 can be used. Instead of driving the tuned circuit with a square wave, a symmetrically clipped sine wave is employed. The clipped sine wave, of course, has less distortion than a square wave and yields a low distortion output when filtered. This circuit is less tolerant of component values than the one in Figure 11-34. To ensure oscillation, it is necessary to apply sufficient signal to the zeners for clipping to occur. Clipping about 20% of the sine wave is usually a good value. The level of clipping must be high enough to ensure oscillation over the entire tuning range. If the clipping is too small, it is possible for the circuit to cease oscillating because of tuning, component aging, or temperature changes. Higher clipping levels increase distortion. As with the circuit in Figure 11-34, this circuit is self-starting.

Table 11-1 lists the component values for the various frequency ranges. Distortion from the circuit in Figure 11-34 ranges from 0.7 to 2%, depending on the setting of R_3. Although greater tuning range can be accomplished by increasing the size of R_3 beyond 1 kΩ, distortion becomes excessive. Decreasing R_3 lower than 50 Ω can make the filter oscillate by itself. The circuit in Figure 11-35 varies between 0.2 and 0.4% distortion for 20% clipping.

About 20 kHz is the highest usable frequency for these oscillators. At higher

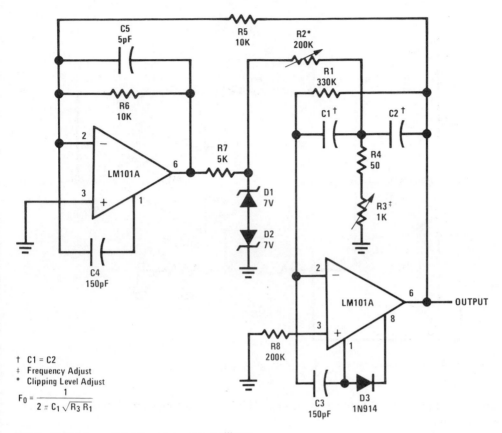

Figure 11-35 Low-distortion sine-wave oscillator.

† C1 = C2
‡ Frequency Adjust
* Clipping Level Adjust

$$F_0 = \dfrac{1}{2 \pi C_1 \sqrt{R_3 R_1}}$$

Table 11-1   Component Values for Various Frequency Ranges.

C_1, C_2	MIN. FREQUENCY	MAX. FREQUENCY
0.47 μF	18 Hz	80 Hz
0.1 μF	80 Hz	380 Hz
.022 μF	380 Hz	1.7 kHz
.0047 μF	1.7 kHz	8 kHz
.002 μF	4.4 kHz	20 kHz

frequencies, the tuned circuit is incapable of providing the high Q bandpass characteristic needed to filter the input into a clean sine wave. The low-frequency end of oscillation is not limited except by capacitor size. In both oscillators, feed-forward compensation is used on the LM101A amplifiers to increase their bandwidth.

Instrumentation Amplifier

The differential-input single-ended output instrumentation amplifier is one of the most versatile signal-processing amplifiers available. It is used for precision

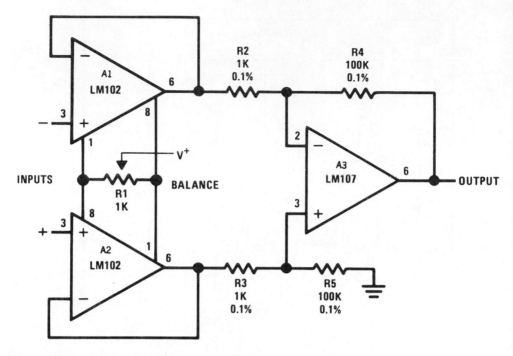

Figure 11-36   Differential-input instrumentation amplifier.

amplification of differential dc or ac signals while rejecting large values of common-mode noise.

Figure 11-36 diagrams a basic instrumentation amplifier which provides a 10-V output for 100-mV input, while rejecting greater than ±11 V of common-mode noise. To obtain good input characteristics, two voltage followers buffer the input signal. Amplifiers A_1, A_2, and A_3 should have a very high input impedance and low input currents. (The LM102 is specifically designed for voltage follower usage and has 10,000-MΩ input impedance with 3-nA input currents.) This high value of input impedance provides two benefits: it allows the instrumentation amplifier to be used with high source resistances and still have low error, and it allows the source resistances to be unbalanced by more than 10,000 Ω with no degradation in common-mode rejection. The followers drive a balanced differential amplifier, as in Figure 11-36, which provides gain and rejects the common-mode voltage. The gain is set by the ratio of R_4 to R_2 and R_5 to R_3. With the values shown, the gain for differential signals is 100. To obtain good CMRR, the ratio of R_4 to R_2 must match the ratio of R_5 to R_3. For example, if these resistors had a total mismatch of 0.1%, the common-mode rejection would be 60 dB times the closed-loop gain, or 100 dB.

For optimum performance, several items should be considered during construction. The R_1, used for zeroing the output, should be a high-resolution, mechanically stable potentiometer, to prevent a zero shift from occurring with mechanical disturbances. Since there are several ICs operating in close proximity, the power supplies should be bypassed with 0.01-μF disc capacitors to ensure stability. The resistors should be of the same type to have the same temperature coefficient.

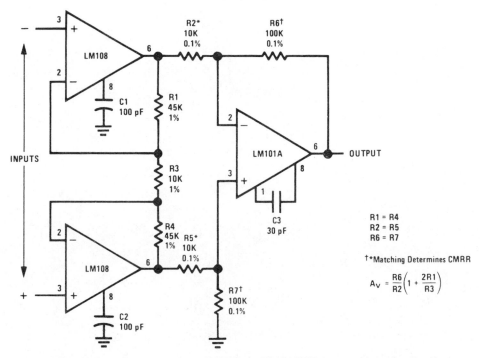

Figure 11-37   Differential-input instrumentation amplifier with high common-mode rejection.

A few applications for a differential instrumentation amplifier are: differential voltage measurements, bridge outputs, strain-gauge outputs, and low-level voltage measurement. Figures 11-37 and 11-38 depict several other instrumentation amplifier connections.

Analog-to-Digital Converter Ladder Network Driver

The use of the LM102 in a switch circuit for driving the ladder network in an analog-to-digital converter is schematized in Figure 11-39. Simple transistor

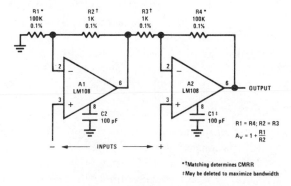

Figure 11-38   High-input-impedance instrumentation amplifier.

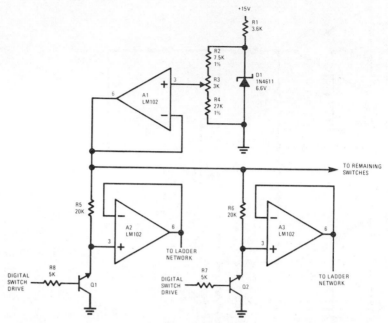

Figure 11-39   Using the LM102 to drive the ladder network in an analog-to-digital converter.

switches, connected in the reverse mode for low saturation voltage, generate the 0 and 5-V levels for the ladder network. The switch output is buffered by A_2 and A_3 to give a low driving impedance in both the high and low states.

The switch transistors can be driven directly from integrated logic circuits. Resistors R_7 and R_8 limit the base drive; the values indicated are for operation with standard TTL and DTL circuits. If necessary, the switching speed can be increased somewhat by bypassing the resistors with 100-pF capacitors.

Even with operation at maximum speed, clamp diodes are not needed on the voltage followers to reduce overshoot. The pull-up resistors on the switches R_5 and R_6 can be made large enough so that the LM102 does not see a positive-going input pulse that is much faster than the output slew rate.

The main advantage of this circuit is that it gives much lower output resistance than push-pull switches. Furthermore, the drive circuitry for these switches is considerably simpler. The LM102 can also be used as a buffer for the temperature-compensated voltage reference as in Figure 11-39. The output of the reference diode is divided down with a resistive divider, and it can be set to the desired value with R_3.

Analog Commutator

In the expandable four-channel analog commutator of Figure 11-40, two 7473 dual flip-flops form a 4-bit static shift register. The parallel outputs drive DM7800 level translators, which convert the TTL logic levels to voltages suitable for driving MOS devices: the MM451 four-channel analog switch. An extra gate on the input of the translator can be used, as shown, to shut off all the analog switches.

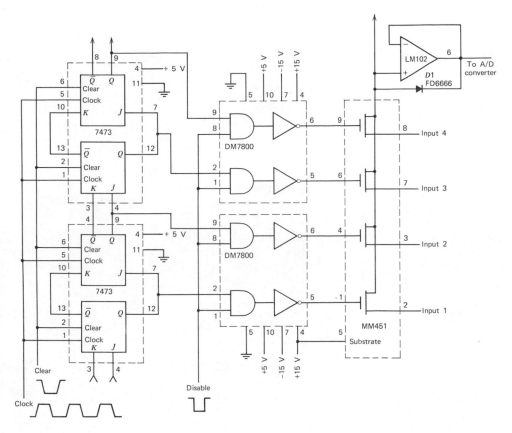

Figure 11-40   Analog commutator with buffered output.

In operation, a bit enters the register and cycles through at the clock frequency, turning ON each analog switch in sequence. The "clear" input is used to reset the register such that all analog switches are OFF. The channel capacity can be expanded by connecting registers in series and hooking the output of additional analog switches to the input of the buffer amplifiers. When the outputs of a large number of MOS switches are connected together, the capacitance on the output node can become high enough to reduce accuracy at a given operating speed. This problem can be avoided, however, by breaking up the total number of channels, buffering these segments with voltage followers, and subcommutating them into the analog-to-digital converter.

Integrated Nanoammeter Amplifier Circuit[1]

The schematic diagram of a nanoammeter amplifier is presented in Figure 11-41. In this circuit, the meter amplifier is a differential current-to-voltage converter with input protection, zeroing and full-scale adjust provisions, and input resistor

[1]Simple IC Meter Amplifier Circuit Measures 100 Nanoamps, Full-Scale, by Marvin Vander Kooi, *EDN/EEE*, April 15, 1972.

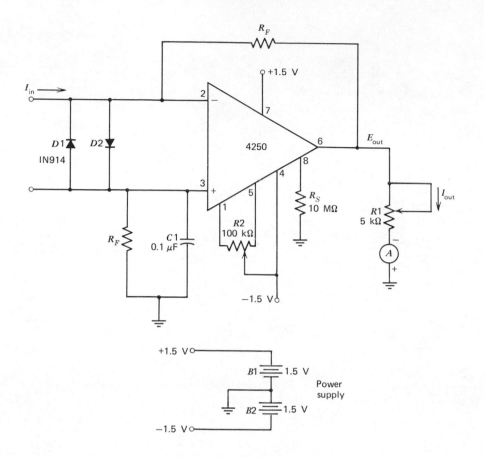

Figure 11-41   Circuit for a dc nanoammeter and microammeter.

Resistance Values for dc Nano-
and Microammeter

I Full Scale	R_F (Ω)	R_F (Ω)
100 nA	1.5 M	1.5 M
500 nA	300 k	300 k
1 μA	300 k	0
5 μA	60 k	0
10 μA	30 k	0
50 μA	6 k	0
100 μA	3 k	0

balancing for minimum offset voltage. Resistor R_F (equal in value to R_F for measurements of less than 1 μA) ensures that the input bias currents for the two input terminals of the amplifier do not contribute significantly to an output error voltage. The output voltage E_{out} for the differential current-to-voltage converter is equal to $-2I_F R_F$, since the floating input current I_{in} must flow through R_F and R'_F; R'_F may be omitted for R_f values of 500 kΩ or less because a resistance of this value contributes an error of less than 0.1% in output voltage. Potentiometer R_2 provides an electrical meter zero by forcing the input offset voltage V_{OS} to zero. Full-scale meter deflection is set by R_1. Both R_1 and R_2 only need to be set once for each op amp and meter combination. For a 50-μA, 2-kΩ meter movement, R_1 should be about 4 kΩ to give full-scale meter deflection in response to a 300-mV output voltage. Diodes D_1 and D_2 provide full input protection for overcurrents up to 75 mA.

With an R_F resistor value of 1.5 MΩ, the circuit in Figure 11-41 becomes a nanoammeter with a full-scale reading capability of 100 nA. The voltage drop across the two input terminals is then equal to the output voltage E_{out} divided by the open-loop gain. Assuming an open-loop gain of 10,000 gives an input voltage drop of 30 μV or less.

A unique feature of the 4250-type programmable operational amplifier is the provision for an external master bias-current setting resistor R_s at pin 8. With this resistor set at 10 MΩ, the total quiescent current drain of the circuit is 0.6 μA for a total power supply drain of 1.8 μW. The input bias current required by the amplifier at this low level of quiescent current is in the range of 600 pA. Notice that this circuit operates on two 1.5-V flashlight batteries and has quiescent power drain so low that an ON–OFF switch (for when the meter is not in use) is not required.

The circuit of Figure 11-41 can be easily modified to handle current readings higher than 100 μA (see Figure 11-42). In the second circuit, resistor R_A develops a voltage drop in response to input current I_A. This voltage is amplified by a factor equal to the ratio of R_F/R_B; R_B must be sufficiently larger than R_A that the input signal will not be loaded.

Electronic Thermometer

A new and unique use for operational amplifiers is the electronic thermometer (Figure 11-43). As shown, the sensor is attached to a bridge whose output goes to both a zero-crossing detector and a sample-and-hold circuit. Upon contacting the mouth, the thermistor in the sensor tip responds, unbalancing the bridge. When the sensor response passes 0 V – the bridge balance point that corresponds to the threshold temperature – the detector switches, turning on the 15-sec timer and connecting the output of the sample-and-hold circuit to the instrument's meter. Meanwhile, the timer's output closes the switch between the bridge and the sample-and-hold circuit. The sensor output then is connected by way of the bridge and the sample-and-hold circuit to the meter. After 15 sec, the timer turns OFF, disconnecting the sample-and-hold circuit from the bridge and turning ON the read lamp, a light-emitting diode. The temperature displayed by the meter equals T_{SS}. The sample-and-hold circuit maintains this reading for at least 7 sec after the read light comes ON.

The battery cutoff circuit continuously monitors the power supply for low output. If either the positive or negative battery voltage drops below a predetermined level at which accuracy could be degraded, the thermometer automatically turns OFF and cannot be used again until the batteries are replaced.

The instrument's sensor consists of a thermistor mounted in the tip along with a positive-temperature-coefficient resistor. This resistor compensates the total

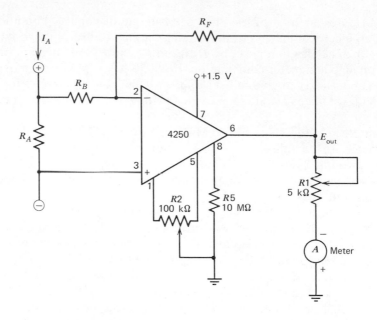

Resistance Values for dc Ammeter

I Full Scale	$R_A (\Omega)$	R_B (kΩ)	R_F(kΩ)
1 mA	3.0	3	300
10 mA	0.3	3	300
100 mA	0.3	30	300
1 A	0.03	30	300
10 A	0.03	30	30

Figure 11-42   Ammeter circuit for measurement of currents from 1 mA to 10 A.

sensor assembly for any lowering of tissue temperature occurring when the thermometer is inserted, as well as the sensor's initial temperature in the thermistor.

The circuit of Figure 11-43 works in the following manner. The bridge consists of a zero-adjustment potentiometer R_1 and resistors R_2, R_3, and R_4. Battery voltage to the bridge is regulated by zener diodes D_1 and D_2. Operational amplifier A_1, located at the bridge's output, provides a gain of 5 as well as high-input impedance, in order to protect the bridge from loading effects. The output of A_1 passes through sample switch Q_1 to charge-hold capacitor C_1 at the input of hold amplifier A_2. Like A_3, A_2 is an op amp, but it is connected in a voltage-follower configuration to give very high input impedance and low output impedance at a voltage gain of 1. Amplifier A_2 drives a series combination of potentiometer R_6, the meter, the meter switch Q_2, and the scaling resistor R_6.

The output of A_1 also goes to A_3, an open-loop amplifier that provides the zero-crossing detection. Since A_3's positive input terminal is referenced to ground, its output is positive when A_1's output is below ground and negative when A_1's output is above ground.

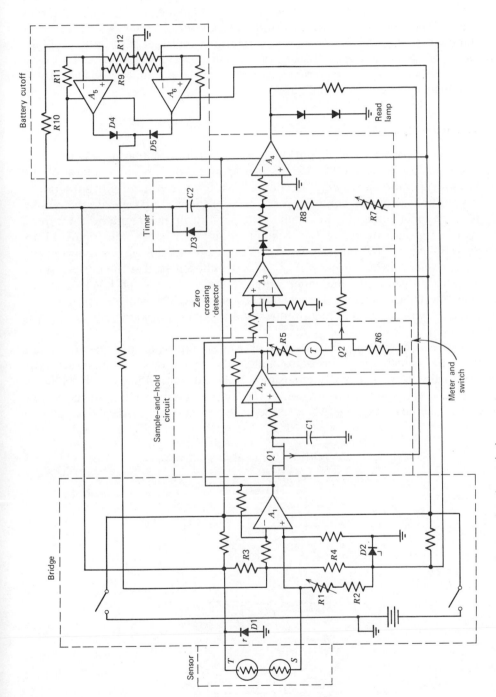

Figure 11-43 Schematic diagram of an electronic thermometer.

When A_3 switches to its negative state, Q_2 turns ON, allowing positive current to flow through the meter. Before it goes negative, A_3's output is more positive than the cathode of diode D_3; therefore, A_3 clamps the voltage across C_2 the timing capacitor. When A_3's output swings negative, C_2 charges through R_7 and R_8. When the junction of C_2 and R_8 reaches the reference voltage of A_4, the amplifier switches to a positive output, lighting the read lamp. The output of A_4 also turns OFF Q_1, maintaining the meter reading; the transistor's charge is sufficient to hold a constant current for a minimum of 7 sec.

Operational-Amplifier Voltage Regulator

The use of an operational amplifier as the precision comparing element in a voltage regulator is widely known. As in Figure 11-44, a basic regulator circuit employs an operational amplifier to compare a reference voltage with a fraction of the output voltage and to control a series-pass element Q_1 to regulate the output; Q_2 is used to increase the load-current capability of the regulator. A typical dissipative regulator using an IC as the error amplifier appears in Figure 11-45. The LM107 has built-in output current limiting as protection against burning out. At high output voltages, however, Q_2 must be chosen so that the current it can draw is held to within the maximum rating of the amplifier. The LM107 op amp also has internal frequency compensation. Capacitor C_4, typically rated in tens of microfarads, will supply the high-frequency components of load demand from its stored energy.

A low-frequency power transistor is chosen as Q_1, using the manufacturer's "safe area" specification curves to match the regulator's maximum current, voltage, and wattage needs. By choosing a transistor with f_t above 100 MHz for Q_2, the total phase-shift contribution of the Darlington pair is minimized at gain crossover, thereby lessening the closed-loop stability problem. Also, since emitter

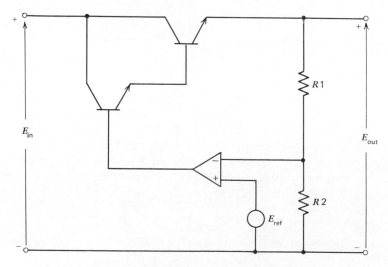

Figure 11-44   Basic operational-amplifier regulator circuit.

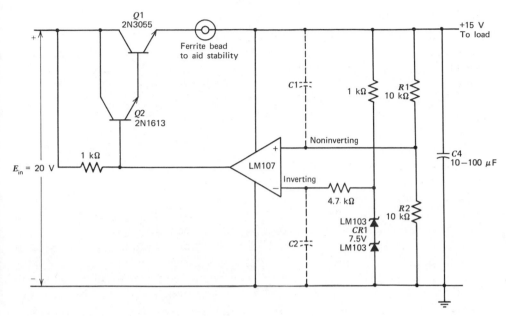

Figure 11-45   A typical dissipative regulator.

followers may oscillate because of the effects of inductance in the base circuit, the base lead lengths should be kept short and, if necessary, a small powered-magnetic bead or toroid can be added to the emitter circuit.

Voltage dividers R_1 and R_2 form a sampling network by selecting a fraction of the output voltage for comparison with the reference voltage. This fractional load voltage sample can be adjusted by a trimmer; the sample appears at the noninverting operational-amplifier terminal. Stable resistors must be used in the divider — the temperature coefficient of the voltage divider is as important as that of the zener diode D_1. The dashed lines in Figure 11-45 indicate optional circuitry — C_2 for noise suppression, C_1 to aid in loop stabilization, and C_1, R_3 and C_3 to minimize load voltage overshoot when driving a fast-switching load.

At its input terminals, the operational amplifier is a differential amplifier operated at an emitter current of about 15 μA, in order to minimize noise and base current and to ensure high input impedance. The impedance of the sampling network is a compromise: the current drawn must not be excessive — 1 mA of "bleeder" current is typically acceptable. But at the same time, it must be large enough to swamp out any rise in operational-amplifier input current with a rise in temperature, thus limiting the regulator's temperature sensitivity. The inverting terminal has a resistance, R_4 which is approximately equal to R_1 in parallel with R_2, inserted in series for the same purpose. Other elements are C_2, a noise filter capacitor, typically 0.01 μF, and C_1, a phase-lead-compensating capacitor, usually between 5 and 20 pF.

Short-circuit load protection for the series-regulating element can be added as fold-back current limiting, a technique in which load voltage is sampled to sense overcurrent and the series-pass transistor is turned OFF.

Dual-Voltage Regulator

A low-cost dual-voltage regulator utilizing the common-mode rejection properties of two operational amplifiers to eliminate the need for a second reference diode may be easily fabricated (see Figure 11-46). With this particular circuit, the total change in output voltage for positive or negative input voltages from 14 to 18 V, and for load currents from 0 to 45 mA, is only 40 mV. The load current changed caused only 2 mV of the 40-mV output voltage change.

A variation of the dual voltage regulator[2] of Figure 11-46 is shown in Figure 11-47. In this circuit, precision positive and negative voltages are supplied, which track each other in absolute value and are stable over wide ranges of temperature, input supply conditions, and output load conditions.

The dual voltage regulator of Figure 11-47 operates in the following manner: A constant current is supplied to the precision, aged, temperature-compensated Zener diode CR_1 by means of transistor Q_1 and associated components; variations of the Zener current due to the effect of temperature fluctuations on the V_{be} of Q_1 are automatically compensated for by changes in the value of the silicon resistor R_1(sensistor). The voltage drop across CR_1 is compared by means of the high-gain monolithic operational amplifier A_1 with a similar voltage obtained by means of a voltage divider across the $+V$ output and the common terminal. The extremely high gain of the amplifier (typically in excess of 150,000) causes the difference between these voltages to be very small; however, any variation between the compared voltages is supplied as an error signal by A_1 to the series transistor Q_2. The sense of the error signal causes the voltage drop across Q_2 to decrease if the output voltage becomes lower or to increase if the output voltage rises.

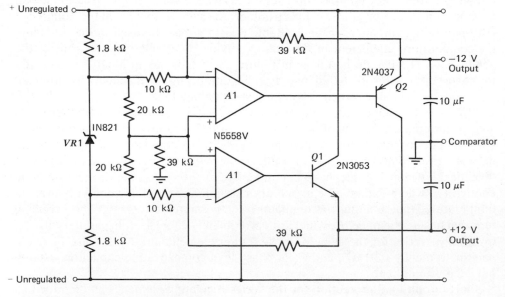

Figure 11-46   A dual-voltage regulator.

[2]NASA Tech Brief B72-10097, *Precision Voltage Regulator*, 1972.

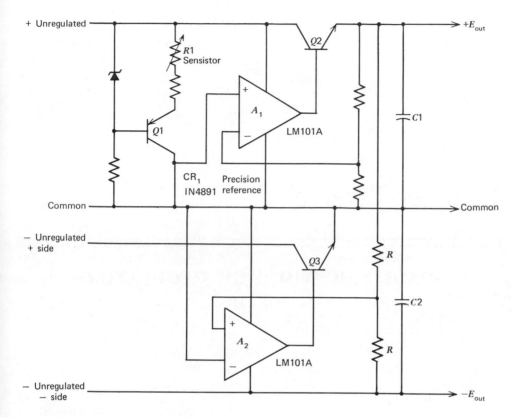

Figure 11-47   A precision dual voltage regulator.

Operational amplifier A_2 is identical with A_1 and operates analogously; it compares the difference between the negative and positive output voltages with the output common point, thus causing the negative output voltage to track the positive output voltage. Error signals from A_2 control the series transistor Q_3 in the same manner that those from A_1 control Q_2.

Capacitors C_1 and C_2 reduce the output impedance at higher frequencies. To obtain maximum performance, all comparison and division resistors must be precision-wire-wound, low-temperature-coefficient types, and location of the voltage sensing points must be carefully controlled.

The foregoing examples were chosen because they are typical of operational-amplifier applications circuits. Of course, because of the many uses of operational amplifiers, this chapter could be expanded into a separate volume.

Chapter Twelve

MONOLITHIC VOLTAGE REGULATORS

Next to the operational amplifier, the voltage regulator is the most common linear IC building block. The advent of the monolithic voltage regulator has added a new dimension to regulation. The monolithic regulator offers the performance to replace discrete assemblies or portions of discrete assemblies in conventional central regulators. In addition, the small size, low cost, and ease of use uniquely fit the monolithic units for point-of-use regulation. With the many options available with IC regulators, it behooves an engineer to evaluate the possibility of employing such devices in his circuit.

The small size and low cost certainly make the monolithic regulator uniquely suited for local regulation. The extremely high-speed capability of these devices means that effective local regulation of communications equipment and computer boards is now practical. Hand-in-hand with these technical and price advantages is a significant reduction in the engineering effort necessary to design regulated power supplies.

Instead of a major engineering project, design of a power supply with ICs is literally a matter of minutes. The user need only determine the current and voltage requirements, in order to select the most economical unit that will meet his requirements. Once the right regulator and package have been chosen, the only remaining design effort involves calculating the appropriate values for the voltage-setting resistors (to provide the required output voltage) and for the current-limiting resistor (to protect against overloads). Conservatively, the first time a designer uses a monolithic regulator, this procedure may take as long as 10 min.

When higher current values are required, or if a switching regulator is desired, the design effort will be necessarily greater to select the external circuitry, but the time needed for the undertaking is still substantially less than the time required to start from scratch with a do-it-yourself regulator.

The primary function of any voltage regulator is to hold the voltage in its output circuit at a predetermined value over the expected range of load currents. Working against the regulator are variations in load current, input voltage, and temperature. The degree to which a regulator can maintain a constant voltage in the face of these variations is the basic figure of merit. Although there is a degree of interaction between these performance-degrading factors, especially between output current and temperature, it is most convenient to consider their effects separately.

DEFINITION OF TERMS

Load Regulation

Load regulation is defined as the percentage change in regulated output voltage for a change in load from minimum load current to the maximum load current specified. The general formula for load regulation is

$$\text{load regulation } (\%) = \frac{E_{\text{out(min)}} - E_{\text{out(max)}}}{E_{\text{out(min)}}} \times 100 \qquad (12\text{-}1)$$

where $E_{\text{out(min)}}$ = output voltage with minimum rated load
$E_{\text{out(max)}}$ = output voltage with maximum rated load

Line Regulation

Line regulation, defined as the percentage change in regulated output voltage for a change in input voltage, is given by

$$\frac{\%}{V} = \left(\frac{\Delta E_{\text{out}}}{\Delta E_{\text{in}} \times E_{\text{out}}}\right) \times 100 \qquad (12\text{-}2)$$

where ΔE_{out} = change in output voltage E_{out} for a change in input voltage ΔE_{in}
ΔE_{in} = change in input voltage
E_{out} = nominal output voltage
A low value of line regulation is desirable.

Ripple Rejection

Ripple rejection is ac line regulation. Whereas line regulation is measured at dc, ripple rejection is defined as the line regulation for ac input signals at or above a given frequency with a specified value of bypass capacitor on the reference bypass terminal.

Standby (Quiescent) Current Drain

Standby current drain is the current that flows into the regulator and through to ground. It does not include any current drawn by the load or by external resistor networks.

Temperature Drift

Temperature drift is the percentage change in output voltage (or reference voltage) for a temperature variation from room temperature to either temperature extreme. Approximately 85 to 90% of this drift occurs in the regulator's internal reference circuitry. The remaining 10 to 15% is due to error amplifier or bias current drift. Temperature drift is the major cause of output voltage change in an IC regulator.

BASIC REGULATOR TYPES

Figure 12-1 depicts a typical circuit of a generalized series-dissipative regulator. An error amplifier converts the difference between the output sample and the reference voltage into an error signal that controls the voltage dropped by the series-pass transistor. In a series-dissipative regulator, the excess voltage is dissipated in the series transistor, which acts as a variable resistor. These losses result in a low circuit efficiency.

DC voltage regulation is accomplished most efficiently by nondissipative or switching regulators. All high-efficiency regulators use a high-frequency switch (the higher the frequency, the smaller the components, but the more filtering required), a choke (energy-storage device), a filter capacitor, and a flyback diode to provide a path for the choke current during the period when the switch is off.

The nondissipative (switching) regulator has higher efficiency and lower weight than the dissipative regulator. It allows a designer to use smaller components and

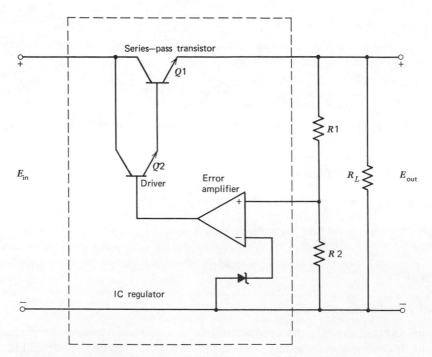

Figure 12-1   Series-dissipative regulator.

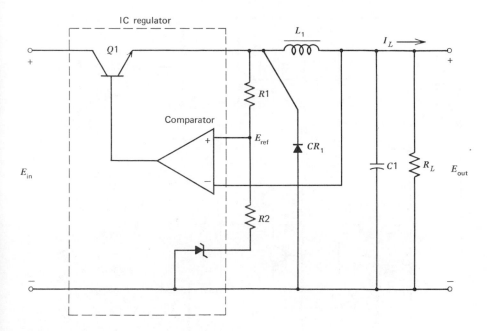

Figure 12-2 Basic switching regulator circuit.

heat sinks to provide a specified output power capability than he would need with a series regulator. The basic configuration of the switching regulator is presented in Figure 12-2.

Regulation occurs by controlling the duty cycle of Q_1. Transistor Q_1 serves as a switch and is either ON (saturated) or OFF, so that power dissipation is at a minimum. Free-wheeling diode CR_1 conducts during the time that Q_1 is cut OFF, thus maintaining current flow through inductor L_1. When Q_1 is ON, CR_1 is reverse biased and does not conduct. The load current I_L through L_1 increases according to the relationship

$$E_{in} - E_{out} = L_1 \frac{(\Delta i_L)}{t_{on}} \qquad (12\text{-}3)$$

and it flows through the load and charges capacitor C_1. When E_{out} reaches E_{ref}, the voltage comparator turns Q_1 OFF. The current through L_1 then decreases until CR_1 is forward biased. At this point the inductor current flows through CR_1 and decreases at a rate given by

$$E_{out} = L_1 \frac{(\Delta i_L)}{(t_{off})} \qquad (12\text{-}4)$$

When the inductor current falls below the load current, the output capacitor begins to discharge and E_{out} decreases. When E_{out} decreases to slightly less than E_{ref}, the voltage comparator turns Q_1 back ON and the cycle repeats itself. The output voltage is given by

$$E_{out} = \frac{E_{in} \cdot t_{on}}{t_{on} + t_{off}} \qquad (12\text{-}5)$$

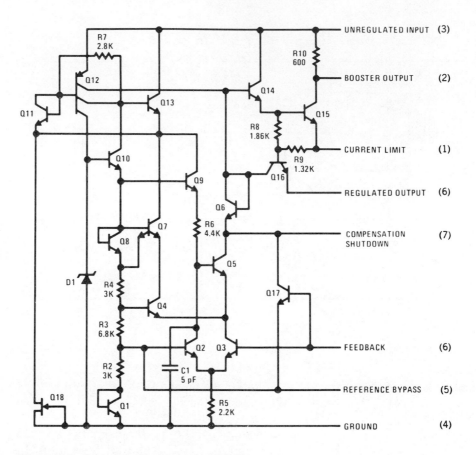

Figure 12-3   Schematic diagram for the LM105.

INTEGRATED CIRCUIT REGULATORS

In Figures 12-1 and 12-2, E_{out} is compared before integration with a reference voltage by an operational amplifier driving the pass transistor. IC regulators include the amplifier and the pass transistor. Most regulators have an internal voltage reference, which may be supplemented in some applications by an external zener diode. In some of the regulator application examples to be presented, the LM105 regulator is used. The schematic of the LM105 is shown in Figure 12-3.

Briefly, the left-hand side of the LM105 is a constant-current source and zener diode with temperature compensation. It provides a constant E_{ref} of 1.8 V to a two-stage, high-gain error amplifier formed in the center of the chip by Q_2–Q_3 and Q_4–Q_5. The amplifier controls the output stage. Maximum output current is 20 mA, which can be boosted to several amperes by driving one or more external transistors with the booster output. A stabilizing capacitor is generally placed between pins 6 and 7 when the IC is used as a linear regulator.

In order to facilitate comprehension of the ease of designing with IC regulators,

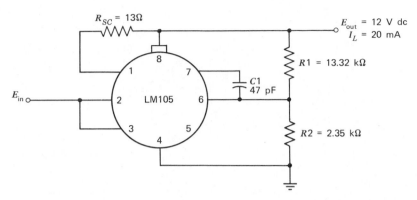

Figure 12-4   Basic LM105 regulator circuit.

a practical example is presented. A linear regulator is desired which has the following specifications:

$$
\begin{array}{rl}
E_{\text{in}}: & \text{22 to 30 V dc} \\
E_{\text{out}}: & \text{12 V dc} \\
I_L: & \text{10 mA} \\
\text{temperature range}: & -55 \text{ to} +125°\text{C} \\
\text{current limit}: & \text{at 25 mA}
\end{array}
$$

These specifications dictate the choice of an LM105 type of IC regulator. The basic regulator circuit is diagrammed in Figure 12-4. In this circuit, the output voltage is set by R_1 and R_2. The resistor values are selected based on a feedback voltage of 1.8 V to pin 6 of the LM105. To keep thermal drift of the output voltage within specifications, the parallel combination of R_1 and R_2 should be about 2 kΩ. The values of R_1 and R_2 are easily found from Figure 12-5. For an output voltage of 12 V dc, we have

$$R_2 \simeq 2.35 \text{ kΩ} \tag{12-6}$$

and

$$R_1 = 1.11 \, E_{\text{out}} = 1.11 \, (12) = 13.32 \text{ kΩ} \tag{12-7}$$

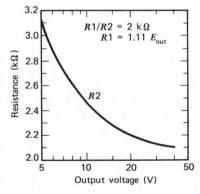

Figure 12-5   Optimum divider resistance values.

If we wish to use current limiting in the circuit, resistor R_{SC} must be added. The value of R_{SC} is computed from the formula

$$R_{SC} = \frac{325}{I_{SC}} \qquad (12\text{-}8)$$

In the present example, $I_{SC} = 25$ mA. Thus we can write

$$R_{SC} = \frac{325}{25} = 13 \ \Omega$$

From these routine calculations, it is seen that designing with the LM105 IC regulator is a matter of minutes.

Figure 12-6 indicates how an external pass transistor is added to the basic regulator circuit of Figure 12-4, to increase the load-current capability of the regulator. In the case of Figure 12-6, the load current capability has been increased from 20 to 500 mA by the addition of external transistor Q_1. In this circuit, capacitor C_2 is required to suppress oscillations in the feedback loop involving the external booster transistor Q_1 and the output of the LM105; C_3 is added if the lead length between the regulator and the power supply is long.

Figures 12-7 and 12-8 are the schematic diagram and the basic connection diagram, respectively, of the 723, another widely accepted IC regulator. Figure 12-9 shows how an external pass transistor is added to the basic 723 regulator circuit of Figure 12-8 to increase its load-current capability. To convert the IC to a switching regulator, E_{ref} and E_{out} (Figure 12-2) are permitted to alternately imbalance the positive and negative feedback so that the operational amplifier goes into oscillation at a high frequency. The optimum switching frequency for switching regulators has been determined to be between 20 and 100 kHz. At lower frequencies, the core becomes unnecessarily large; and at higher frequencies, switching losses in Q_1 and D_1 become excessive. It is important, in this respect, to minimize switching losses by assuring that both Q_1 and D_1 are fast-switching devices.

To obtain a switching regulator with the same specifications as before, simply modify the circuit of Figure 12-6 by the addition of a choke L_1, and a free-wheeling diode, D_1. The circuit that would result is shown in Figure 12-10. In this circuit, R_3 determines the base drive for the switch transistor Q_1, providing enough drive to saturate it with maximum load current. Capacitor C_2 minimizes output ripple,

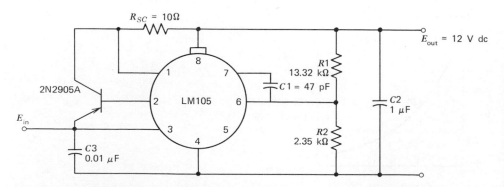

Figure 12-6   A 500-mA regulator circuit.

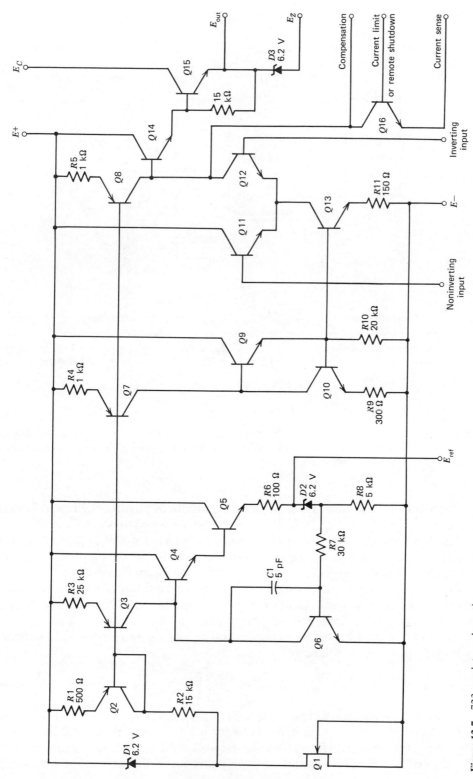

Figure 12-7 723 regulator schematic.

337

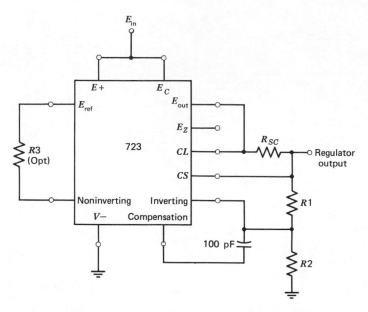

Typical Performance		Design Guidelines	
Regulator output voltage	15 V	E_{out} range	7–37 V
Line regulator	1.5 mV	E_{in} range	9.5–40 V
($\Delta E_{in} = 3$ V)		$E_{out} = [(R_1 + R_2)/R_2] \times 7$ V;	
Load regulator	4.5 mV	$R_3 = R_1 R_2/(R_1 + R_2)$ for minimum drift.	
($\Delta I_L = 50$ mA)			

Figure 12-8 Medium-current 723 regular configuration for higher output voltages. Resistor attenuator regulates levels about internal reference voltage. Feedback capacitor in frequency-compensation network uses breakpoint provided by R_1.

C_3 removes fast-rise time transients, and R_4 produces positive feedback. Losses are kept low by Q_1 and D_1, which are both fast-switching devices.

The following equations are used to calculate the values of R_4, L_1, and C_1, given the switching frequency and the output ripple. These equations provide a starting point for designing a switching regulator to fit a given application.

The output ripple of the regulator at the switching frequency is mainly determined by R_4. It should be evident from the description of circuit operation that the peak-to-peak output ripple will be nearly equal to the peak-to-peak voltage fed back to pin 5 of the LM105. Since the resistance looking into pin 5 is approximately 1000 Ω, this voltage will be

$$\Delta E_{ref} \simeq \frac{1000\, E_{in}}{R_4} \tag{12-9}$$

In practice, the ripple will be somewhat larger than this. When the switch transistor shuts OFF, the current in the inductor will be greater than the load current; thus the output voltage will continue to rise above the value required to shut OFF the regulator. An important consideration in choosing the value of the inductor is

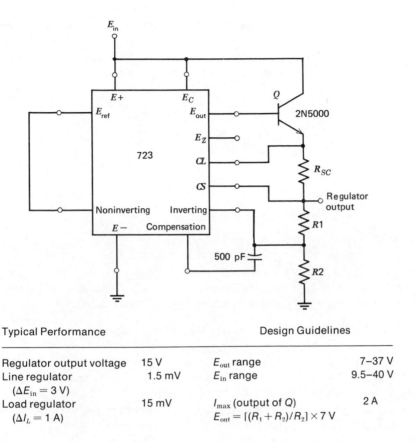

Typical Performance		Design Guidelines	
Regulator output voltage	15 V	E_{out} range	7–37 V
Line regulator	1.5 mV	E_{in} range	9.5–40 V
($\Delta E_{in} = 3$ V)			
Load regulator	15 mV	I_{max} (output of Q)	2 A
($\Delta I_L = 1$ A)		$E_{out} = [(R_1 + R_2)/R_2] \times 7$ V	

Figure 12-9 In this high-current 723 regulator, an external *npn* pass transistor boosts power capabilities and accommodates ampere outputs.

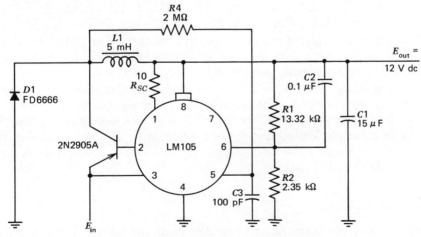

Figure 12-10 A 500-mA switching regulator using the LM105.

that it be large enough that the current through it does not change drastically during the switching cycle. If it does, the switch transistor and the catch diode must be able to handle peak currents which are significantly larger than the load current. The change in inductor current can be written as

$$\Delta I_L \simeq \frac{E_{out} \cdot t_{off}}{L} \qquad (12\text{-}10)$$

In order for the peak current to be about 1.2 times the maximum load current, we must have

$$L_1 = \frac{2.5 E_{out} \cdot t_{off}}{I_{out(max)}} \qquad (12\text{-}11)$$

A value for t_{OFF} can be estimated from

$$t_{off} = \frac{1}{f}\left(1 - \frac{E_{out}}{E_{in}}\right) \qquad (12\text{-}12)$$

where f is the desired switching frequency and E_{in} is the nominal input voltage. The size of the output capacitor can now be determined from

$$C_1 = \frac{E_{in} - E_{out}}{2 L_1 \, \Delta E_{out}}\left(\frac{E_{out}}{f E_{in}}\right)^2 \qquad (12\text{-}13)$$

where ΔE_{out} is the peak-to-peak output ripple and E_{in} is the nominal input voltage. The values of R_1 and R_2 are calculated as before.

Having these basic IC voltage regulator circuits in mind, we can use them in a myriad of applications. Some of these are now presented.

INTEGRATED CIRCUIT REGULATOR APPLICATIONS

Constant-Current Source

The circuit in Figure 12-11 utilizes the LM105 as a 1-A constant-current source. Here the LM105 regulates the emitter current of a Darlington-connected transistor, and the output current is taken from the collectors. The use of a Darlington connection for Q_1 and Q_2 improves the accuracy of the circuit by minimizing the base current error between the emitter and collector current.

The output of the LM105, which drives the control transistors, must be protected against short circuits with R_6 to limit the current when Q_2 saturates. The minimum load current for the IC is provided by R_7; D_1 is included to absorb the kickback of inductive loads when power is shut OFF. The output current of the circuit is adjusted with R_2. When used as a current source, the LM105 is capable of driving servo control motors.

Temperature Controller

A circuit for an oven-temperature controller using the LM105 is given in Figure 12-12. Temperature changes in the oven are sensed by a thermistor. This signal

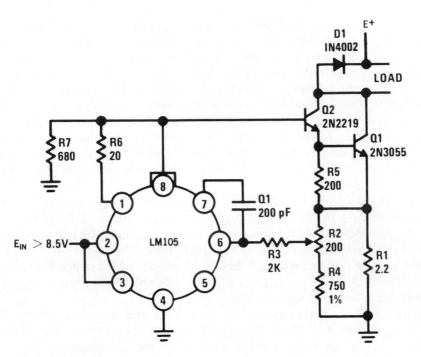

Figure 12-11 A 1-A current source using the LM105.

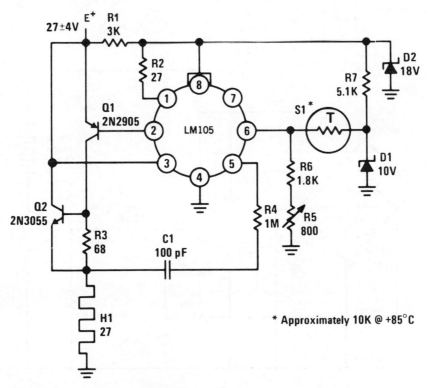

Figure 12-12 Switching-temperature controller.

341

is fed to the LM105, which controls power to the heater by switching the series-pass transistor Q_2 ON and OFF. Since the pass transistor will be nearly saturated in the ON condition, its power dissipation is minimized. In operation, if the oven temperature should try to increase, the thermistor resistance will drop, increasing the voltage on the feedback terminal of the regulator. This action shuts off power to the heater. The opposite would be true if the temperature dropped.

Variable-duty-cycle switching action is obtained by applying positive feedback around the regulator from the output to the reference bypass terminal (which is also the noninverting input to the error amplifier) through C_1 and R_4. When the circuit switches ON or OFF, it will remain in that state for a time determined by this RC time constant.

Additional details of the circuit are that base drive to Q_1 is limited, to a value determined by R_2, by the internal current-limiting circuitry of the LM105. Moreover, D_2 provides a roughly regulated supply for D_1 in addition to fixing the output level of the LM105 at a level which properly biases the internal transistors. The reference diode for the thermistor sensor D_1 need not be a temperature-compensated device as long as it is put in the oven with the thermistor. Finally, the temperature is adjusted with R_5.

Figure 12-13 is another circuit for an oven-temperature controller using the 723 voltage regulator. This circuit will control the temperature of an oven within $\pm 1°C$ over an ambient swing of $-50°C$ to the temperature of the oven. In addition, if we select R_2 to provide adequate drive to Q_1, the input voltage can vary from 10 to 37 V. Because transistor Q_1 is saturated when ON, no heat sink is required.

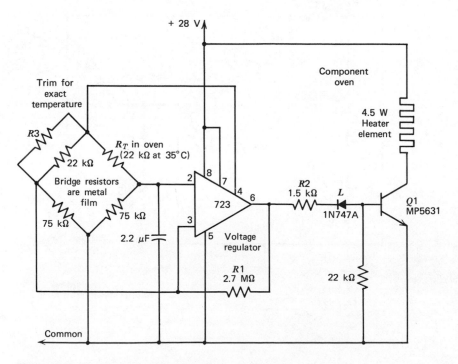

Figure 12-13   Oven-temperature controller.

Resistor R_1 sets the trip-point hystersis at 2°C with the circuit values indicated. If the heater current is less than 150 mA, it can be supplied directly from the 723 by connecting the element between pin 6 and the common line.

Power Amplifier

In Figure 12-14 the LM105 is used as a high-gain amplifier and connected to a quasi-complementary power-output stage. Feedback around the entire circuit stabilizes the gain and reduces distortion. In addition, the regulation character-istics of the LM105 are used to stabilize the quiescent output voltage and to minimize ripple feedthrough from the power supply.

The LM105 drives the output transistors Q_5 and Q_6 for positive-going output signals, whereas Q_1, operating as a current source from the 1.8 V on the reference terminal of the LM105, supplies base drive to Q_3 and Q_4 for negative-going sig-nals. Transistor Q_2 eliminates the dead zone of the class-B output stage, and it is bypassed by C_5 to present a lower driving impedance to Q_3 at high frequencies. The voltage drop across Q_2 will be a multiple of its emitter–base voltage, deter-mined by R_9 and R_{10}. These resistors can therefore be selected to give the desired quiescent current in Q_4 and Q_6. It is important that Q_2 be mounted on the heat sink with the output and driver transistors to prevent thermal runaway.

Output-current limiting is obtained with D_2 and D_3. Diode D_2 clamps the base drive of Q_3 when the voltage drop across R_6 exceeds one diode drop, and D_3 clamps the base of Q_5 when the voltage across R_7 becomes greater than two diode

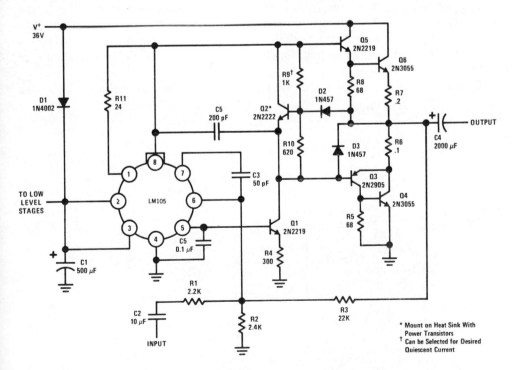

Figure 12-14 Power amplifier with current limiting.

drops. When D_3 becomes forward biased, R_{11} is needed to limit the output current of the LM105. The power-supply ripple is peak detected by D_1 and C_1 to obtain increased positive output swing by operating the LM105 at a higher voltage than Q_5 and Q_6 during the troughs of the ripple. This also reduces the ripple seen by the LM105. Capacitor C_5 bypasses any zener noise on the reference terminal of the LM105 that would otherwise be seen on the output.

The quiescent output voltage is set with R_2 and R_3 in the same way as with a voltage regulator. The ac voltage gain is determined by the ratio of R_1 and R_3, since the circuit is connected as a summing amplifier.

High-Voltage Regulator

IC regulators were designed primarily for applications with output voltages below 80 V. However, they can be used as high-voltage regulators under certain circumstances. An example of this, a circuit regulating the output of a 2-kV supply, appears in Figure 12-15. The LM105 senses the output of the high-voltage supply through a resistive divider and varies the input to a dc/dc converter, which generates the high voltage. Hence, the circuit regulates without having any high voltages impressed across it.

Under ordinary circumstances, the feedback terminal of the LM105 wants to operate from a 2-kΩ divider impedance. Satisfying this condition on a 2-kV regulator would require that about 2 W be dissipated in the divider. This, however, is reduced to 40 mW by the addition of Q_1, which acts as a buffer for a high-impedance divider, operating the LM105 from the proper source resistance. The other half of the transistor Q_2, compensates for the temperature drift in the emitter–base voltage of Q_1, so that it is not multiplied by the divider ratio. The circuit does have an uncompensated drift of 2 mV/°C; but since this is added directly to the output, and not multiplied by the divider ratio, it will be insignificant with a 2-kV regulator.

A variation of the high-voltage regulator is diagrammed in Figure 12-16. In

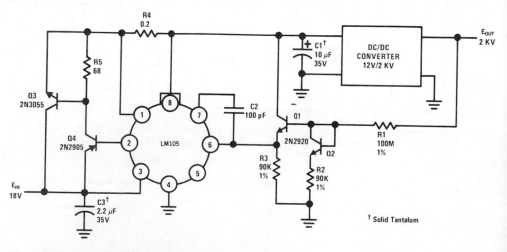

Figure 12-15   The LM105 used as a high-voltage regulator.

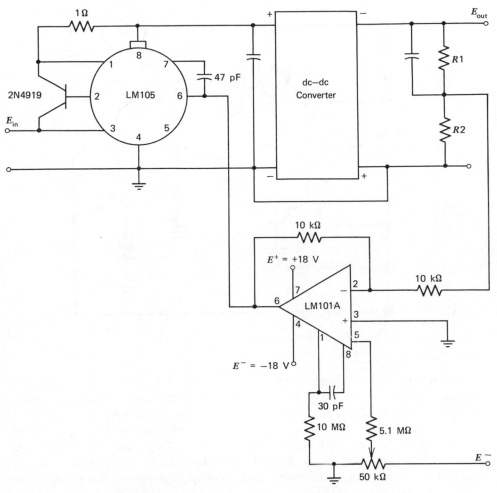

Figure 12-16   A high-voltage regulator using the LM101A and LM105.

this circuit the LM101A compensates for temperature drifts in the dc/dc converter and in the LM105; it also improves temperature stability. The output voltage
is set by the values of R_1 and R_2. If we desired an adjustable output voltage, a
potentiometer could be inserted between R_1 and R_2. If a 10-kΩ potentiometer
were inserted, the output could be adjusted from about 400 to about 1500 V. For
an output voltage of -500 V, the line and load regulation are better than 0.1% for
line voltage from 10 to 24 V and load currents to 10 mA. Temperature stability
is 0.2%/°C over the temperature range -55 to $+125$°C.

Light-Intensity Regulator

Figure 12-17 gives the circuit for a light-intensity regulator using the LM105. A
phototransistor senses the light level and drives the feedback terminal of the
LM105 to control current flow into an incandescent bulb; R_1 serves to limit the
inrush current to the bulb when the circuit is first turned on.

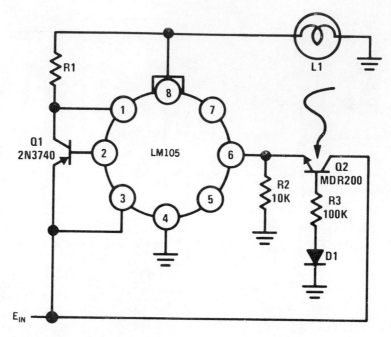

Figure 12-17 Light-intensity regulator.

The current gain of the phototransistor Q_2 is fixed at 10 (to make it less temperature sensitive) by R_3 and the temperature-compensating diode D_1. A photodiode, such as the IN2175, could be substituted for the phototransistor if it had sufficient light sensitivity, and R_3 and D_1 could be eliminated. The input voltage does not have to be regulated because the sensitivity of a phototransistor or photodiode is not greatly affected by the voltage drop across it. A photoconductor can also be used in place of the phototransistor, except that input voltage would have to be regulated.

Tracking Regulators

In systems that have more than one regulated-output voltage, it is desirable to adjust all supplies with a single potentiometer. In Figure 12-18 a single potentiometer controls both outputs of the 5- and 15-V regulators. To ensure that both regulators operate with the same reference voltage, their internal reference voltages are tied together (pin 5). Resistors R_2 and R_{2_1} are fixed at 2 kΩ. Resistor R_1 is calculated for an output voltage using 1.6 V as a reference voltage. The following expression is used to determine either R_1 or R_{1_1}:

$$(R_1)(R_{1_2}) = \frac{(E_{\text{out}} - 1.6)\,2000}{1.6} \qquad (12\text{-}14)$$

Potentiometer R_5 adjusts both regulators within 2% of the desired output for reference variations from 1.6 to 2 V. If the reference is 2 V, R_5 is 324 Ω and the output voltages are 5.1 and 14.9 V. If the reference is approximately 1.8 V, both outputs are within 1% nominal.

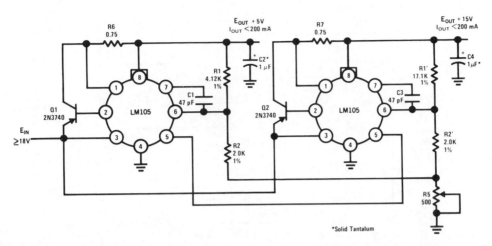

Figure 12-18   A tracking regulator.

Potentiometer R_5, connected between the lower section of the output voltage divider network and ground, compensates for any variations from the nominal 1.8-V reference of the regulators. Note that the wiper of R_5 is connected to one side of the potentiometer. If a rheostat connection were used, the arm might open-circuit during adjustment, causing large transients on the output.

If 1% resistors are used, there is an additional worst-case error of 2% of each regulator. Resistor errors are inherent in any type of tracking regulator system, even if the adjustment is proven to be exact theoretically.

Negative Regulators

If a negative voltage is the primary voltage source rather than a positive voltage, the 723 may be connected as in Figure 12-19. The other option is to replace the LM105 by the LM104 negative regulator in the previous applications.

Figure 12-20 depicts a power supply that uses both the LM105 and LM104 to supply ±15 V dc for driving operational amplifiers. These two regulators can be made to track each other, in a method similar to that of Figure 12-18.[1]

Low-Cost Switching Regulator

Figure 12-21 is a circuit schematic diagram of a high-efficiency low-cost IC switching regulator designed for portable equipment using the LM376 regulator. This switching regulator combines high efficiency with low cost. Normally, switching regulators require fast-switching diodes and transistors and a core (choke) with a sharp saturation characteristic to minimize losses. The circuit of Figure 12-21 works at relatively high efficiency (87%) yet does not require fast components or sharp saturation cores. Instead, almost any diode, transistor, and choke can be used efficiently. In fact, an inexpensive Japanese transformer

[1]*Linear Applications*, National Semiconductor Corporation, January 1972.

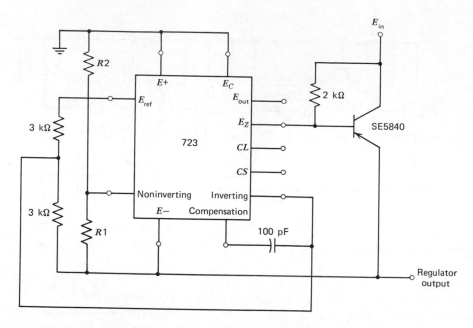

Typical Performance		Design guidelines	
Regulated output voltage:	-15 V	$E_{out(min)}$ range:	-15 V
Line regulation:	1.0 mV	$E_{in(min)}$ range:	-20 V
($\Delta E_{in} = 3$ V)			
Load regulation:	2.0 mV	$I_{out(max)}$:	1 A
($\Delta I_L = 100$ mA)		$E_{out} = 3.5(R_2 + R_1)R_1$	

Figure 12-19 Using the 723 as a negative regulator.

secondary can act as the choke, which points up the circuit's low cost, versatility, and performance.

Also, the high switching frequency (50 to 100 kHz) generates transients that could be troublesome. Because of the inherent design of the LM376N (for voltage-breakdown ratings of the integrated transistors) and the selective filtering employed, however, spikes are not a problem.

Almost any type of choke can be used for L_1, which gives a significant cost savings. Diode CR_1 can be any diode (even a low-frequency type) that can handle the actual load current, and transistor Q_1 can be any variety of transistor that is capable of handling the load current and remaining within its maximum ratings. Even though it was previously recommended that a fast-switching diode be used, a low-frequency diode will work in the circuit of Figure 12-21. A switching frequency of 33 kHz was selected for this regulator.

Low-r-f-Noise Switching Regulator

Figure 12-22 is the complete schematic diagram of a low–r-f-noise switching regulator. This circuit is more complex than the previous switching regulator

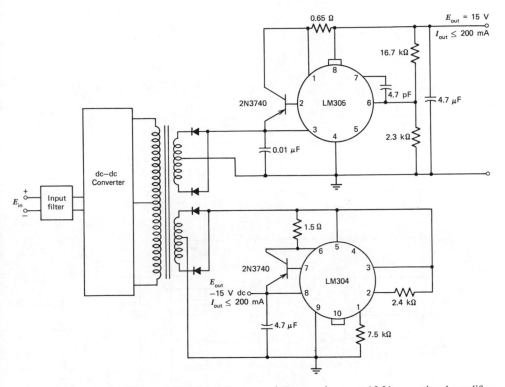

Figure 12-20 The LM104 and LM105 voltage regulators used on a ±15-V operational-amplifier power supply.

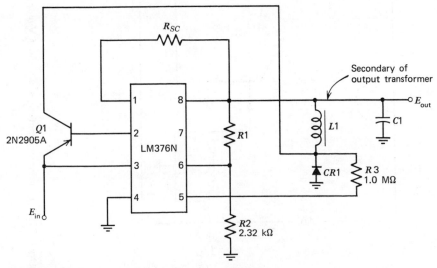

Figure 12-21 Low-cost switching-regulator schematic.

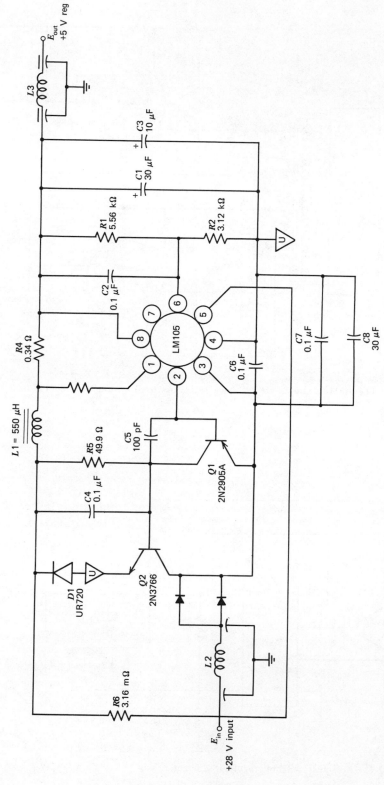

Figure 12-22 Low-noise switching regulator with 5 V output regulated to 1% by LM105 IC regulator.

circuits because of the stringent r-f requirements. Also, because of the added components and r-f requirements, the performance has degraded somewhat from that of a conventional switching regulator. Capacitors C_4 and C_5 were added to the circuit of Figure 12-10 to reduce high-frequency switching transients by slowing down the switching speeds of the output transistors. They also protect the semiconductor devices from transients. Capacitor C_4 projects the emitter–base junction of Q_2, whereas C_5 protects both the IC and the collector–base junction of Q_1 by controlling the rise times of the switching voltages.

In addition, capacitors C_6 and C_7 suppress transients on the unregulated input, C_8 minimizes the input impedance seen by the regulator, and C_3 improves filtering of the switching noise in the regulated output. The other capacitors have the same function as before, except that C_1 is much larger. This circuit oscillates in the range of 20 to 40 kHz. Resistors R_1 and R_2 still determine the output voltage level, but R_6 was added to ensure sufficient positive feedback. The feedback is now mainly through R_6.

Inductor L_1 has the same function as before, but the lower E_{out} and the added inductive resistance provided by R_4 allow L_1 in this case to be only $550\,\mu\text{H}$; L_2 and L_3 are high-frequency input and output filters. The diodes on the collector of Q_2 are used to prevent reverse-input voltage polarity and heavy negative-going input transients from damaging the regulator.

Although I_{max} is only 500 mA, the higher dissipation of this slower-switching circuit and the need to operate the regulator at relatively high temperatures required the use of two external transistors. Pass transistor Q_2 is an *npn* type for the same reason that Q_1 is a *pnp* — so that cascade connections on the IC booster output could be used. Finally, R_5 was added to terminate Q_2 and ensure low leakage in the OFF state.

Development of a 5-V, 1-A Monolithic Regulator

The trend in IC voltage regulators is for more foolproof, high-current circuits with a minimum of external components, whether they be active or passive. One of the drawbacks of earlier regulators was the considerable number of external parts they required, which led the user to think twice about changing from discretes to ICs. Additionally, because of the high current requirements of digital systems, single-point regulation creates many problems. Heavy power buses must be used to distribute the regulated voltage. With low voltages and currents of many amperes, voltage drops in connectors and conductors can cause an appreciable change in the load voltage. This situation is further complicated because TTL draws transient currents many times the steady-state current when it switches. These current transients can cause false operation unless large bypass capacitors are used.

These problems have led to the development of the LM109 on-card monolithic regulator. It is quite simple to use in that it requires no external components. The IC has three active leads — input, output, and ground — and can be supplied in standard transistor power packages. Output currents in excess of 1 A can be obtained. Furthermore, no adjustments are required to set up the output voltage, and with the overload protection provided, it is virtually impossible to destroy the regulator. The simplicity of the regulator, coupled with low-cost fabrication and improved reliability of monolithic circuits, now makes the regulator quite attractive.

The design of the LM109 regulator is based on several breakthroughs (in the absence of which progress previously had been retarded) in the development of high-current IC regulators: On-chip thermal shutdown in conjunction with current limiting; on-chip series-pass transistor; and a stable internal voltage reference. Because of their uniqueness, each item is discussed.

Thermal shutdown limits the maximum junction temperature and protects the regulator regardless of input voltage, type of overload, or degree of heat sinking. With an external pass transistor, there is no convenient way to sense junction temperature; thus it is much more difficult to provide thermal limiting. Thermal protection is, in itself, a very good reason for putting the pass transistor on the chip.

When a regulator is protected by current limiting alone, it is necessary to limit the output current to a value substantially lower than is dictated by dissipation under normal operating conditions, in order to prevent excessive heating when a fault occurs. Thermal limiting provides virtually absolute protection for any overload condition. Hence the maximum output current under normal operating conditions can be increased. This tends to make up for the limitation that an IC has a lower maximum junction temperature than a discrete transistor.

When a regulator works with a relatively low voltage across the IC, the internal circuitry can be operated at comparatively high currents without causing excessive dissipation. Both the low voltage and the larger internal currents permit higher junction temperatures. This can also reduce the heat sinking required — especially for commercial-temperature-range parts. Figure 12-23 is the schematic diagram of the LM109 1.5-A, 5-V monolithic voltage regulator, which utilizes thermal shutdown and current-limiting techniques.

Thermal protection of the LM109 limits the maximum chip temperature to 175°C, thereby protecting the regulator from overheating, regardless of input voltage, type of overload, or degree of heat sinking. Having the temperature sensor on the chip, close to the series-pass transistor, enables the regulator to shut down in a matter of milliseconds, should an overload occur. Once the overload is removed, and the chip cools to 165°C, the regulator safely turns back on, resuming normal operation.

The output current of the regulator is limited when the voltage across R_{14} becomes large enough to turn on Q_{14}. This ensures that the output current cannot get high enough to cause the pass transistor to go into secondary breakdown or to damage the aluminum conductors on the chip. Furthermore, when the voltage across the pass transistor exceeds 7 V, the current through R_{15} and D_3 reduces the limiting current, again to minimize the chance of secondary breakdown. The performance of this protective circuitry is illustrated in Figure 12-24.

Even though the current is limited, excessive dissipation can cause the chip to overheat. In fact, the dominant failure mechanism of solid-state regulators is excessive heating of the semiconductors, particularly the pass transistor. Thermal protection attacks the problem directly by putting a temperature regulator on the IC chip. Normally, this regulator is biased below its activation threshold and does not affect circuit operation. If for any reason, however, the chip approaches its maximum operating temperature, the temperature regulator turns on and reduces

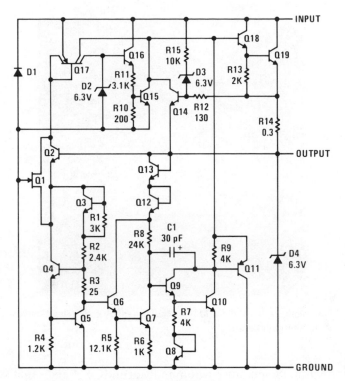

Figure 12-23   Detailed schematic of the LM109 regulator.

internal dissipation to prevent any further increase in chip temperature.

The thermal protection circuitry develops its reference voltage with a conventional zener diode D_2. Transistor Q_{16} is a buffer that feeds a voltage divider, delivering about 300 mV to the base of Q_{15} at 175°C. The emitter–base voltage Q_{15} is the actual temperature sensor because, with a constant voltage applied across the junction, the collector current rises rapidly with increasing temperature.

Although some form of thermal protection can be incorporated in a discrete regulator, ICs have a distinct advantage: the temperature-sensing device detects

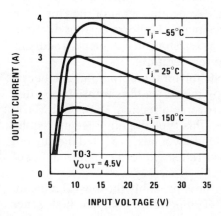

Figure 12-24   Current-limiting characteristics.

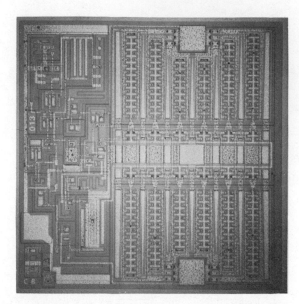

Figure 12-25   Photomicrograph of the LM109 regulator shows that high-current-pass transistor takes more area than control circuitry.

increases in junction temperature within milliseconds. Schemes that sense case or heat-sink temperature take several seconds, or longer. With the longer response times, the pass transistor usually blows out before thermal limiting has come into effect.

Another protective feature of the regulator is the crowbar clamp on the output. If the output voltage tries to rise, D_4 will break down and limit the voltage to a safe value. If the rise is caused by failure of the pass transistor such that the current is not limited, the aluminum conductors on the chip will fuze, disconnecting the load. Although this destroys the regulator, it does protect the load from damage. The regulator is also designed so that it is not damaged if the unregulated input is shorted to ground when there is a large capacitor on the output. Furthermore, if the input voltage tries to reverse, D_1 will clamp this for currents up to 1 A.

Figure 12-25 is a photomicrograph of the regulator chip. It can be seen that the pass transistors, which must handle more than 1 A, occupy most of the chip area. The output transistor is actually broken into segments. Uniform current distribution is ensured by also breaking the current-limit resistor into segments and using them to equalize the current and thus the chip temperature.

With many regulators that have a large capacitor on the output, an accidental short on the input of the regulator could destroy the series-pass transistor by discharging the output capacitor through it in the reverse direction. With the LM109 operating in the fixed 5-V output configuration, this type of failure is impossible. Also, the regulator is protected from input voltage reversals, provided the current is limited to some value less than 1 A. Larger currents will result in the melting of the aluminum interconnections on the chip itself.

The internal voltage reference for this regulator is probably the most significant departure from standard design techniques. Temperature-compensated zener diodes are normally used for the reference. However, these have breakdown voltages between 7 and 9 V, which puts a lower limit on the input voltage to the regulator. For low-voltage operation, a different kind of reference is needed.

The reference in the LM109 does not use a zener diode. Instead, it is developed from the highly predictable emitter–base voltage of the transistors. In its simplest form, the reference developed is equal to the energy-band-gap voltage of the semiconductor material. For silicon, this is 1.205 V; thus the reference need not impose minimum input voltage limitations on the regulator. An added advantage of this reference is that the output voltage is well determined in a production environment, eliminating the need for individual adjustment of the regulators.

A simplified version of this reference is shown in Figure 12-26. In this circuit, Q_1 is operated at a relatively high current density. The current density of Q_2 is about 10 times lower, and the emitter–base voltage differential (ΔV_{BE}) between the two devices appears across R_3. If the transistors have high current gains, the voltage across R_2 will also be proportional to ΔV_{BE}. Here Q_3 is a gain stage that will regulate the output at a voltage equal to its emitter–base voltage plus the drop across R_2. The emitter–base voltage of Q_3 has a negative temperature coefficient, whereas the ΔV_{BE} component across R_2 has a positive temperature coefficient. It will be shown that the output voltage is temperature compensated when the sum of the two voltages is equal to the energy-band-gap voltage.

Conditions for temperature compensation can be derived starting with the equation for the emitter–base voltage of a transistor, which is

$$V_{BE} = V_{g0}\left(1 - \frac{T}{T_0}\right) + V_{BEO}\left(\frac{T}{T_0}\right) + \frac{nkT}{q}\log_e\frac{T_0}{T} + \frac{kT}{q}\log_e\frac{I_C}{I_{CO}} \qquad (12\text{-}15)$$

where V_{g0} is the extrapolated energy-band-gap voltage for the semiconductor material at absolute zero, q is the charge of an electron, n is a constant depending on how the transistor is made (ca. 1.5 for double-diffused npn transistors), k is Boltzmann's constant, T is absolute temperature, I_C is collector current, and V_{BEO} is the emitter–base voltage at T_0 and I_{CO}.

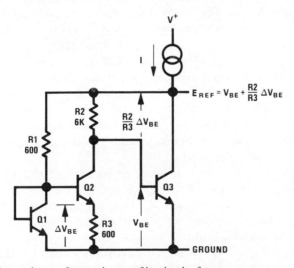

Figure 12-26 The low-voltage reference in one of its simpler forms.

The emitter–base voltage differential between two transistors operated at different current densities is given by

$$\Delta V_{BE} = \frac{kT}{q} \log_e \frac{J_1}{J_2} \tag{12-16}$$

where J is current density.

Referring to Equation 12-15, we see that the last two terms are quite small and are made even smaller by making I_C vary as absolute temperature. At any rate, they can be ignored for now because they are of the same order as errors caused by nontheoretical behavior of the transistors that must be determined empirically.

If the reference is composed of V_{BE} plus a voltage proportional to ΔV_{BE}, the output voltage is obtained by adding Equation 12-15 in its simplified form to Equation 12-16.

$$V_{\text{ref}} = V_{g0}\left(1 - \frac{T}{T_0}\right) + V_{BEO}\left(\frac{T}{T_0}\right) + \frac{kT}{q} \log_e \frac{J_1}{J_2} \tag{12-17}$$

Differentiating with respect to temperature yields

$$\frac{\partial V_{\text{ref}}}{\partial T} = -\frac{V_{g0}}{T_0} + \frac{V_{BEO}}{T_0} + \frac{J_1}{J_2} \tag{12-18}$$

For zero temperature drift, this quantity should equal zero, giving

$$V_{g0} = V_{BEO} + \frac{kT_0}{q} \log_e \frac{J_1}{J_2} \tag{12-19}$$

The first term on the right is the initial emitter–base voltage; the second is the component proportional to emitter–base voltage differential. Hence, if the sum of the two is equal to the energy-band-gap voltage of the semiconductor, the reference will be temperature compensated.

In Figure 12-23 the ΔV_{BE} component of the output voltage is developed across R_8 by the collector current of Q_7. The emitter–base voltage differential is produced by operating Q_4 and Q_5 at high current densities while operating Q_6 and Q_7 at much lower current levels. The extra transistors improve tolerances by making the emitter–base voltage differential larger. In addition, R_3 serves to compensate the transconductance of Q_5, so that the ΔV_{BE} component is not affected by changes in the regulator output voltage or the absolute value of components.

Thus the LM109 performs a complete regulation function on a single silicon chip, requiring no external components. It was basically designed to provide a $+5$-V supply for TTL and DTL circuits. However, the design breakthroughs achieved in the 109 have led to the widespread development of many high-current IC regulators by all the major IC manufacturers. (National's LM120 and LM140 series, Fairchild's 7800 series, and so on).

Figures 12-27 and 12-28 depict the ease of use of the LM109 and the minimal number of external components required to operate the device. The LM109 is very simple to use in that it requires no external components and it has three active leads (input, output, and ground); output currents in excess of 1 A can be obtained, no adjustments are required to set up the output voltage, and the device has internal overload and thermal shutdown protection—making it virtually blow-out proof. The internal frequency compensation of the regulator permits it to

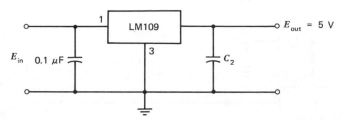

Figure 12-27   Fixed 5-V regulator: 0.1 pF required, as shown, if regulator is located an appreciable distance from the power-supply filter; although C_2 is not required, it gives improved transient response if this is desired.

operate with or without a bypass capacitor on the output. However, an output capacitor improves the transient response and reduces the high-frequency output impedance.

Precision Regulators

Precision regulators can be built using an IC operational amplifier as the control amplifier and a discrete zener as a reference, where the performance is determined by the reference. Figure 12-29 presents the circuit schematic of a simple regulator. It is capable of providing better than 0.01% regulation for worst-case changes of line, load, and temperature. Typically, the line rejection is 120 dB to 1 kHz; and the load regulation is better than 10 μV for a 1-A change. Temperature is the worst source of error; however, it is possible to achieve less than a 0.01% change in the output voltage over a −55 to +125°C range.

The operation of the regulator is straightforward. An internal voltage reference is provided by a high-stability zener diode. The LM108A operational amplifier compares a fraction of the output voltage with reference. The output of the amplifier controls the ground terminal of an LM109 regulator through source follower Q_1. Frequency compensation for the regulator is provided by both the R_1–C_2 combination and output capacitor C_3.

The use of an LM109 instead of discrete power transistors has several advantages. First, the LM109 contains all the biasing and current-limit circuitry needed to supply a 1-A load. This simplifies the regulator. Second, and probably most important, the LM109 has thermal overload protection, making the regulator virtually burn-out proof.

Although the regulator is relatively simple, some precautions must be taken to eliminate possible problems. A solid tantalum output capacitor must be used.

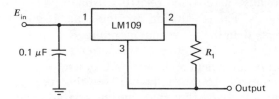

Figure 12-28   Current regulator, where R_1 determines the output current.

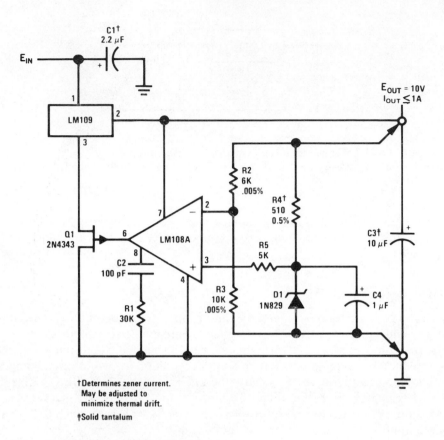

Figure 12-29   High-stability positive regulator.

Unlike electrolytics, solid tantalum capacitors have low internal impedance at high frequencies. Low impedance is needed both for frequency compensation and to eliminate possible minor loop oscillations.

Some unusual problems are encountered in the construction of a high-stability regulator. Component choice is most important, since resistors, amplifier, and zener can contribute to temperature drift. Also, good circuit layout is needed to eliminate the effect of lead drops, pickup, and thermal gradients.

The resistors must be low-temperature-coefficient wirewound or precision metal film. Ordinary 1% carbon film, tin oxide, or metal film units are not suitable because they may drift as much as 0.5% over temperature. The resistor accuracy need not be 0.005% as in Figure 12-29, however, they should track better than 1 ppm/°C. Additionally, wirewound resistors usually have lower thermoelectric effects than film types. The resistor driving the zener is not quite as critical, but it should change less than 0.2% over temperature.

The IN829 diode is representative of the better zeners available. However, it still has a temperature coefficient of 0.0005%/°C or a maximum drift of 0.05% over a −55 to 125°C temperature range. The drift of the zener is usually linear with temperature and may be varied by changing the operating current from its nominal value of 7.5 mA. The temperature coefficient changes by about 50 μV/°C for a

15% change in operating current. Therefore, the temperature drift of the regulator may be minimized by adjusting the zener current.

Good construction techniques are important. It is necessary to use remote sensing at the load, as shown in the schematic. Even an inch of wire will degrade the load regulation. The voltage setting resistors, the zener, and the amplifier should also be shielded. Board leakages or stray capacitance can easily introduce 100 μV of ripple or dc error into the regulator. Generally, short wire length and single-point grounding are helpful in obtaining proper operation.

Chapter Thirteen

THIN–FILM HYBRID CIRCUITS

THIN–FILM CIRCUIT DESIGN

The development of thin-film circuits somewhat parallels semiconductor technology, for it was not until transistors became available that the potential rewards of thin-film techniques became apparent. The applications that appeared to be possible for thin films involved resistors, capacitors, and interconnections.

The performance advantages of thin-film resistors and capacitors are compared with those of equivalent diffused elements in Table 13-1. Chiefly, they are:

1. Reduction of parasitic capacitance and series resistance (higher Q).
2. Larger range of component values.
3. Lower temperature coefficient.
4. Tighter tolerance control.

Additionally, thin-film components require less area both because of their higher resistivity or dielectric capabilities and isolation (they require no isolation junctions). Because of the many thin-film processes, the circuit design is always oriented toward the process required to achieve the desired electrical characteristics of the passive components. In general, the design evolves from a discrete-component prototype and is adjusted to compensate for the process. The layout, however, may look nothing like the breadboard. It must be scaled to substrate size, taking into account such problems as power dissipation, lack of component packaging, and the necessity to provide bonding pads and circuit pins at the substrate periphery.

Table 13-1   Comparison of Thin-Film and Diffused Monolithic Circuits

Thin-Film Components	Diffused Monolithic Components
• Capacitors, resistors, and conductors are readily available; inductors, diodes, and active devices have been demonstrated	• Transistors, diodes, resistors, and capacitors are readily available
• Availability of inductors is limited	
• Multitude of materials is available	• Essentially limited to silicon as basic material, plus diffusants, metallization, and isolation dielectric
• Circuits can be adjusted and components replaced before packaging	•
• More complete assortment of component parameter values and circuit types than with silicon	• Limited component values
• Additional processing steps and thus increased cost	
• High-power passive components and interconnections are practical	• Very small circuits with limited power handling capability
• More radiation resistant	• Large parasitics, directly proportional to both area and voltage
• Low-temperature packaging — chips can be cemented to substrate with conductive epoxy	
• Can be deposited on silicon substrate to improve capabilities of monolithic devices	• Looser resistor tolerance: ±20%
• Monolithic silicon chips can be mounted in thin-film structure to provide active devices	
• Smaller parasitics which are directly proportional to area only	
• Tighter resistor tolerances: ±15%	
• Thin-film resistor is narrower than diffused resistor	
• Large circuits	• Smaller circuits
• Higher sheet resistivity, typically 400 $\Omega/\square$	• Lower sheet resistivity, typically 200 $\Omega/\square$
• Lower temperature coefficient: 50 to 250 ppm/°C	• Higher temperature coefficient, up to 2000 ppm/°C

General design considerations include:

- Circuits must be analyzed and tested very thoroughly to assure that they meet specifications before fabrication is begun. The breadboard may not be able to simulate temperature coefficient or the thermal tracking of the various components, thus computer simulation is often advantageous.

- Passive components should be used whenever feasible in preference to active components. An example might be resistive or capacitive coupling in digital circuitry rather than diodes. Active components added on after the circuit is deposited are more expensive.

- Use a minimum number of different types of circuits. The fabrication costs depend more on the number of different types of circuits than on the quantity of each circuit type. It is generally more economical, for example, to use a high-speed flip-flop in a low-speed condition than to design a special low-speed flip-flop. Another technique is to design a general-purpose circuit that can be used for more than one function (e.g., a flip-flop and shift register). Extra (spare) passive components can be deposited at nominal cost. Also, the types and placement of the active components can be changed without added cost.

- Circuits should be designed to use ratios of component values rather than absolute component values whenever possible. Ratios of values can be held to much closer tolerances (by perhaps twice) than absolute values of components. This is because the deposition thicknesses of the individual components fabricated on a circuit substrate are the same, owing to the fabrication process. The temperature tracking of these components is also better, by at least an order of magnitude, than the absolute temperature coefficient given for the component type.

- Tunable components should be eliminated, if at all possible. It is better to increase complexity than to employ a tunable component. Components may be trimmed to better than 0.5% to bring a value in line during fabrication; but once the circuit is packaged, a film component cannot be adjusted.

- Minimize power dissipation. With a good thermal heat sink, it is possible to dissipate as much as 1 W on a 1-in.^{2} substrate packaged in a system. The volume dissipation should not exceed 5 W/in.^{3}.

- Operate at low voltages. Although voltage ratings as high as 25 or 50 V can be obtained, preferred levels are below approximately 15 V.

- Make the circuit substrate as large as practical so that a large amount of circuitry can be put on one substrate. As the circuit substrate becomes larger the number of interconnections between substrates drops drastically. Interconnections between substrates represent volume and weight and are the most unreliable part of a microelectronic assembly. The maximum practical size is determined basically by the cost of fabrication and the cost of parts of the thin-film module. As many as four or five circuits may be fabricated at a reasonable cost on a single substrate.

THIN VERSUS THICK FILMS

The terms "thin film" and "thick film" have come to mean the processes of manufacturing hybrid ICs as well as the thickness of the films employed. Most thin-film circuits are made by vacuum-deposition processes and patterned by deposition masks or photoetching. Thick films are almost invariably made by screen printing and firing inks or pastes. Important differences are outlined in Table 13-2.

Thin films are generally less than 1 μ (10,000 Å) thick, but thicknesses may range up to tens of microns. Very high pattern resolution, comparable to resolution of patterns in monolithic ICs, can be achieved with photolithography. Line definition as good as 0.1 mil and lines as narrow as 0.3 mil can be formed. This is particularly important for tight component tolerances, and it also allows the assembly of flip-chip semiconductor components (see Chapter 15).

Table 13-2   Comparison of Thin- and Thick-Film Hybrid Circuits

Thin-Film	Thick-Film
• Vacuum process	• Silk-screen-printing–firing process done in tape furnace
• Subtractive process	• Additive process
• Uses glazed substrate	• Uses unglazed substrate
• Poor power dissipation	• Good power dissipation
• Capacitors, resistors, and conductors are readily available; inductors, diodes, and active devices have been demonstrated	• Uses resistive inks and air-abrasive resistor trimming; also laser trimming
• Many more materials and techniques for handling these materials are available for thin films than for thick films	• Resistors and conductors are readily available; most capacitors used are in chip form
• Thin films allow higher resolution in the devices	• No chance to print active device structure
• Smaller value resistors are made with thin-film techniques	• Higher-temperature processes employed in fabrication might produce higher reliability than for thin films
• Highest unit cost for both small and large quantities	• Lower tooling cost than thin films
• High circuit complexity	• Somewhat lower unit cost for both small and large quantities
• Thin-film frequency limit: 10,000 MHz	
• Capacitors are linear and have smaller temperature coefficients	• Capable of high power dissipation
• Limited power dissipation	• High circuit complexity
	• Flexible technique in which a variety of patterns can be achieved with little process variation
• Maximum sheet resistance: 5000 $\Omega/\square$	• Maximum sheet resistivity: 1 M$\Omega/\square$
• Typical coefficient of resistance: 50 ppm/°C	• Resistance tolerances without trimming: ±10%
	• Typical temperature coefficient of resistance: < 100 ppm/°C
• Typical thin-film resistor characteristics:	• Typical thick-film resistor characteristics:
Design tolerance: 0.1–2%	Design tolerance: 1–10%
Temperature coefficient: 1–100 ppm/°C	Temperature coefficient: 100 ppm/°C
Resistivity: 0.1 $\Omega/\square$–5 k$\Omega/\square$	Resistivity: 10 $\Omega/\square$–1 M$\Omega/\square$
Resistance: 1 Ω–500 kΩ	Resistance: 10 Ω–20,000 MΩ
Power-handling capability: 20 W/in.^{2}	Power-handling capability: 20 W/in.^{2}
Film thickness: 75–20,000 Å	Film thickness: 0.5–2.5 mils
Material: metal or metal oxide on ceramic substrate	Material: metal-bearing ink or paste on ceramic substrate
Minimum line/space: 1 mil	Minimum/space width: 3 mils
	• Maximum achievable capacitor dielectric constant: ≈ 1200–1500
• Typical vacuum-evaporated capacitance: 0.6 μF/in.^{2}	• Maximum capacitance/area = 100,000–200,000 pF/in.^{2}
• Dielectric constant (SiO): 6.0	

The most popular thin-film materials are nichrome for resistors and gold for conductors and chip-bonding pads; but a wide variety of conductive, resistive, and capacitive materials can be used. Basic vacuum evaporation can be combined with co-deposition of different metals, chemical decomposition, sputtering of compounds, and postdeposition chemical processing to obtain virtually the entire range of materials available to both monolithic IC and discrete-component fabricators. Finished film properties can be quite different from those of the bulk materials. Resistivity and temperature coefficients, for instance, may vary with thickness, deposition rate, and substrate temperature and composition. Thick films are typically a fraction of a mil thick—an order of magnitude thicker than most thin films—and lines are several mils wide. The resolution is less because the inks are printed through patterns on stainless-steel meshes or screens with 200 to 300 wires per inch. Then the inks are fired, to bond them to the substrate and to the final electrical properties of the materials.

The range of thick-film materials is limited compared with thin films. Most thick-film circuits have only printed conductors and resistors, although capacitor inks are available. The principal advantage of thick films over thin films is lower cost. Capacitors and semiconductor devices are generally added as wire-bonded chips; flip-chips with wide terminal spacings can be used, however.

Hybrid ICs are not necessarily thin- or thick-film circuits. Many circuits are "chips and wire" constructions. Semiconductor chips, including ICs, transistors and other active devices, MOS capacitors, diffused resistors, and ceramic capacitor chips, are bonded to a substrate with heat-conducting materials and interconnected with wire bonds. In some cases, chips are interconnected by way of conductive patterns on substates that are the equivalent of microminiature PC cards. The resistors and other passive components are not fabricated on the substrate. The substrate may have several layers of conductors, and it may be the equivalent of a multilayer PC board.

FILM CIRCUIT SUBSTRATES

Another major difference between thin-film and thick-film circuits is the choice of film substrate. As indicated in Figure 13-1, the thick-film substrate is usually alumina ceramic. Thin-film substrates are typically glass, a smoother grade of alumina, silicon oxide (in the case of monolithic hybrids), or crystals such as sapphire. The substrate of a monolithic hybrid is called "active" to distinguish it from the conventional substrate of passive networks.

The substrate must be a good insulator, since in many cases it also dielectrically isolates the active components attached to it, and it should be a good heat conductor. In power circuits, the substrate often sits on a metal heat sink or thermal conductor. Roughly surfaced substrates are preferred for thick films, because the printed pastes will bond better after firing. In fact, high-volume thick-film circuits are often printed on long tapes of "green" ceramic which are hardened during the postprinting firing process. Beryllia is sometimes chosen for high-power circuits because of its excellent heat conduction. Microwave ICs are built on a variety of dielectrics because the dielectric, being part of the circuit, must be chosen for its microwave properties.

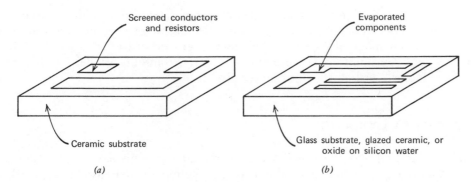

Screened conductors
and resistors

Evaporated
components

Ceramic substrate

Glass substrate, glazed ceramic, or
oxide on silicon water

(a)

(b)

Figure 13-1 Comparison of thick- and thin-film circuits: (*a*) thick-film substrate, (*b*) thin-film substrate.

Some types of dielectrically isolated semiconductor circuits are made as silicon islands on sapphire crystals. Single-crystal substrates are needed because the silicon devices are formed in an epitaxial layer of silicon crystal which can be grown only on a compatible crystal. Sapphire has also been used as the substrate of very high quality thin-film passive networks. Sapphire is basically single crystal aluminum oxide.

Alumina is the standard substrate because of its low cost and good all-around mechanical and electrical properties. A typical alumina substrate size is 1.5 in.^{2} and 15 mils (0.015 in.) thick. It is nominally 99.5% Al_2O_3. The substrate plates are pressed, extruded, and fired. Alumina for thin films is glazed when the process calls for a very smooth surface.

Thin-film tolerances are generally tighter, and the warping and dimensional changes that occur during firing cannot be allowed. Thin-film substrates must be flat, particularly if the metal patterns are deposited through mechanical masks. Any significant warp would cause poor registration and line definition.

THIN–FILM PROCESSES

There is no universal thin-film process, just as there is none for monolithic circuits. The major steps are vacuum deposition, masking, etching, bonding chip devices, trimming or adjusting component values, and packaging. In principle, deposition, masking, and etching are the same as the steps used in forming the thin-film intraconnections of monolithic ICs, but in practice they are much more variable. Often, the processing sequence changes with the materials used.

Thin-film processing is more an art than monolithic metallization because of the greater diversity of materials involved in the former. Personal preference and experience weigh heavily in the process used. It is a batch operation, but each batch generally contains fewer circuits than monolithic IC batches.

A decade ago, several companies built very large automatic processing systems for in-line or sequential deposition of various materials, and hundreds of thin-film substrates could be covered with successive layers of different materials. Today, except for the mass production of resistor networks, automatic thin-film systems are rare. Small-batch processing in bell-jar vacuum systems prevails.

The tricks of the thin-film trade are legion. They range from sputtering to clean substrates between depositions (essentially the reverse of vacuum deposition) in the vacuum chamber all the way to judging the temperature of the vacuum chamber by feel. Certainly, every practitioner of the thin-film art has his favorite "finagle factor," just as IC diffusion experts do.

VACUUM EVAPORATION

Although plating and other methods of forming metal films have applications in hybrid circuits, vacuum evaporation is by far the most widely used process. It is almost invariably combined with photolithography in one form or another. Patterns are obtained in the deposited film by photoetching after film deposition, by a photoresist mask on the substrate, by a thin-film mask developed on the substrate with photoetching, or by etching a mechanical mask through which the film is deposited.

There are several vacuum-evaporation processes. The main ones are conventional evaporation as described in Chapter 2, electron-beam evaporation, and sputtering. Sputtering is used primarily to deposit refractory metals, capacitor dielectrics, and other compounds that would break down if vaporized. In sputtering, ions in a glow discharge dislodge molecular particles from a target of the material to be deposited. It is not a true vacuum process, since a small amount of gas must be introduced into the chamber to support the glow discharge.

The basic "bell-jar" type of deposition is illustrated in Figure 13-2. In this example, the substrate is masked mechanically and only one crucible, or single-

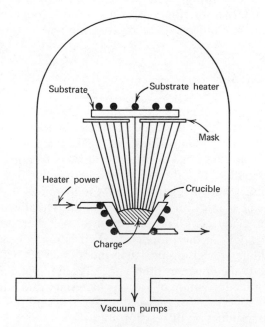

Figure 13-2   Vacuum evaporation process.

layer deposition, appears. The crucible with a charge of the material to be deposited is heated to vaporize the charge. The vapor atoms travel in straight lines. The vapor passes through the mask openings to deposit on the exposed substrate, forming thin films of the required area at the required location. All the required films are deposited through a succession of masks to build up the thin-film circuit. After removal from the chamber, the required "add-ons," principally transistors and diodes, are attached in the spaces left for them.

PROCESS MONITORING

Deposition of conductive films can be monitored by the same method employed for monolithic IC metallization, namely, observing the frequency change of an oscillator crystal while the material is deposited on the crystal. The frequency changes with thickness. For resistive materials, a monitor slide can be placed in the vacuum chamber and ohms per square can be measured directly.

The monitor slide has conductive material such as aluminum on both ends of the glass slide and no aluminum in the center (see Figure 13-3). As the evaporation proceeds, the resistive material bridges the clean, nonconductive glass and makes the contact to the ohmmeter. Since $L = W$ on the monitor slide, the aspect ratio of the resistor is one to one or one square. When the desired resistivity in ohms per square is achieved, the evaporation process is discontinued.

THIN–FILM PATTERNING

Patterning of the thin films can be accomplished before, during, and after deposition. To pattern before deposition, a mask of photoresist or nichrome is usually prepared on the substrate. Patterning during deposition is done with a mechanical mask that is actually a high-resolution stencil prepared by photoetching a thin plate of metal. After deposition, the film patterns are developed with photolithographic techniques similar to those used to pattern thin-film metallization on

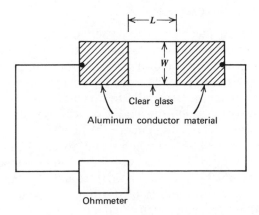

Figure 13-3   Conductive-film resistivity measurement.

monolithic ICs. Selective etching after deposition is much more common than masking on the substrate itself.

One of the process options is to deposit all the materials to be used in sequence without removing the substrate from the vacuum chamber. The advantage is that the interfaces between materials are not contaminated by being exposed to the atmosphere or by handling. After the series of depositions is complete, the different materials are patterned by etching with a series of chemical etching operations that successively pattern the different film layers. The patterns are controlled by photoresist.

This method is frequently used just for nichrome and gold. Nichrome, as mentioned before, has excellent adhesion to substrate materials and is also a good resistor material. Although gold has poor adhesion, it is an excellent conductor; it makes a good surface for chip and wire bonding, and it resists corrosion. Thus the nichrome–gold system can be used to make an entire conductor–resistor network. Where a conductor or bonding pad is wanted, a strip of gold on nichrome is left. Where a resistor is wanted, the gold is etched off and a nichrome stripe is left between gold-on-nichrome electrodes. A typical process sequence for such a network follows.

1. Substrate inspection.
2. Substrate cleaning.
3. Nichrome deposition.
4. Gold deposition.
5. Gold adhesion.
6. Precoat bake.
7. Photoresist spin.
8. Postcoat bake.
9. Align and expose.
10. Develop photoresist.
11. Inspect development.
12. Postdevelop bake.
13. Gold etch for conductors.
14. Nichrome etch for conductors and resistors.
15. Photoresist removal.
16. Wire bond, test, and package.

Deposition through a mechanical mask eliminates the need for photoresist operations on the substrate. It is particularly inexpensive when fine-line resolutions are not required or when two or more film patterns do not have to be aligned. For instance, it may serve to provide resistor networks in a chip-and-wire hybrid. Since the chips, which can be cemented to the substrates, are interconnected by wire bonds, only one deposition is needed. Conversely, a conductor pattern alone may be deposited if only chip components are needed.

Mechanical masks are also used when materials cannot be selectively etched. For example, silicon monoxide (SiO) capacitors can be shaped. This is virtually impossible for most substrates because the etchants that attack SiO also attack glass and would undercut the edges of the film components. Moreover, if the materials deposited are precious, such as gold or some of the resistor metals, it is much easier to recapture the excess from a mask than when the materials have become part of the etching chemicals.

Masking on the substrate is usually done with nichrome, which adheres tightly to glass, allowing nonadherent materials to be stripped from the remainder of the substrate. This approach is often combined with resist masking. Photoresist is developed on the substrate in the manner used to define oxide windows on mono-lithic ICs. However, the windows are located where the designer wants the film to adhere to the substrate. After deposition, the resist is dissolved. Since the un-wanted film no longer has any support, it is readily removed from the substrate (and, if precious, easily recovered). The effect of selective etching of, for example, gold and nichrome, can be achieved without etching by patterning the nichrome and developing a new pattern that masks the nichrome resistor areas.

THIN–FILM MATERIALS

Thin-film materials are discussed in context in the following sections for thin-film components. However, a brief review of some of the more popular materials is useful to set the stage.

The nichrome–gold system is most popular for several reasons. First, nichrome (nickel–chromium alloy) adheres tightly to glass or glazed ceramic substrates because stable oxides form at the glass–metal interface. Good adherence is ab-solutely essential to reliable thin-film circuits. Second, nichrome is an excellent resistor material.

Gold adheres poorly to glass because it does not form a stable oxide. Since it does form intermetallics with nichrome, however, the two metals bond well. Equally important, gold and nichrome can be selectively etched. Gold is excellent both as a conductor and as a bonding surface. Typically, 140 Å of nichrome – a good thickness for resistor stripes – is deposited for the underlay of the conductor lines and the resistors. The gold is deposited at a thickness of about 5000 Å. After etching, the resistor electrodes are simply the ends of the gold conductor lines. The current preferentially travels through the thick gold rather than through the highly resistive nichrome, going everywhere under the conductor pattern except through the bare nichrome resistor lines.

Chromium is often used instead of nichrome. Other materials of major interest include aluminum, tantalum, and cermets. Aluminum and tantalum can be ano-dized to vary film thickness after patterning, thus adjusting component values. Aluminum provides both conductors and low-value resistors.

Cermets are combinations of ceramic or dielectric material and metals. They are of great interest to the IC designer because of their ability to produce high values of resistance with low temperature coefficients. Cermet films can be deposited by vacuum evaporation similiar to nichrome except that multiple sources must be controlled. Flash evaporation, whereby a controlled stream of a mixture of powdered cermet material is fed onto a heated tantalum strip, also gives good results. Substrates on which the film is evaporated are usually heated to 350 to 400°C.

Tantalum films are popular because of their stability, high annealing temper-ature, anodic adjustment of resistors, high dielectric strength and dielectric con-stant for capacitor use, uniformity, and reproducibility. The deposited films can be anodized in an oxygen or nitrogen atmosphere to form a thin passivating layer and to make minor adjustments in resistor value. Tantalum must be deposited by cathodic sputtering because of its high melting point. The sputtered films adhere better than evaporated films and are more uniform. Sputtering is more economical because the deposit is localized. However, the deposition is difficult to monitor and to mask for geometry control.

Sputtering is a vapor-deposition process in which an electrical discharge is set up between two plates in the presence of a low-pressure inert gas such as argon. The ionized gas atoms, accelerated by the high electric field to the tantalum cathode, release their kinetic energy, thus knocking off tantalum atoms (a few of which may become ionized). These are then free to diffuse to the glass substrate on the anode. The resistors and capacitors are patterned by photoetching. Tantalum films usually have a sheet resistance of the order of 100 to 500 $\Omega/\square$. The temperature coefficient is less than 200 ppm/°C.

Certain combinations of resistor and capacitor thin-film technologies are naturally compatible for production of RC circuits. These are as follows:

Resistors	Capacitors	Compatibility Feature
Nichrome	SiO	All vacuum
Conducting glaze or SnO	Dielectric glaze	No vacuum
Ta_2N	Ta_2N	All tantalum

THIN–FILM CONDUCTOR DESIGN

In a typical thin-film conductor and pad design sequence, the following procedures would be observed:

1. Conductors and pads:
   Minimum conductor and resistor width: 2 mils (ceramic substrates)
   Minimum spacing between traces: 2 mils
   Minimum clear area on edge of substrate: 5 mils

2. Long, narrow conductor lines should be avoided whenever possible, especially in high-current paths such as V_{CC} and ground, because of the finite resistance value of the conductor traces. Where conductor length may be questionable, and if temperature coefficient of resistance can be ignored, conductors can be treated as resistors and potentials can be calculated. For such calculations, use $R_S = 0.1 \, \Omega/\square$.

3. A minimum of 10 mils in each dimension must be added to maximum dice dimensions when laying out pads for die attach (Figure 13-4).

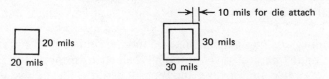

Figure 13-4   Diagram depicting required area around bonding pads for die attach.

4. If a lead bond is to be made from a chip to its own pad, a minimum of 15 mils in addition must be allowed for lead-bond tool clearance (Figure 13-5).

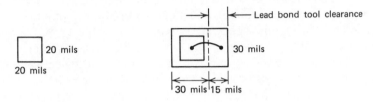

Figure 13-5 Diagram depicting amount of additional clearance required for lead bond tool clearance when a lead bond is made from a chip to its own pad.

5. Die-attach pads must be designed to assure that no die can be placed within a 25-mil distance from a lead-bond area along the line of the lead bond, nor closer than 5 mils perpendicular to this line if ultrasonic bonding is to be employed (Figure 13-6).

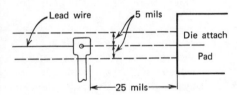

Figure 13-6 Diagram depicting typical minimum dimensions when ultrasonic bonding is used.

6. For thermocompression bonding the dimensions in item 5 reduce to 5 mils and 5 mils, respectively. Die-attach pads must be placed no less than 10 mils apart to assure die-attach–tool clearance. Minimum dimension of lead-bond pads is 5 mils^2.

7. Wire-bonded leads should never be longer than about 150 mils. Leads crossing over film traces should be avoided whenever possible. If they are necessary, a deposition of SiO insulation must cover the traces crossed over.

8. Leads crossing over leads are not allowed.

9. Sufficient space must be allowed at lead-bonding pads for bonding tool clearances.

10. When using thermocompression bonding, calculations for package height should include 15-mil lead height from the bond level (Figure 13-7).

Figure 13-7 Diagram depicting proper lead height for thermocompression bonding.

THIN–FILM RESISTOR PROPERTIES

Thin-film resistors satisfy performance requirements difficult or impossible to achieve with diffused resistors. Commonly used materials are chromium, nichrome, rhenium, tantalum, titanium, and a cermet composed of a mixture of silicon monoxide and chromium. These are compared in Table 13-3.

Nichrome thin-film resistors deposited on monolithic circuits can provide higher resistor values, tighter tolerances, and tighter resistance ratios between resistors. Compared with diffused resistors, nichrome has lower parasitic capacitance, is less affected by frequency changes, and is less affected by temperature changes. Thin resistive films several times higher in resistance than the typical 200-$\Omega/\square$ base diffusion of monolithic ICs are readily available.

Cermets may have sheet resistivities up to several thousand ohms per square. Table 13-4 provides a comparison of diffused and thin-film resistors. The properties of tantalum thin-film resistors are summarized in Table 13-5.

A sheet-resistivity versus thickness curve for 80–20 nichrome, deposited at

Table 13-3   Comparison of Thin-Film Resistors

Nichrome Resistors	Chromium Resistors	Cermet Resistors	Rhenium Resistors
• Most common material • Adhere well to substrate • Have small thermal coefficients: −50 to 600 ppm/°C • Have large values of sheet resistance up to 400 $\Omega/\square$ depending on film thickness • High-temperature fabrication process	• Used often • Adhere well to ceramic • High thermal coefficients of resistance: −50 to 100 ppm/°C depending on thickness • Sheet resistivity of 100 to 300 $\Omega/\square$ • No serious vapor-source problems	• Fabrication in two-step process: Lay film down; Anneal film • Temperature coefficient: ±100 ppm/°C • Sheet resistivity up to 1 k$\Omega/\square$ • 1-MΩ resistor can be made • Difficulty of depositing a well-controlled film • Substrate affects resistivity, thermal coefficient, and stability during annealing cycle • Resistors can be trimmed to ±1%	• Usable at high temperatures and high power levels • High-temperature fabrication process • Temperature coefficient: 250–300 ppm/°C for 10-k$\Omega/\square$ films • Sheet resistivity of up to 10 k$\Omega/\square$ • Deposition rate important in fabrication • Possess excellent age stability • Low variation of resistivity with film thickness

350°C on a borosilicate glass, is plotted in Figure 13-8. A typical sheet resistivity is 40 to 400 $\Omega/\square$. Films with resistivity up to 200 $\Omega/\square$ are reasonably stable and can withstand TO-5 fabrication process with only a slight change. At 400 $\Omega/\square$, films require passivation before any of the standard assembly processes can be used, in order to prevent significant resistance changes. Passivated resistors up to 400 $\Omega/\square$ have proved to be stable under IC assembly temperatures and storage and operating life tests.

Nichrome and cermet resistors have considerably lower temperature coefficients than diffused resistors. They range from 50 to 250 ppm/°C. The temperature coefficient of nickel–chromium films depends on the rate of deposition, the nature of the substrate, the composition of the film, and the residual gas pressure during evaporation. Temperature coefficients of bulk alloys can vary between 100 and 600 ppm/°C. Temperature coefficients of 80–20 nichrome films have been reported between −50 and 600 ppm/°C. Rate of deposition, the condition of the substrate, the vacuum pressure, and the source composition have to be carefully checked. Good systems should yield a temperature coefficient less than 150 ppm/°C. In

Aluminum Resistors	Anodized Tantalum Resistors	Anodized Titanium Resistors
• Suitable for small-valued resistors — on the order of 10 Ω • Best-known evaporant • Good adhesion to glass • Temperature coefficient: 1500 ppm/°C	• Resistor quality independent of line width, film thickness, and anodizing voltage • Sheet resistivity: 5–75 Ω • Temperature coefficient: −50 to 100 ppm/°C • Use narrower lines • Preferred substrates are sapphire, glazed alumina, and glass • Can be made with ±0.1% initial tolerance with 20-year stability	• Sheet resistivity of 1–2 k$\Omega/\square$ • Tolerances of ±1% possible • Temperature coefficient: 50 ppm/°C for resistivity of 1 k$\Omega/\square$ • Voltage coefficient: 425 ppm/°C

Table 13-4   Typical Diffused and Thin-Film Resistor Comparisons

Variable	Diffused Resistors		Thin-Film Resistor
	Monolithic	Multichip	Nichrome
Resistivity ($\Omega/\square$)	2.5 (emitter) 100–300 (base)	2.5–300	40–400
Resistance range (Ω)	100–30,000 (base) 10–1000 (emitter)	–	20–50,000
Temperature coefficient (ppm/°C)	500–2000 100 (emitter)	500–2000	36–60
Maximum power (W)	0.1	0.25	0.1
Tolerance (%)	± 10	± 5	± 5
Range of values (Ω)	15–30,000	15–30,000	40–100,000

Table 13-5   Properties of Tantalum Thin-Film Resistors

Property	Typical	Possible
Resistance value (Ω–MΩ)	10–0.25	1–5
Precision (%)		
Individual	0.1	0.02
Substrate	3.0	–
Aging (end of life – %)	1	0.05
Line Width (mils)	2	0.2
Sheet resistance ($\Omega/\square$)	10–100	to 3000
Temperature coefficient resistance	-70 ± 30 ppm/°C	Zero (or as required)
Power density (W/in.²)	20	100

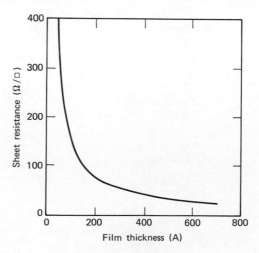

Figure 13-8   Sheet resistance as a function of thickness for a nichrome film.

contrast, the temperature coefficient of diffused resistors depends on the resistivity profile. The more heavily doped the material, the lower the temperature coefficient. A diffused resistor of $200 \, \Omega/\square$ may have a temperature coefficient up to 2000 ppm/°C.

RESISTOR PARASITICS

Thin-film parasitic materials are directly proportional to area, whereas diffused-component parasitics are both area and voltage dependent. The junction area includes not only the surface of the resistor area, but also the sidewalls and the ends. The primary effect of parasitic capacitance of resistors is to limit their frequency response. The magnitude of this effect is seen in Table 13-6. The parasitics in thin-film components are much smaller than those associated with diffused components and are not voltage dependent. Except for coupling with adjacent components, they are almost negligible in passive-substrate circuits.

The *contact resistance* of thin-film resistors is an important, two-part factor, especially for the lower valued components. The first part is the resistance of the thin film underneath the contact through which the current must flow. The second part is the interface resistance resulting from the alloy, mixture, or compound formed between the resistor surface and metallized contact. Both should be minimized by making the contact as large as possible. The contact resistance of a 1-mil² contact can be of the order of 2 to $10 \, \Omega$ – depending on the thickness of the thin film and the particular process steps used. Occasionally, much higher contact resistance (up to $1 \, k\Omega$) can be found. This results from a high interface–contact resistance, probably due to an oxide formation or other contamination. Such severe cases can be observed as a nonlinearity on a curve tracer. Many times, these can be "burned in" by increasing the voltage across the resistors.

Table 13-6   Cut-off Frequency and Parasitic Capacitance of Integrated Resistors

Resistor Type	Reverse Bias Voltage (V)	Capacity (pF/kΩ)		Cut-off Frequency (MHz), 5000-Ω Resistor	
		1-Mil Width	2-Mil Width	1-Mil Width	2-Mil Width
Diffused	1	0.66	2.7	19.3	4.7
$200 \, \Omega/\square$	5	0.41	1.5	31.0	5.7
(0.5 Ω-cm substrate)	10	0.32	1.1	39.8	11.6
Diffused	1	1.35	4.7	9.4	2.7
$200 \, \Omega/\square$	5	0.93	3.3	13.7	3.9
(0.1-$\Omega/\square$ substrate)	10	0.64	2.25	19.9	5.7
Thin film (0.022 pF/mil²)					
$200 \, \Omega/\square$	—	0.11	0.44	115	28.7
$400 \, \Omega/\square$	—	0.055	0.22	230	57.4

Careful processing and high-temperature annealing tend to minimize contact resistance problems.

THIN–FILM RESISTOR DESIGN

Several rules for film resistor design are as follows.

1. The largest practical tolerance should be allowed.
2. The resistor should be designed for the lowest values possible.
3. Large contact areas should be used.
4. Spacing between contacts (the resistor length) should be more than 1 mil.
5. Bends and corners should be avoided. Where they are required, their effects must be considered.
6. Subsequent process temperatures should be held to a minimum.

In simplified form, a thin-film resistor is a metallic film of thickness t, length L, and width W. Its resistance R is given from Figure 13-9 as

$$R = \frac{\rho L}{tW} = \frac{\rho}{t} \cdot \frac{L}{W} = \text{sheet resistance} \times \text{aspect ratio} = R_S \frac{L}{W} \qquad (13\text{-}1)$$

where ρ is the material resistivity.

The ratio ρ/t is known as the sheet resistivity and is given in ohms per square. The ratio L/W is known as the aspect ratio. A widely used resistivity R_S is $100\ \Omega/\square$. Whenever possible, all resistors in a device should be designed with the same width. Thin-film sheet resistivities have values up to $5\ \text{K}\Omega/\square$. If ρ and t are fixed by processing, resistance can be designed by adjusting the aspect ratio L/W usually specified as $L/W\ \square$ of dimension W (Figure 13-9).

Where a resistor turns a sharp corner, the corner square should be counted as only $0.4\ \square$ in design, owing to current crowding along the inside edge of the corner (Figure 13-10). Resistors are affected by bending, which is quite common, particularly with high values of resistance. At present, the best way to handle these problem areas is to minimize them. Whenever possible, long straight resistors (particularly for resistivity $> 100\ \Omega/\square$) should be used. If the resistor becomes too long for convenient layout, a corner will have to be made. In practice, all corners are rounded off somewhat. Thus it is recommended that the corners be drawn square and that the practical approximations of Figure 13-11 be used as the equivalent number of squares.

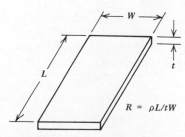

Figure 13-9 Thin-film resistor defined.

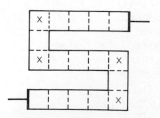

Figure 13-10 Example of thin-film resistor turning a sharp corner.

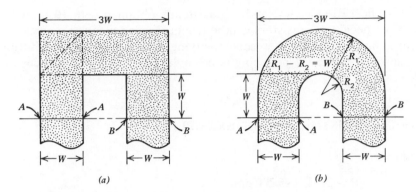

Figure 13-11   Typical thin-film resistor corners:

(a)   Square corners:

calculated
$$R_{AA} - R_{BB}/\rho_S = 4.10 \, \square;$$

practical
$$R_{AA} - R_{BB}/\rho_S = 4.0 \, \square.$$

(b)   Round corners:

$$R_{AA} - R_{BB}/\rho_S = (\pi/\ln R_1/R_2) + 1$$
$$= 3.86 \, \square.$$

Customary resistor design practice dictates that resistors terminate on conductors with the two ends facing in opposite directions, leaving a minimum of 0.2 mil overlap to ensure freedom from mask-registration errors.

Various resistor layout techniques are presented in Figure 13-12.

Resistors must cover enough substrate area to provide efficient heat conduction away from the resistor film. A given packaging configuration and the resistor materials determines the effective power density for resistors in units of power

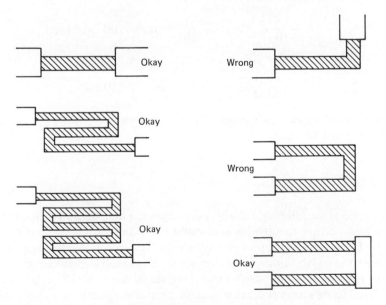

Figure 13-12   Resistor layout techniques.

per area. Deposited resistors can be consistently obtained with 10 to 20% tolerances. When necessary, resistors can be trimmed using electronic or mechanical techniques; however, there is an accompanying decrease in yield. With trimming, $\pm 0.5\%$ absolute and $\pm 1\%$ matching is possible, but 2 to 5% tolerances are preferred.

THIN–FILM RESISTOR DESIGN EXAMPLE

REQUIRED:

$$R = 2.5 \text{ k}\Omega, \pm 200 \text{ ppm/}°\text{C}$$

$$P_D = 40 \text{ mW at 50\% duty cycle} = P_{D,\text{area}} = 20 \text{ mW}$$

DESIGN:

$$P_{D,\text{area(min)}} = \frac{20 \text{ mW}}{0.01 \text{ mW/mil}^2} = 2000 \text{ mil}^2$$

Then

$$L \times W = 200 \text{ mil}^2$$

Assume: 10-mil resistor width at $100 \ \Omega/\square$. Then

$$\frac{L}{W} = 2000 \text{ mil}^2 = \frac{L}{10} = 200 \text{ mil} \qquad \text{or} \qquad 20 \ \square$$

or

$$10 \times 10 \text{ mil} = \frac{100 \text{ mil}^2}{100 \ \Omega/\square \text{ area}} = 1 \text{ mW/}\square = 20 \ \square \text{ min}$$

Resistor length L is derived from

$$R = \frac{\rho L}{tW} \qquad \text{and} \qquad \frac{\rho}{tW} = R_S = 100 \ \Omega/\square \text{ area}$$

Then

$$L = \frac{2500 \ \Omega}{100 \ \Omega/\square} = 25 \ \square \qquad \text{or} \qquad 250 \text{ mils}$$

If this value is not equal to or greater than that obtained for the P_D calculation, a larger square must be used.

THIN–FILM CAPACITORS

There are several varieties of thin-film capacitors. The true thin-film capacitor has metal top and bottom electrodes separated by a dielectric. The entire capacitor (Figure 13-13) is built on top of the substrate. It can be constructed on a silicon chip on top of the passivating silicon dioxide layer, as well as on a ceramic substrate.

Another type of capacitor is the metal–oxide–silicon (MOS) capacitor. It may be considered to be either a semiconductor chip component or a compatible thin-film component. This easily fabricated capacitor is particularly useful and is

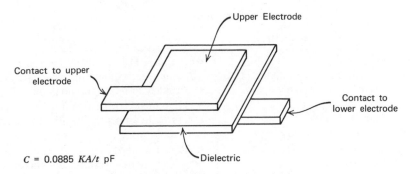

$C = 0.0885\, KA/t$ pF

Figure 13-13   Construction of typical thin-film capacitor.

finding wide application. The silicon substrate serves as the lower conductor. The substrate area under the capacitor usually contains an emitter-type diffusion for the purpose of lowering the sheet resistance. Thermally grown silicon dioxide or a thin film of deposited glass can act as the dielectric. The upper electrode is the same metallization as that used for the IC metallization.

The capacity of a thin film capacitor is given by,

$$C = \frac{0.0885 KA}{t} \text{ pF} \tag{13-2}$$

where K is the dielectric constant, A the area (cm²), and t the dielectric film thickness (cm).

Table 13-7 summarizes the characteristics of commonly used thin-film capacitors. Table 13-8 and Figures 13-14 and 13-15 compare vacuum-evaporated capacitors with anodized capacitors. Figures 13-14 and 13-15 depict capacitance versus temperature for a typical evaporated capacitor and a typical anodized tantalum capacitor, respectively. Diffused and thin-film (oxide) capacitors are compared in Table 13-9.

As indicated previously, the dielectric may be vacuum deposited as a dielectric or it may be chemically developed after deposition and patterning by anodizing. Since dielectric compound materials are subject to break down in a conventional vacuum-deposition process, the materials are generally sputtered. The dielectric may also be thermally grown on silicon, for MOS capacitors. Very-high-valued capacitors are sometimes made by depositing electrodes on opposite sides of the

Table 13-7   Typical Thin-Film Capacitor Data

Material	Dielectric Constant	Temperature Coefficient (ppm/°C)	Capacitance/ mil² (pF)	Film Thickness (Å)
Silicon monoxide	4.4–6.3	+90–200	0.25–0.36	1000
Silicon dioxide	5.8	+200	0.33	1000
Tantalum oxide	20–25	+150–3000	1.1–2.3	1000–6000
Boroaluminum–silca	6.5	+200	0.37	1000
Aluminum oxide	9	<+400	1.2	400

Table 13-8   Comparison of Thin-Film Capacitors

Vacuum-Evaporated Capacitors	Anodized Capacitors
<ul><li>Dielectric is usually SiO</li><li>Normally used counterelectrode materials are aluminum or chromium plus copper</li><li>Typical characteristics Capacitance = 0.6 μF/in.^{2} Dielectric constant (SiO) = 6.0 Temperature coefficient of capacity: 110 ppm/°C Breakdown potential at 25°C: 50 V</li><li>Problem of shorting of counter-electrodes due to pinholes in dielectric</li><li>A structural weak point can occur if the edges of the films are sharp</li><li>Leakage of evaporated capacitors may be reduced by using two evaporated dielectrics</li></ul>	<ul><li>Anodized titanium, aluminum, and tantalum capacitors are available</li><li>Primary film is tantalum pentoxide</li><li>High dielectric constant, which yields smaller areas than equivalent silicon monolithic capacitors</li><li>Poor frequency performance above 10 kHz due to series resistance of tantalum film</li><li>Temperature coefficient of tantalum oxide: 220 ppm/°C</li><li>May be trimmed to desired magnitude</li><li>For second counterelectrode, metals having good adhesion allow lower operating voltages (Ta and Al) and metals with poor adhesion have large values of breakdown voltages (Au and Pd)</li><li>Good design guide: max operating voltage × capacity/cm^2 = 5</li><li>Not polarization sensitive</li><li>Anodized aluminum capacitors have been used for high-frequency applications</li><li>Temperature coefficient of aluminum same as tantalum, but dielectric constant is about 10 for aluminum oxide; thus twice as much area is needed as for tantalum</li></ul>

substrate, but attached ceramic-chip capacitors are generally much less expensive when high values are needed. Attaching the chips avoids the necessity of double processing in the deposition chamber.

Film-capacitor fabrication methods evolved from discrete-capacitor techniques. Deposited dielectrics basically replace the glass or ceramics of multilayer-type glass or ceramic capacitors. Since it is not very practical to deposit many layers of dielectrics and electrode materials, the number of layers is usually limited to one and sometimes two. Therefore, deposited capacitors tend to have small values.

Anodized tantalum capacitors are very popular in film circuits, principally because the same deposition can be used for resistors. Quite high capacitance can be achieved in a small area. However, as Figure 13-15 indicates, a rapid increase with frequency in the dissipation factor for tantalum thin-film capacitors limits their use in high-frequency circuits. To extend this frequency, a relatively thick layer of tantalum is deposited over a conductive metal such as aluminum. Capacitors having a Q of 50 to 5 MHz may be fabricated by underlaying the tantalum film with aluminum and by using thick counterelectrodes of copper.

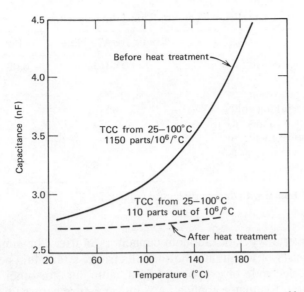

Figure 13-14   Evaporated capacitor capacitance as a function of temperature and heat treatment.

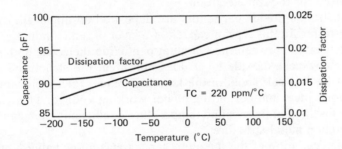

Figure 13-15   Capacitance and dissipation factor versus temperature for an anodized tantalum capacitor.

Table 13-9   Typical Diffused and Thin-Film (Oxide) Capacitor Data

| Variable | Diffused Capacitors | | Thin-Film Capacitor |
	Monolithic	Multichip	Oxide
Maximum Capacitance (pF/mil²)	0.2 for collector–base junction	1	0.25–0.40
Breakdown voltage (V dc)	5–20	5–50	50
Q (at 10 MHz)	1–10	10–50	10–1000
Voltage Coefficient	$c = kV^{-1/2}$	$c = kV^{-1/2}$	0
Tolerance (%)	±20	±10	±10

Table 13-10   Properties of Tantalum Thin-Film Capacitors

Property	Routinely Available	Possible
Capacitance density (μF/cm²)	0.05–0.1	0.002–0.5[a]
Dielectric constant	21–22	5–6
Forward working voltage (V)	8	15[a]
Capacitance values (μF)	to 0.05	to 1[a]
Temperature coefficient of capacitance (ppm/°C)	200	+500 to 1000
Stability (%)	1	0.1

[a]Value obtained with duplex dielectric capacitors.

There are several types of tantalum capacitors in use. The simplest tantalum–metal type consists of a tantalum bottom electrode, a tantalum pentoxide dielectric, and a top electrode of evaporated gold, aluminum, or other metal. This capacitor is suitable for many applications, but moisture can readily penetrate the electrodes. However, this susceptibility to degradation from humidity can be reduced by putting a layer of nickel–chromium or titanium over the metal (gold or aluminum) used for the top electrode.

Some tantalum pentoxide capacitors are deposited on unglazed, rough surface ceramic substrates. In this type of capacitor, a layer of semiconducting manganese dioxide is deposited between the tantalum pentoxide dielectric and the top electrode. The manganese dioxide layer does not affect the structure's capacitance value, but it heals weak spots in the tantalum oxide, thus increasing yield, improving performance, and permitting higher working voltage. This structure provides higher dissipation factors and temperature coefficients than can be obtained with the tantalum metal structure.

Still another variation of the tantalum capacitor is a low value type that uses a silicon monoxide layer between the dielectric and the top electrode. The capacitance value is determined almost entirely by the thickness of the silicon monoxide layer because of the material's low dielectric constant of 6 (cf. 22 for tantalum pentoxide). Capacitance values to a few picofarads can be obtained. Table 13-10 summarizes the properties of tantalum thin-film capacitors.

THIN–FILM INDUCTORS

Inductors are often made in thin-film r-f circuits when a low-valued inductor is needed and high Q is not required. They are simply a round or square spiral pattern, like those in Figures 13-16 and 13-17.

Thin-film designers generally reserve the use of thin-film inductors to applications in which it is possible to completely surround the conductors with high-quality dielectric, such as in a microwave hybrid IC with carefully selected substrate materials. These applications require very exhaustive design analysis, since the coils operate as lumped components rather than simple coils.

Monolithic IC fabricators have had poor luck with inductors deposited on the

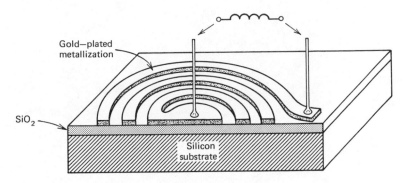

Figure 13-16   Round spiral pattern for an inductor.

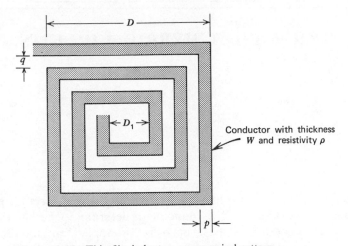

Figure 13-17   Thin-film inductor, square spiral pattern.

oxide coating of silicon wafers. Since values of only 5 μH or less, with low Q, are obtainable, it is clear that monolithic technology favors the use of active filters.

The inductance of a thin film inductor is given by

$$L = 8.5 \times 10^{-3} DN^{5/3} \, \mu\text{H} \tag{13-3}$$

where

$$N = \text{the number of turns} = D/2(q+p)$$
$$\text{spiral length} = d = 2DN \text{ (Figure 13-17)}$$

Semiconductor techniques have not successfully demonstrated methods of substitutionally replacing inductors. The transistor turn-on delay has a negligible inductive characteristic at low and moderate frequencies. At sufficiently high frequencies, the input impedance of the transistor can appear inductive because the turn-on delay is approximately 90°C. The engineering response to this problem has been to avoid the use of inductors in IC design, except for gyrator–capacitor networks, which are practical alternatives.

Chapter Fourteen

THICK–FILM HYBRID CIRCUITS

Each of the advantages of thin film over monolithic circuits could, perhaps, be made for thick-film structures. It is difficult to determine whether the thick- or thin-film approach is better. At best, each hybrid application should be investigated separately to find out which is the most feasible approach.

Like thin films, thick films allow a designer to come closer to converting his breadboard to a more reproducible package than he could obtain with a fully monolithic approach. However, thin films can be used to form capacitors, resistors, conductors, inductors, and even diodes and transistors, whereas thick films are usually restricted to resistors and conductors. Resistors of several megohms are practical. The screen-and-fire technology can be extended to capacitors by using glass dielectrics. Devices as large as several hundred picofarads appear to be practical.

Thick films are generally several microns thick and are less precisely controlled than thin films. Silk screening is commonly used to define the film pattern. The silk-screen process is relatively simple, but it is incapable of directly achieving tight tolerances. Close control of characteristics can be gained by using special geometric control techniques in conjunction with silk screening (e.g., "sandblasting" resistors after firing). Line widths and spacings greater than 10 mils must be used. The resolution is ±5.0 mils. The process is straightforward and additive, although the films must be "fired" after they have been deposited. The thick film process is the most economical, requiring the smallest outlay in capital equipment of any of the film processes.

Thick-film structures might be expected to be more reliable owing to the high-temperature processes employed in their fabrication, but techniques have been developed for many more materials with thin films than with thick films. Also, there is apparently no foreseeable chance to obtain an active thick-film device structure; this is not the case with thin films. If size is important, thin films allow higher resolution in the structures.

Generally speaking, thick-film resistor tolerances should not be better than $\pm 1\%$ and the circuit should not contain any resistors larger than several megohms. Aside from processing methods and layout dimensions, the design rules are similar to those given previously for thin-film circuits.

DESIGNING A THICK–FILM CIRCUIT

The first step in deciding whether a given circuit can be built in thick-film form within the necessary confines of volume is to list everything that constitutes the circuit: All the transistors, diodes, resistors, capacitors, and other components are written down. This helps to decide what size substrate the designer can mount the circuit on, which in turn determines the package size and the number of packages required for the implementation of the original circuit. For this purpose, we need the best information available about the actual size of the components. As a first approximation, however, it is customary to employ an average value for the size of each of the various kinds of components. The designer should assume the following component sizes for a first approximation: A typical thick-film resistor would not exceed 40×40 mils in area. A capacitor would seldom be more than 80×50 mils. A typical diode die would be 20×20 mils. A transistor die might be 30×30 mils. Most linear IC's are 60×60 mils, whereas RTL or DTL die generally run about 40×40 mils; on the average, TTL die have a 50×50 mil area. The use of MSI yields a die size of about 80×80 mils; a 140×140 mil LSI die would be perhaps the largest single component to be included in a thick-film hybrid circuit.

Then we apply the following rule of thumb: The total area of these components should not be greater than 20% of the area of the substrate size. The remaining 80% of the substrate is required for the conductors, the pads, and the space between all these. Of course, if a designer wants to cram more components in and still use thick-film technology, he can resort to such methods as multilevel metallizations, or he can use the back side of a substrate and have holes through the substrate. But in general these methods should be avoided because they complicate fabrication and reduce the yield. Knowing the substrate area needed for a circuit, the number of different packages required to implement the circuit design can be obtained once a definite package size has been selected. Alternatively, the substrate area tells how big a package may have to be used in order to fit everything into one package.

Power Dissipation

The next step in confirming that a circuit can be made in thick-film hybrid form is to look at the power dissipation requirements. It is necessary to determine how much power each component will be dissipating. Then the total power of all these components is summed to determine the ability of the package to dissipate the total power. The important package parameters are θ_{jc} (the thermal-resistance junction to case) and θ_{ja} (the thermal-resistance, junction to ambient, where the ambient is free air). For example, an eight-lead TO-99 package would be rated at about 40°C/W, as the junction-to-case thermal resistance; and θ_{ja} would be about 250°C/W going to free air. A $\frac{3}{8} \times \frac{3}{8}$ in. flatpack with 14 leads having a substrate size of 245 mil² would be rated at about 40°C/W in junction-to-case thermal resistance but only 125°C/W for the junction-to-free-air thermal resistance. These values for a 30-lead 1×1 in. flatpack with a $\frac{3}{4}$-in.² substrate would be 20°C/W for the θ_{jc} and 60°C/W for the thermal resistance between the junction and the ambient. The values of thermal resistance from junction to case can vary at least 50%, depending on the type of dice used and how they are packaged together. The values of the thermal resistance from junction to free air can be reduced by proper heat sinking. These thermal resistance values are just rough guides, but they can help to avoid the error of trying to place a hybrid circuit in a package that cannot dissipate the required power.

The power-dissipation problem requires a consideration of substrates. Up till now there has been the implicit assumption that the hybrid circuit substrate would be 96% alumina, which is used in the vast majority of cases. However, beryllia (BeO) substrates sometimes serve where higher thermal conductivity is a definite requirement. The thermal conductivity of beryllia is about 0.38 cal/(sec)(cm²)(°C) (cm), whereas the 250°C value of thermal conductivity for 96% alumina is about 0.084 cal/(sec)(cm²)(°C)(cm). Since beryllia is much more expensive than alumina, and beryllia dust presents certain toxicity problems, alumina is used in most instances.

There seems to be general agreement that the maximum power dissipation that a 96% alumina substrate is capable of handling is about 10 W/in.², if proper heat sinking is provided. Since most substrates are between 15 and 25 mils thick, the heat generated on the top surface must be conducted through that thickness of alumina and into the package. One final checkpoint before getting down to the artwork stage is verifying that the number of pins on the package intended for use is equal to or greater than the number of terminations on the circuit schematic. This is seldom a problem with the wide variety of multilead packages available today.

Laying Out the Circuit

Having concluded that the circuit can be made in thick-film form, the designer must "lay out" the circuit so that it can be fabricated. This means drawing a layout to scale at least 10 times the final size of the circuit which can be used for preparation of the rubylith artwork. The rubyliths are then given to the photolithography group for preparation of the thick-film screens.

Thick-film hybrid circuit layout begins with a consideration of the design rules related to the metallized pattern to which various components are to be attached and by which they will be electrically interconnected. (These rules are spelled out at the end of the chapter.) Of course, the designer and the builder of the circuit must first have agreed on the kind of metallization scheme to be used on the substrate. This choice is made from the standpoint of compatibility with all the processes to be used in circuit assembly and in light of the end-use requirements.

In preparing the artwork for the metallization screen, the designer should plan to have a minimum width of 10 mils for each of the conducting traces on the substrate. Remember that the artwork must be at least 10 times larger than the finished size on the substrate. The 10-mil minimum-conductor-width design rule is a result of a compromise between resolution and ease of screening.

If very fine-mesh screens are used, it is difficult to force the ink through the screens, and uneven traces are obtained. Too large a mesh, on the other hand, leads to lack of definition, and adjacent traces must be spaced further apart. Size 200 mesh screens appear to be a reasonable compromise between resolution and ease of printing. With the 200 mesh screen, a good rule of thumb is to have a minimum width of 10 mils and a very minimum spacing of 10 mils between adjacent conductors. It would be much better to use 15 to 20 mils between adjacent conductors, because a width in this range would make the yield at the screening step that much better. If the design calls for external lead pads, they would probably be about 50×75 mils in area.

THICK–FILM CONDUCTOR DESIGN

In general, the conductor traces should be kept as wide as possible to minimize the resistance and hence the power dissipation. The circuit should be designed such that all the metallization is on one side of the substrate. If it becomes necessary to interconnect top and bottom surfaces, holes should be used rather than running a conducting trace around the edge of the substrate. Holes, if used, should have a diameter at least as large as two-thirds of the substrate thickness so that electrical continuity through the holes can be established in the screening process.

In the conductor layout, cross-overs are to be avoided because of the complication of having to put down a glass intermediate layer. Another rule is never to have wires crossing each other. There is no good way to insulate such wires, and it is too risky simply to rely on their remaining apart.

The components have to be put down on conducting pads. The back sides of the transistors make electrical contact through the bulk silicon down into that pad. Thus as a conservative design rule, a minimum of about 5 mils spacing should be maintained around the die on every side. This is necessary to allow the person who has to put the die down to center it on the pad and still see some metallization around the outside, which will make it possible to obtain a fillet, to assure a reliable die attach. If a wire bond to the pad is required directly adjacent to a transistor die, we will need room to get the bonding tool down on the pad, and a minimum of 15 mils between the edge of the chip and the edge of the pad should be allowed. No matter what kind of bond is being used—whether a wedge bond, a ball bond, or an ultrasonic bond—a minimum area of 10×10 mils on the bonding pads should be provided.

The general rule for wire lengths is to avoid any wire longer than 100 mils. Obviously, wires should be kept as short as is feasible, without being bent at such a sharp angle that they represent an undue stress on the bonding. There is another rule of thumb which might at first glance seem a little odd: Never bond directly

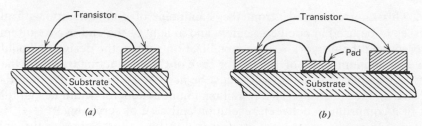

Figure 14-1   Thick-film transistor connection guidelines: (*a*) not recommended, (*b*) recommended.

between semiconductor chips. If, for example, we want to connect the emitter of one transistor to the base of another, it might appear that a direct wire between the two chips would be the proper procedure. However, this is not so. It is recommended that an intermediate pad be used to which wires from both transistors would be bonded (see Figure 14-1). At first, the intermediate pad sounds bad because it requires twice as many bonds. However, it is frequently necessary to replace one or the other component, and a component can be removed much more easily if it is not directly connected to another component. Removing a wire from a metal pad on the substrate leaves plenty of area for another bond, but removing a wire from the small contact pad on a transistor chip frequently leaves too little undisturbed area to permit the making of a reliable replacement bond.

Another good design rule is to avoid having components or conducting traces near the edge of the substrate. The reason for this, of course, is to prevent handling damage. A pair of tweezers can seriously damage gold traces. Ten mils is about the minimum allowable edge clearance, but 20 or 25 mils is better if that much area is at the designer's disposal.

A component should not be mounted across a conductor. The eutectic material used in attaching the component could flow and short two conductors without being visible.

THICK–FILM RESISTOR DESIGN

The resistors are layed out next. The range of resistor values should be limited so that they can be fabricated with a minimum number of screenings. A rule of thumb suggests a maximum of three screenings. Any one ink can usually be counted on to cover a range of resistor values anywhere from 30 to 1 to perhaps 100 to 1. Still, in certain situations even three screenings are insufficient. In such cases, ink combinations should be chosen that can make most of the resistors with the film-screening technique. Then the others are put down as resistor chips.

As previously discussed, the resistor value is the product of the sheet resistivity (which is the bulk resistivity divided by the thickness) and the aspect ratio (which is the length divided by the width, as given earlier). It is assumed that the resistors are rectangular sheets. With that assumption, and given a certain value of a resistor to be fabricated by screening, how do we calculate the physical size and shape to be provided on the screen? First, allowance is made for trimming so that the resistor values can be adjusted to a $\pm1\%$ tolerance. Although the as-fired resistors probably could be held to 20% or maybe $\pm10\%$ inaccuracy, frequently that is not good enough. The standard procedure for trimming, then, calls for an air-abrasive unit. Or, more recently, a laser is used. Ordinarily the resistor layout size planned should represent about 85% of the final design value. For example, if a designer wanted to have a 25-kΩ resistor in his circuit, he would want the screened value to be 85% of that, or 21.3 kΩ.

In using a 12.5-kΩ/□ ink, the number of squares, which is equal to L divided by W, is 21.3/12.5, or 1.7 □. This is still indefinite because only a ratio of length to width has been defined. To decide on a definite size, we arbitrarily choose the width to be some value, say 50 mils; then the length will be 50×1.7. Thus the area of the resistor between the metal contact pads should be 85 mil². The actual hole in the screen would have to be somewhat larger to allow for the overlap, which is discussed below. A rule of thumb is to make the resistors as large as possible, depending of course on the available area. This permits greater freedom when trimming, it makes the power dissipation problems simpler, and it also means that slight errors or tolerances in the artwork will be insignificant. The L/W aspect ratio should not, in general, exceed 10:1. Good conservative design requires that L/W is less than 5:1 for the case where the length will be greater than the width. The reverse situation can also occur (i.e., width > length); then the aspect ratio should be 1:2 or greater.

Already mentioned were the desirability of using the minimum number of inks and the minimum number of screenings. For example, a circuit containing resistors having values of 100, 50, 10, and 0.5 kΩ could be made in two screenings if there were available an ink of 37.5-kΩ/□ sheet resistivity and a second ink with 375-Ω/□ sheet resistivity. Obviously many other ink combinations would work just as well. If several combinations of inks are available, all requiring the same number of screenings to make the resistor values called for in the circuit, the actual areas needed with the different ink combinations can be calculated and a selection can be made based on the optimum utilization of space on the substrate.

RESISTOR POWER DENSITY

Earlier, power dissipation of the total package was mentioned. There is also the matter of power dissipation of resistors, which frequently are the circuit elements scheduled to be dissipating the most power. Typically, thick-film resistors are rated at 35 W/in.² of resistor area at 125°C.

An example of using this kind of information in a layout is to calculate whether the resistors are large enough to dissipate the necessary power. For example, a 30×60 mil resistor occupies 0.0018 in.², and at 35 W/in.² can dissipate 0.063 W. If the voltage drop across this resistor and the current that must pass through it will produce less than 0.063 W, then the resistor will not overheat and there is no reliability problem. But if the anticipated power dissipation in the resistor exceeds 0.063 W, the resistor must be made larger by using a different ink.

A general rule of thumb for resistor layout is to first lay out all the resistors parallel to the substrate edges. This makes it possible to trim all the resistors with a minimum amount of problems. Also it is desirable to be able to work in $X–Y$ directions and not to have to rotate the piece all the time. Again from the standpoint of trimming, resistor loops should be avoided where alternate current paths make it impossible to trim the resistors in the very elementary and simple rectangular configuration. Try to avoid zig-zags where the corners tend to overheat. Given a very high aspect ratio, in the 5 : 1 or 10 : 1 range, it is best to use the hat configuration (discussed in the layout rules at the end of the chapter). Again, the desirability of having 20 mils minimum clearance between any resistor and the edge of the substrate is stressed. Resistors should be made as large as is practical. A frequently quoted minimum resistor size is 20×20 mils. Under no circumstances should a resistor be smaller than 20×20. Also, the overlap of the conductor, around the ends of a resistor, should be a minimum of 10 mils, to make sure that there is good contact around the entire resistor. The spacing between the resistor and the pads or the conductors should be 20 mils minimum. In addition it should be kept in mind that at least 25 mils of access space is required for the air-abrasive trimmer. In addition to screened thick-film resistors, discrete chip resistors are available. Chip resistors are discussed at the end of this section.

All foregoing rules have been two-dimensional. There is also, of course, the third dimension of height. A hybrid circuit containing large capacitors that are much higher than the other chips can be very awkward to trim, even though the above-cited rules have been followed to the letter. It should be certain in the thick-film layout stage that enough room exists for the required tools to finish the job when all the components are in place.

CHIP RESISTORS

Although the thick- and thin-film deposition process fabricates resistors directly on substrates, at times it is more efficient to use discrete components. Chip resistors are used for critical values to diminish substrate yield losses and, in general, to simplify the process when only one to two resistors are required. This condition occurs when the desired values would call for a different type of paste for each resistor. Thick-film discrete resistors also serve for attaining higher values on a thin-film hybrid circuit, or when a resistor of higher wattage rating is needed. Discrete chips can offer advantages in achieving the most efficient area for circuit layout, since they can be strategically placed to bridge conductor runs. However, care should be taken in this respect because of the potential effects of induced capacitance.

Chip resistors are available in three designs: thick film, thin film, and solid cermet. The film resistors are usually deposited on an alumina chip utilizing the same techniques as used for normal thick- and thin-film deposition. (Glass may also be used as the base chip material.)

The resistive thin films are vapor deposited by way of either aperture mask or selective etch techniques and are usually fabricated of chromium or nickel–chromium with resistivities up to $500 \, \Omega/\square$. High resistivities tend to be unstable, and for purposes of achieving a balance between stability and size, the normal working resistivity is $100 \, \Omega/\square$.

The thick-film chip-resistor element is screen-printed and fired using a noble metal resistive paste, resulting in resistivities of up to $1 \, M\Omega/\square$. In direct contrast to thin film, the higher resistivity formulations of thick-film resistors tend to be more stable because the lower resistivity pastes have less glass binder acting as an encapsulant. If the resistivity is too high, however, reproducibility becomes a problem. High-value chip resistors can be fabricated using the thick-film process, with some sacrifice in noise level and stability as compared with thin-film resistors. Most film chip resistors are designed for a single resistance value, but tapped resistors are available in up to four taps per chip.

Solid-cermet resistors are fabricated of a material similar to thick-film resistors, but without a base substrate or chip. Cermet is a mixture of metal, metal oxide, and powdered glass combined with organic binders and solvents to form a paste. Upon forming and firing at elevated temperature, the organic binders and solvents vaporize, and the melting of the powdered glass securely binds the metal and oxide powders to form a homogeneous mass.

Thin-film resistor chips are available with a temperature coefficient of resistance (TCR) of better than 75 ppm/°C and a tracking TCR of from 2 to 5 ppm. Resistance values range up to approximately 1 MΩ, with a standard tolerance of from 5 to 10%; lower tolerances may be obtained on special order. Temperature drift is often specified as 0.05% maximum at 125°C for 1000 hr.

Thick-film ceramic-chip resistors and solid-cermet resistors are offered with a TCR of between 30 and 300 ppm/°C and a tracking TCR of from 25 to 100 ppm. The range of values of from 10 Ω to 15 MΩ is available in a standard tolerance of 5 to 10%; lower tolerances require trimming. Thick-film chips may be obtained with a temperature drift of 0.5% at 150°C for 1000 hr.

There is no size standarization for chip resistors. Sizes range from less than $0.050 \times 0.050 \times 0.012$ in. to any reasonable size necessary to accommodate the desired resistor value. Pellet solid-cermet resistors are available as small as 0.050 in. in diameter and 0.030 in. thick.

Discrete-chip film resistors can be supplied with a gold backing for eutectic bonding to the substrate, the interconnection being accomplished by wire bonding to the resistor terminating conductor pads on the top surface. Metallized edges may be specified as an alternate scheme for mounting between two land areas. Microdiscrete chip resistors, also available with beam-lead termination, are attached by thermocompression bonding. Solid-cermet resistors are supplied with metallized edges, similar to the chip monolithic capacitors with the same choice of edge materials. Tables 14-1 and 14-2 provide data on silicon chip resistors and cermet chip resistors, respectively.

Table 14-1   Typical Silicon Chip Resistor Data (Hybrid)

Property	Value
Available resistance	47–47,000 Ω
Temperature coefficients	700–1200 ppm/°C
Tolerance	
Below 100 Ω	±20%
100 Ω or above	±10%
Matched pairs	± 5%
Dissipation	250 mW
Breakdown voltages	25 V

Table 14-2   Typical Cermet Chip Resistor Data

Property	Value
Temperature coefficient ($\Omega/\square$)	
250 at	+ 500 ppm/°C
1000 at	+ 400 ppm/°C
5000 at	+ 250 ppm/°C
12,000 at	+ 150 ppm/°C
50,000 at	0 ppm/°C
330,000 at	− 150 ppm/°C
Resistor size	0.1–10 $\square$
Tolerance	± 10% → ± 1%
Dissipation	0.5 mW/mil^2
Breakdown voltage	150–350 V

THICK–FILM CAPACITORS

Up to now, little has been said of thick-film capacitors from a design standpoint. Most capacitors for use in thick-film circuits are obtained as individual chips. However, screened capacitors are also used. With thick-film capacitors of a ceramic material having a dielectric constant as high as 1000 values of capacitance as high as 15,500 pF/(cm^2)(μ) of capacitor thickness can be obtained. This would provide a breakdown strength of up to 300 V with a thickness of 25 to 50 μ. Thus the total capacity might be 15,500 pF/cm^2 with a thick-film capacitor. Chip capacitors are not limited to certain capacity values except by size considerations. They have a wide range of values and they are quite stable. The use of screened capacitors has little to recommend it except perhaps lower cost, and maybe some simplicity in assembly, which minimizes the number of attachments. But as far as control over values and range of values are concerned, chips appear to be better.

One major problem associated with thick-film capacitors is the difference in thermal expansion between the titanate materials used and the alumina substrate. To reduce the problem, glass with a high softening temperature is used, thus bringing down the average expansion coefficient. Also, the titanate must have fine particles to overcome the stresses at the glass–titanate interface.

Another important item is the conductor material of the electrodes — such materials are seldom interchangeable. Any variations are clearly attributable to the differences in concentration and composition of glass frits and other fluxes used in the conductors, rather than to interaction between the metals themselves.

CHIP CAPACITORS

Planar-deposited capacitors in general require a multiple screening or deposition procedure and consequently use considerable substrate area. Discrete capacitors are thus in demand and have, accordingly, a wide range of selection on the passive chip component market. Table 14-3 provides typical MOS capacitor chip data for various chip sizes.

Table 14-3   Typical MOS Chip Capacitor Data

Chip Size (mils)	Capacitance Available (pF)	Working Voltage (V dc)	Tolerance (%)
25 × 25	1.0–24	~ 50	20–10
30 × 30	13–56	~ 50	10–5
35 × 35	30–110	~ 50	10–5
46 × 46	62–220	~ 50	10–5
61 × 61	120–430	~ 50	10–5
Miscellaneous			

Temperature coefficient: 50 ppm/°C max
Q at 1 MHz: 1200–1500 for 1–430 pF

Capacitance is directly proportional to electrode area as given by

$$C \simeq K\frac{A}{t} \tag{14-1}$$

where C = capacitance
   K = dielectric constant
   A = electrode area
   t = dielectric thickness

Therefore, compressing the size of a capacitor results in an attempt to offset the corresponding decrease in area by increasing the number of electrodes, maximizing the dielectric constant, and minimizing the dielectric thickness. The most logical choices of dielectric materials for this application are drawn from the ceramics, whose dielectric constant can be adjusted during fabrication by modification of the constituents during the slurry stage. As an example, barium titanate ceramics can be modified to get dielectric constants ranging from 20 to more than 8000. The ability to tailor the dielectric constant is a distinct advantage, providing flexibility in design for the capacitor manufacturer. The additional qualities of mechanical strength and thermal tolerance during soldering strengthen the selection of a ceramic as the dielectric material.

The process of minimizing the dielectric thickness, while attempting to maintain electrode area, results in a multiple-electrode–dielectric stacking arrangement. This configuration achieves maximum capacitance for a given size package. The design arrangement provides greater electrode area at the expense of increased height, and when the component is attached it encroaches on less substrate area than a planar-deposited capacitor.

A multilayer capacitor design requiring minimum height is a feature of monolithic construction. This process consists of casting the unfired ceramic dielectric into thin flexible strips and screening with a noble metal paste. The strips are stacked, compressed, and fired, resulting in minimum-thickness electrodes and dielectric. The electrode thickness attained by this method is approximately 0.0001 in., with the dielectric capable of a minimum thickness approaching 0.001 in.

For applications requiring a high voltage rating at low values of capacitance, a two-electrode capacitor design allows increased dielectric thickness and results in a higher capacitor breakdown voltage. An important factor in the selection of a capacitor is stability. Capacitor stability, as a function of temperature, is primarily dependent on the dielectric material: The higher the dielectric constant, the more unstable the capacitor. For this reason, chip capacitors may be grouped according to their temperature coefficient characteristics, temperature stability for a given size package being achieved at the expense of decreased capacitance.

COMPARISON OF THIN– AND THICK–FILM HYBRID CIRCUITS

One of the reasons for going to thin-film as opposed to thick-film hybrid circuits is the need to put as many components in a circuit as possible. This means using much narrower line widths, perhaps down to 0.5 mil if possible (although values of 2 or 3 mils are probably regarded as being good design procedure in laying out conducting traces). Up to the very recent past, the thin-film resistors have typically had much lower values than thick-film resistors: $100\ \Omega/\square$ was perhaps typical of nichrome resistors, as opposed to megohms per square for thick-film resistors. But the new development of silicon–chrome resistors permits the fabrication of high-value resistors in hybrid circuits. The thin-film design rules, in general, are about the same as for thick films with the proviso that everything is going to be squeezed together by a factor of 2 or 3. Thin-film conductors are about 5000 Å thick instead of 0.5 mil thick, as in thick-film hybrids. Generally, smaller value resistors are made with thin-film techniques. (Table 13-2 compared the main features of thin-film and thick-film hybrid circuits.)

THICK–FILM PROCESSING

The following list presents a typical order of thick-film substrate processing steps. In this sequence, the less critical steps are performed first (i.e., conductor firing is a less critical step than resistor firing). The screening, firing, and trimming of the resistors are the most critical steps of the thick-film process. Steps 6 and 7 are omitted if chip capacitors are used in the circuit.

1. Clean substrate.
2. Screen conductors.
3. Fire conductors.
4. Dielectric screen.
5. Fire dielectric.
6. Screen capacitor counterelectrodes.
7. Fire capacitor counterelectrodes.
8. Screen resistors.
9. Fire resistors.
10. Trim and test resistors.
11. Clean substrate.
12. Inspection.

Processing a thick-film circuit involves a series of stainless-steel screens, each containing a resistor, a conductor, or a dielectric pattern. The screens are placed on top of a substrate, and the resistive, conductive, or dielectric material (depending on which pattern is being applied) is wiped across the screen. The pastes are transferred through the pattern on the screen to the substrate.

Typical inks available for conductor screening are gold, platinum–gold, or palladium–silver compositions. Other metallizations can be screened, but these three are most commonly used. The sheet resistivity of the gold film is about 0.005 $\Omega/\square$; that of the platinum–gold is about 0.1 $\Omega/\square$, and the palladium–silver has an intermediate sheet resistivity of 0.04 $\Omega/\square$. These are approximate figures for sheet resistivities, but obviously gold has the best conductivity. However, since gold films are not compatible with soldering, palladium–silver or another suitable metallization must be used if components are to be attached to the substrate with lead–tin solder.

A platinum–gold composition has become widely used for screening conductors on the substrate because of its excellent solderability and high adhesion to the substrate. A disadvantage is its high cost.

Palladium–gold conductors are becoming more widespread because this composition possesses essentially the performance characteristics of platinum–gold while having the added advantage of costing less. Other materials that have been used are palladium–silver and pure gold and silver.

Thick-film resistor inks are basically noble metal cermets, commercially available in a number of proprietary formulations. The inks are a mixture of oxides, metal, glass powder, and binders. They are specified in ohms per square for a nominal thickness, but this generally varies with firing temperatures and times. The firing process drives out the binders and solvents. Unlike thin-film resistor materials, it is not the conductivity of the basic metals that determines conductivity (or resistivity) of the fired traces. The finished composition and its reasons for being resistive are quite complex and could be termed an almost infinite number of tiny "bad connections," making the total trace highly resistive.

Firing, which develops the resistivity, varies with the number of screen-and-fire steps. Recommended practice is to screen the most heat-sensitive composition last and to calculate the dimensions of all inks used in terms of the total number of firings of each ink. Because of the glass and oxide content, the fired lines bond tightly to the substrate and develop a glassy surface as a rule. The glassy surface is not necessarily impervious to moisture and contaminants – some manufacturers have found it possible to make minor adjustments in resistor values by wetting the resistors when a direct current is applied across the resistors. Some compositions increase and others decrease in resistance. Before firing, the pattern on the wafer is air dried to drive out part of the binder. Then each separate deposit is fired separately, so that the entire process is repeated for each type of paste. Each paste requires a different firing temperature, and the firings are usually in order of decreasing temperature. Total firing time is typically one hour, although the firing temperatures and times vary with the materials used.

The next step in manufacturing a thick-film circuit is trimming the resistors. Resistors can be screened and fired to an accuracy of about ±10% and then trimmed to about ±1% with air-abrasive jets or laser methods.

In both abrasive and laser trimming, material is removed to increase the resistor value. Abrasive trimming equipment consists of a set of nozzles of any size or shape, positioned over the resistor to be trimmed. Aluminum oxide powder, usually 27-μ size, is propelled at the resistor and the material is removed. The operation is continuously monitored until the proper value is attained; then the abrasive flow is stopped.

Tolerances as low as ±0.5% can be realized with abrasive trimming. In this process, the tools never come in contact with the substrate; thus no shock, vibration, or heat is introduced. This process can be easily automated, and high production rates are practical. There is no damage to the substrate, because the pressure of the abrasive can be adjusted to prevent the removal of substrate material. The problem of removing the abrasive material from the substrate is readily overcome with a vacuum-cleaner setup. Another problem is overspray of the abrasive powder that could hit a neighboring resistor. This can be controlled only by careful layout of the resistors on a substrate.

When a focused laser beam is used to accomplish the task of trimming, resistor material in the path of the beam is vaporized, leaving an exposed substrate. Thus the resistor area that is not required is merely isolated from the conductive path. There are a variety of laser systems—the most common types are yttrium–aluminum–garnet (YAG) and carbon dioxide.

Laser trimming offers some advantages over abrasive techniques because the trimming can be performed on glazed resistors without altering the glaze. In laser trimming, narrow trim paths are possible. Laser-trimmed resistors seem to maintain characteristics identical to those of the basic resistive material, and abrasive units exhibit higher drift and noise parameters. However, the capital investment for a laser system is large and can only be justified for large production requirements.

Finally, the circuit is hermetically packaged or dipped in some type of conformal or epoxy coating. Encapsulation plays an important part in the characteristics of thick-film circuits from the standpoint of protecting the resistors and the active devices from further changes. It is sometimes accomplished using various plastic materials internally and, in some cases, the circuit may be placed in an additional container for hermetic sealing. Good encapsulating material prevents moisture (vapor or liquid) from reaching the resistor film or active devices. Silicone resins have been useful as protective coatings for thick film substrates. Figure 14-2 depicts the steps in the fabrication or a typical thick-film circuit.

THICK–FILM CIRCUIT LAYOUT

Many of the thick-film design and layout rules are identical to those for thin-film circuits. However, dimensional tolerances are quite different:

    Minimum conductor width: 10 mils
    Minimum resistor width: 30 mils
        (resistors should be as large as practical)
    Minimum spacing between conductor traces: 10 mils
    Minimum spacing between resistor traces: 25 mils
    Minimum clearance between edge of substrate and conductor: 10 mils

To assist in layouts, all dimensions should be in 5-mil multiples. Multiple circuits should be laid out symmetrically.

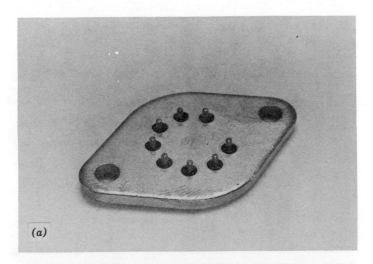

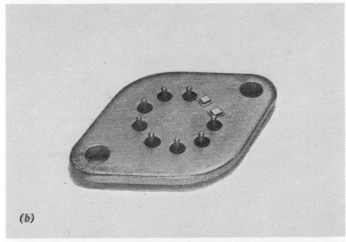

Figure 14-2 Example of thick-film fabrication procedure: (*a*) base of package; (*b*) power transistors are mounted; (*c*) alumina substrate showing screening and fixing stages of resistors, capacitors, and conductors.

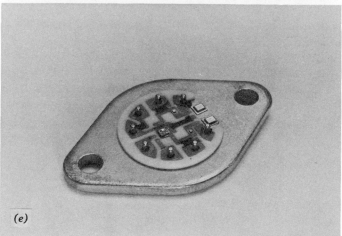

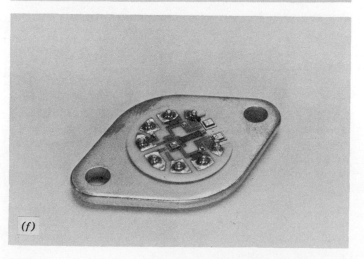

Figure 14-2 (continued) (d) add chip resistors, capacitors, and transistors to substrate; (e) attach substrate to base of package; (f) wires are bonded from package post to substrate bonding pads.

Figure 14-2 (continued) (*g*) thick-film hybrid package is sealed.

Thick-Film Conductor Layout

Long, narrow conductor traces should be avoided whenever possible, especially in high-current paths such as V_{OL} and ground, because of the finite resistance value of the conductor traces. Where conductor length may be questionable, conductors can be treated as resistors and potentials can be calculated. For such calculations, use

$$R_S = 0.01 \ \Omega/\square$$

A minimum of 10 mils in each dimension must be added to maximum dice dimensions when laying out pads for die attach (see Figure 13-4). If a lead bond is to be made from a chip to its own pad, a minimum of 15 mils must be allowed for lead-bond tool clearance (see Figure 13-5). Die-attach pads must be designed to ensure that no die can be placed within 25 mils of a lead-bond area along the line of the lead bond, nor closer than 5 mils perpendicular to this line if ultrasonic bonding is to be employed (see Figure 13-6).

Die-attach pads must be placed no less than 10 mils apart to ensure die-attach tool clearance.

Minimum dimension of lead-bond pads is 10 mil^2.

Conductors must extend 10 mils on each side of resistor film traces along the terminations to the resistors (Figure 14-3).

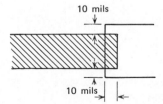

Figure 14-3 Diagram showing required dimensions of conductor resistor termination.

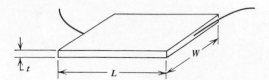

Figure 14-4   Thick film resistor dimensions.

Thick-Film Resistor Layout

Figure 14-4 depicts the germane dimensions of a thick-film resistor. A specific resistor material is specified by a bulk resistivity ρ. The deposition process may also control the film thickness t. Resistance is specified by the well-known equation

$$\frac{\rho}{t(L/W)} = R_S L/W$$

If ρ and t are fixed by processing, resistance can be designed by adjusting the aspect ratio L/W. If L/W is given the dummy dimension of "square," then ρ/t has the dimensions "ohms per square." This quantity is constant for a given process and is termed sheet resistivity R_S. Note that in ultrahigh-frequency circuits, skin-depth effects modify the foregoing description.

The two squares in Figure 14-5, composed of the same material, have the same sheet resistivity and the same ratio $L/W = 1$. Consequently they have the same resistance. The larger resistor has a higher power rating, since it occupies more area and can dissipate more heat, assuming equal thermal conductivities for the two substrates on which the resistors are deposited.

Thick-film sheet resistivities range in values from 10 $\Omega/\square$ to 1 M$\Omega/\square$.

Usual resistor design practice dictates that resistors terminate on conductors with the two ends facing in opposite directions with a minimum of 10 mils overlap to ensure screen registration tolerances. Various resistor layout techniques are presented in Figure 14-6.

Aspect ratios L/W for a given resistor screening should not exceed 5 or be less than 0.5. If all resistors cannot be deposited conforming to this rule with one screening, a second resistor ink is called for.

For high-aspect-ratio resistors a "hat"-shaped geometry (Figure 14-7) should be used to eliminate sharp corners and to reduce the probability of hot spots. Resistors must cover enough substrate area to provide efficient heat conduction away from the resistor film. A given packaging configuration and specified materials determine the effective power density for resistors in units of power per area.

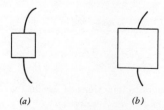

(a)   (b)

Figure 14-5   Illustration of the aspect ratio. Since $R_S = 100\ \Omega/\square$ for (a) and (b), they have the same resistance.

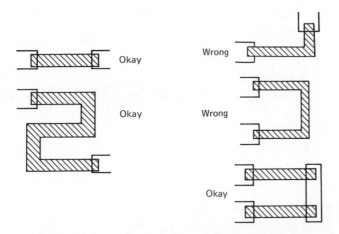

Figure 14-6   Examples of acceptable and unacceptable resistor layout techniques.

Conservative design with resistors covering a small percentage of the total sub-strate area calls for no more than 25 W/in.² (0.025 mW/mil²). For short-pulse applications, present estimates indicate that 40 W/in.² can be used for peak power ratings for pulse widths of less than 500 msec, where average power remains below 25 W/in.².

Deposited resistors can be consistently obtained with 10 to 20% tolerances. When necessary, resistors can be trimmed using air-abrasive trimming or laser trimming to 1% absolute and 1% ratio tolerances. Nonetheless, 2 to 5% tolerances are preferred. When trimming, the following procedure is recommended:

1. After determining the specified geometry for a power dissipation 20% larger than necessary, increase the width by 40%. The resultant geometry ensures that the resistor can be trimmed to the desired value without exceeding the rated power density.

2. Avoid closed electrical paths with resistors in them. These resistors cannot be trimmed.

3. Typically, designs should be made using TCRs of ±200 ppm/°C. Lower TCRs may be obtained, but caution should be exercised.

Thick-Film Resistor Design Examples

DESIRED:

$$10 \text{ k}\Omega \pm 1\% \text{ } 200 \text{ ppm/°C}$$

$$P_{D,\text{av}} = 4 \text{ mW}$$

Figure 14-7   The so-called hat-shaped thick film resistor geometry.

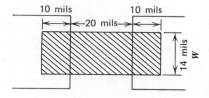

Figure 14-8   Thick film resistor design example.

DESIGN:

$$A_{min} \cong \frac{(1.2)(4 \text{ mW})}{0.025 \text{ mW/mil}^2} = 200 \text{ mil}^2$$

Choose 5-kΩ/$\square$ resistor ink.
Then

$$\frac{L}{W} = \frac{10 \text{ k}\Omega}{5 \text{ k}\Omega/\square} = 2$$

$$LW = 200 \text{ mil}^2$$

$$\frac{L}{W} = 2$$

$$W^2 = 100 \text{ mil}^2$$

$$W = 10 \text{ mils}$$

then

$$L = 20 \text{ mils}$$

To allow trim, increase W by 40% (Figure 14-8).
The following general rules should be noted.

1. Lead lengths should never exceed 150 mils.
2. Leads crossing over film traces should be avoided whenever possible. If they are necessary, a nonconductive insulating film should be deposited over the film traces to prevent unintentional shorts.
3. Leads crossing over leads are not allowed.
4. Sufficient space must be allowed at lead-bonding pads for bonding tool clearances.
5. When using thermocompression bonding, calculations for package height should include 15-mil lead height from the bond level (Figure 14-9).

Figures 14-10 through 14-13 depict several typical examples of thick-film hybrid ICs.

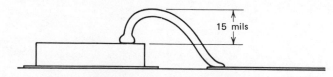

Figure 14-9   Diagram depicting proper lead height for thermocompression bonding.

Figure 14-10 Examples of thick-film hybrid circuits: (*a*) Simple circuit on TO-5 header (*b*) circuits on special Alumina substrates.

Figure 14-11 Examples of thick-film hybrid circuits: (*a*) JFET analog switch, (*b*) MOS analog switch.

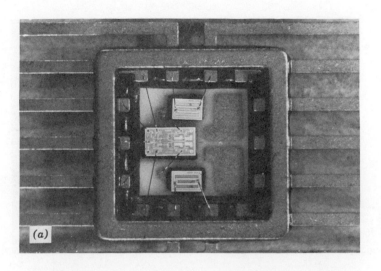

Figure 14-12 Examples of thick-film hybrid circuits: (*a*) JFET analog switch, (*b*) MOS clock driver.

Figure 14-13 (*a*) Various operational amplifiers, (*b*) single operational amplifier with stacked substrates.

Chapter Fifteen

INTEGRATED CIRCUIT PACKAGING AND INTERCONNECTION TECHNIQUES

The ultimate requirements of the final system greatly influence the circuit package chosen. The package determines the system's form factor, its component density, and its ability to perform reliably in a particular operating environment and temperature range.

A package has four basic functions:

1. To protect the sensitive semiconductor device from any hostile environment that could, in time, degrade circuit performance.
2. To provide adequate mechanical protection.
3. To provide a convenient means for interconnecting many individually packaged circuits.
4. To act as a path for heat resulting from power dissipation in the IC; the package must dissipate this heat to the surrounding air or conduct it to a heat sink.

The type of package to use for a particular application is generally evaluated on the basis of five factors: environmental capability, comparative interconnecting cost, comparative package cost, component density per unit volume, and comparative system size.

Generally, IC package types are classified as either hermetic or nonhermetic. At present, the majority of packages are hermetic, especially for military applications, although the nonhermetic or plastic package is widely used in commercial equipment because of its economy, simplicity of assembly, and ruggedness. Hermetic packages have historically provided better chip protection under certain environments. This factor becomes significant for complex and costly chips or for devices such as MOS ICs, which are very sensitive to contaminants. The quality of a hermetic seal is expressed in terms of a maximum allowable leakage rate, which necessarily applies only to package types possessing an internal cavity. Noncavity packages cannot "leak," but if not properly constructed, they can allow contaminants to reach the silicon chip.

Additionally, packages are also classified by the materials that are used in their manufacture (ceramic, metal, glass, and plastic) and by the package configuration itself: TO-5, flat-pack, and dual-in-line (DIP) packages. The TO-5 is a modified version of the conventional "transistor-outline" package, with more leads. A flat pack is a miniature package, usually about 0.25 in.². A DIP (both cavity type and plastic or noncavity type) is a larger package designed for dip-soldering to a PC board.

The TO-5 package, the flat-pack and the cavity DIP are hermetically sealed packages capable of operating reliably over a wide temperature range. Plastic DIPs are designed to operate at lower temperatures, typically to a maximum ambient temperature of 70°C.

The DIPs are designed to accommodate mass assembly techniques in applications where size and space are not a premium consideration, such as commercial computers. Insertion into the PC boards can be either manual or automatic. Lower assembly costs result because of the 100-mil grid spacing, which permits relaxed tolerances and wider PC lines. These are commensurate with the widely employed discrete-component mounting standards. Even though the cavity DIP is expensive, its use may permit overall savings in the total systems assembly.

Flat-pack leads are thin, flexible, and only 50 mils apart, which makes them difficult to insert into PC boards. They are generally inserted by hand into board holes for dip soldering, or welded to or soldered to wiring on top of the board.

Most TO packages are inserted by hand and then soldered. The methods are similar to those used for transistor circuits, except that three or four times as many leads must be formed and inserted.

PACKAGING MATERIALS

Packaging materials are chosen for chemical inertness and compatibility with silicon assembly temperatures. Matched coefficients of thermal expansion are also desirable. The four main package materials are metal, ceramic, glass, and plastic. All but plastic are considered hermetically sealable.

Most metal packages have ASTM F-15 alloy, commonly known as Kovar, as the structural material. Kovar is a nickel–iron alloy particularly useful in sealing glass to metal. Kovar has a temperature coefficient of expansion close enough to that of glass to prevent the glass-to-metal seal from cracking over a wide range of temperatures. Oxidized Kovar forms a good bond between itself and the glass. However, the metal is easily oxidized, and the external leads on an IC package become plated, usually with gold. The gold prevents oxides from forming and interferring with soldering and welding.

Kovar's disadvantages are low electrical and thermal conductivity (for a metal). Clad or laminated high-conductivity materials with glass-compatible expansion coefficients are sometimes used instead of Kovar. Cladding also allows precious metal to be placed only at the chip attachment area, eliminating the usual method of plating the entire lead frame. As a rule, all the package leads are mechanically connected by a frame and cut apart after the packaging is completed. A new technique of selectively plating lead frames allows strips of lead frames for plastic packages to be spot gold-plated in the desired chip-bonding area.

Solid copper is also used for hermetic IC packages, but with Kovar eyelet seals. Copper cannot be directly sealed to normal sealing glass because the coefficients of expansion of the two materials are different. Copper packages provide high thermal dissipation, and the lid can be sealed by a simple cold-welding process.

Ceramic is a good material for hermetic packages. Its thermal conductivity is high, and its expansion coefficient is compatible with glass. Typical ceramic materials are alumina (85 to 99% Al_2O_3) and beryllia. Beryllia, which has a much higher heat conductivity than alumina, is favored for high-power circuits.

Although glass packages provide a good seal and, by virtue of their simplicity, reduce fabrication costs by more than 50%, their thermal conductivity and heat dissipation are poor. This has limited the glass package to low-power, less-heat-sensitive devices. Alternative designs include ceramic bases for better heat conduction and an all-glass ring for lead sealing. In addition, glass plates serve as substrates for many hybrid film circuits. The flatness and uniformity specifications are satisfied at surprisingly low cost. Special glass and plastic materials are often used to coat, protect, and seal completed IC chips.

Plastic or molded packages are made by assembling and interconnecting the chip on a substrate or lead frame and molding the entire structure in plastic (except for lead ends) to form the package body. Epoxy, phenolic, and silicone resins are the most commonly used plastics. Molded packages are not considered hermetic because the resin–lead interface does not form a true seal. The materials usually have unmatched coefficients of thermal expansion, and some atmospheric constituents can diffuse through resins. In addition, the absence of an internal cavity precludes any means of hermetic testing.

Epoxy is probably the most desirable plastic molding material. It is low in cost and chemically stable, and it exhibits excellent characteristics (similar to those of hermetic devices) when exposed to hostile environments. Silicone compounds are structurally weaker than epoxy and are attacked by some important classes of chemicals, namely, salts. Thus silicone does not perform well in a salt atmosphere. Silicone is somewhat more expensive than epoxy. Phenolics are strong and economical but can contain some harmful chemical impurities.

PACKAGE TYPES

TO-Style Packages

Besides the TO-5, TO-style metal packages include the TO-3, TO-99 and TO-100, which are also related to the original top-hat transistor styles. TO-5 packages have 8, 10, 12, and 14 pins coming out of the bottom of the base or header. The base is usually a gold-plated Kovar header which provides for hermetic glass-to-metal seals around the pins. A monolithic circuit die is usually soldered directly to the

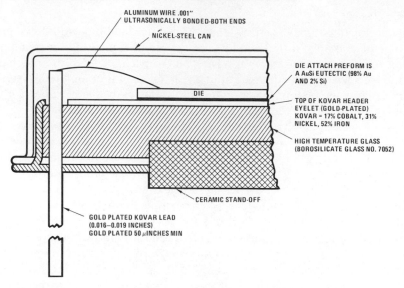

Figure 15-1　Cross-sectional view of a monolithic IC header package.

header with a gold–germanium eutectic solder preform. If the package is to contain several chips, the chips are usually isolated from one another and from the header by mounting them on a ceramic insulator. The cross section of a typical TO package appears in Figure 15-1.

Most TO packages are hermetically sealed by welding a Kovar cap to the header flange. Figures 15-2, 15-3, and 15-4 depict typical 8-, 10-, and 12-pin TO-5 pack-

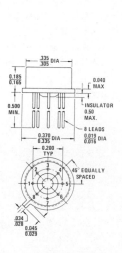

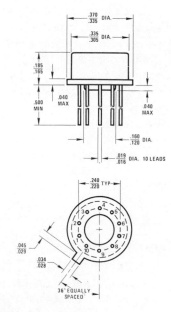

Figure 15-2　An eight-lead metal can (TO) package.

Figure 15-3　A 10-lead metal can (TO) package.

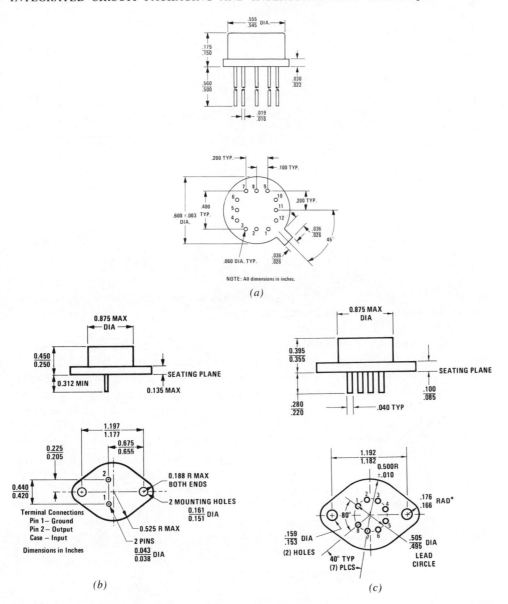

Figure 15-4 (*a*) A 12-lead metal can (TO) package. (*b*) A 2-lead TO-3 metal can package used for high current voltage regulators. (*c*) A 8-lead TO-3 metal can package.

ages, respectively. Figures 15-4b and c depict 2 and 8 lead TO-3 metal can packages which are specifically used for high current voltage regulators, operational amplifiers, and audio amplifiers.

Flat Packs

Several types of flat packages are presently available. A typical flat pack has gold-plated leads embedded in a glass frame. The die is bonded in the same manner

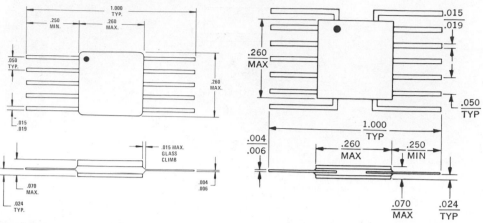

Figure 15-5   A 10-lead flat package.   **Figure 15-6**   A 14-lead flat package.

as in the TO-5 package. This type of package is sealed by attaching a ceramic Kovar lid with a low-temperature glass frit. Figures 15-5 and 15-6 show typical 10-lead and 14-lead flat packages, respectively.

The Kovar parts in a metal pack are the base, the side structure or ring, the lid, and the leads, which extend through a glass between the ring and base.

Glass packages consist of a one-piece base and a ring in which the Kovar leads are sealed. The glass is either a borosilicate or Pyroceram (which becomes ceramiclike after processing). Both have excellent resistance to thermal and mechanical shock. The chip is mounted on the glass base and wire bond connections are made to the protruding leads. The large volume of glass around the Kovar leads ensures a reliable hermetic seal. Lids are attached by a layer of low-temperature sealing glass between the ring and lid. The lid material may be glass, ceramic, or metal. Figure 15-7 is a cross-sectional view of a glass–metal flat pack.

Ceramic flat packs are constructed similar to the metal type with base, ring, and lid of either alumina or beryllia. The cross section of a typical ceramic flat pack is shown in Figure 15-8.

Cavity Dual-in-Line Packages (DIP)

The cavity DIP has rapidly been increasing in popularity. A glass frit is used to bond together the ceramic substrate and cap and to provide the seal for the leads. In this type of package, the glass frit also serves to bond the circuit die.

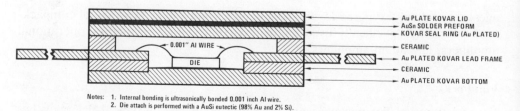

Notes: 1. Internal bonding is ultrasonically bonded 0.001 inch Al wire.
         2. Die attach is performed with a AuSi eutectic (98% Au and 2% Si).

Figure 15-7   Typical glass–metal flatpack, in cross section.

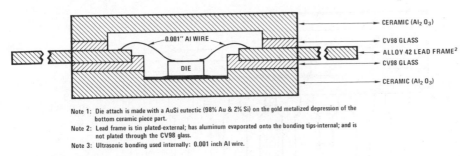

Note 1: Die attach is made with a AuSi eutectic (98% Au & 2% Si) on the gold metalized depression of the bottom ceramic piece part.

Note 2: Lead frame is tin plated-external; has aluminum evaporated onto the bonding tips-internal; and is not plated through the CV98 glass.

Note 3: Ultrasonic bonding used internally: 0.001 inch Al wire.

Figure 15-8   Typical ceramic flatpack, in cross section.

A popular ceramic DIP is a multilayered structure with thick-film conductors on an internal layer and another layer cut out to form a cavity in which the chip is placed. Conductors extend from the chip compartment to exposed terminating pads at the outer edge. The pads are brazed to a lead frame. The package is sealed by a brazed or solder-sealed ceramic lid enclosing the chip compartment. To accommodate this type of cover seal, the ceramic must be initially metallized over the seal area. Alternatively, a glass frit is often used as a sealant between the ceramic base and a metal or ceramic cover. Another variation is a Kovar ring attached to the ceramic by means of a glass frit and a lid sealed to the ring either by solder or welding.

Figure 15-9 shows the cross section of a typical multilayer glass-to-metal DIP package with 14 leads. The multilayer package, which does not depend on glass to provide a seal, eliminates any lead sealing problems because the seal is accomplished by the ceramic multilayered structure and is remote from the external lead. Another design is a pair of ceramic rectangles with a Kover lead frame sandwiched between. The bottom ceramic plate is cemented to the frame. A cutout in the frame provides room for attaching the circuit die to the ceramic by a glass frit. The top ceramic plate is bonded over the opening with a glass frit.

Figure 15-10 is a cross section of a ceramic DIP package. The *C* DIP seal is made by sealing the leads in glass at low temperature, adding the ceramic cover, and sealing it to the glass by firing at a higher temperature until the glass is devitrified.

Figures 15-11, 15-12, and 15-13 are typical 14-, 16-, and 24-lead cavity DIP package drawings, respectively.

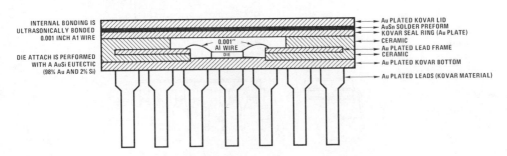

Figure 15-9   Typical glass–metal DIP, in cross section.

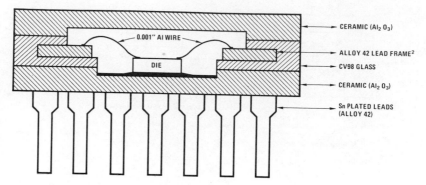

CERAMIC (Al$_2$O$_3$)

ALLOY 42 LEAD FRAME[2]

CV98 GLASS

CERAMIC (Al$_2$O$_3$)

Sn PLATED LEADS (ALLOY 42)

Note 1: Die attach is made with a AuSi eutectic (98% Au & 2% Si) on the gold metalized depression of the bottom ceramic piece part.

Note 2: Lead frame is gold plated-external; has aluminum evaporated onto the bonding tips-internal; and is not plated through the CV98 glass.

Note 3: Ultrasonic bonding used internally: 0.001 inch Al wire.

Figure 15-10   Typical ceramic DIP, in cross section.

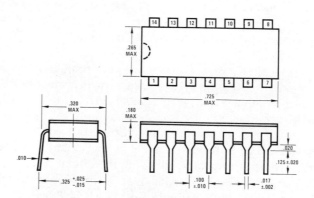

Figure 15-11   A 14-lead cavity DIP.

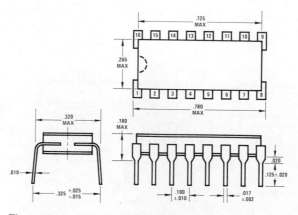

Figure 15-12   A 16-lead cavity DIP.

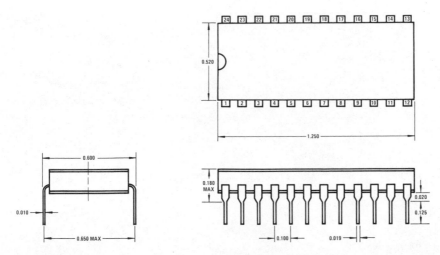

Figure 15-13   A 24-lead cavity DIP.

Noncavity DIPs

Plastic or molded DIPs are popular because their manufacturing costs are low. The most formidable problems posed by these types of packages are improving resistance to moisture and contamination and preventing mechanical damage during thermal cycling. The main savings are that the molding process can be adapted to high production rates and the user can assemble them with automatic insertion and PC board soldering equipment.

Plastic packages do not have the heat dissipation capability of other types but they do possess the capability of meeting the wide temperature operating limits demanded of military equipment. However, they are usually specified for 0 to 70°C, whereas the military range is −55 to +125°C or −25 to 85°C. At reduced temperatures, as encountered in industrial and commercial equipment, plastic encapsulation is adequate. Additional chip processing, such as the deposition of a thick glass protective layer on top of the normal silicon dioxide passivating film, makes the IC chip itself more impervious to moisture and contaminants; thus the plastic encapsulation need provide only mechanical protection. Therefore, even though ICs in plastic packages lack a true hermetic seal, they are receiving widespread acceptance because of their low cost. Figure 15-14 depicts the molded DIP immediately after encapsulation (molding) with the lead frames still attached.

To protect the die from a hostile environment, the molded package should completely surround the die and the internal wiring with materials that are impervious to penetration by gases and liquids. To be reliable the package must be molded of an inert material that keeps the IC chip, lead wires, wire bonds and lead frame in compression throughout the worst-case operating and junction temperature range. Furthermore, the molding compound should provide a thermal coefficient expansion match between itself and the IC chip, should possess high dimensional stability to a high temperature and should possess high tensile strength. The outline dimensions for 8-, 10-, 14-, and 16-pin molded plastic DIPs appear in Figures 15-15, 15-16, 15-17, and 15-18, respectively.

Figure 15-14 A molded DIP with lead frames attached.

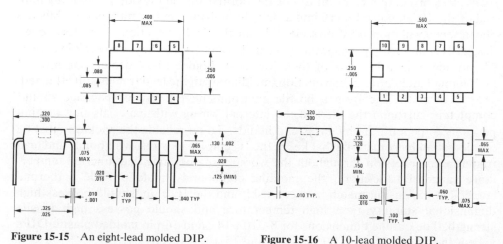

Figure 15-15 An eight-lead molded DIP.

Figure 15-16 A 10-lead molded DIP.

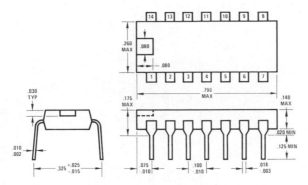

Figure 15-17   A 14-lead molded DIP.

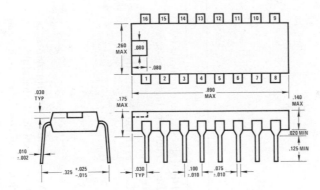

Figure 15-18   A 16-lead molded DIP.

Hybrid Packages

All the previously mentioned package types are used for hybrid circuits, as well as a great many types of larger specialized packages constructed by similar techniques. At present, no single package is best for hybrid circuits from cost and reliability viewpoints (see Table 15-1). Also, the need for hermetically sealing the flip-chip package is yet unresolved by the industry. In some commercial and consumer electronics equipment, the chips and substrate are simply covered by a conformal coating of plastic. Table 15-1 compares the various available Hybrid packaging options and Figure 15-19 depicts these options photographically. Note that Figures 15-4a, and c shown previously, as well as Figures 14-2 and Figures 14-10 through 14-13 also depict various hybrid packages.

LSI (LARGE–SCALE INTEGRATION) PACKAGE PROBLEMS

Packages for LSI circuits are rapidly becoming a more serious problem than the chips themselves. Array complexity runs to hundreds of logic circuits and dozens of package pins. Many pinouts mean high assembly costs and low assembly yield. The packages themselves are costly; yet they crack, break, and warp. Their seals

Table 15-1 Hybrid packaging options

Description	Typical Examples	Advantages	Disadvantages
Hermetic package with passivated or sealed devices.		1. Highest resistance to moisture 2. Best chance for high, long-term reliability 3. Easy to shield electrically	1. Highest packaging cost 2. Double sealing is redundant 3. Sealed semiconductor packages require large overall size and weight
Hermetic package with unsealed devices.		1. Meets military environments 2. Easy to shield electrically 3. Semiconductor devices relatively low in cost 4. Smallest package size	1. "Chip-and-wire" techniques costly in assembly time, with long learning curve 2. Chip probing before assembly difficult unless entire wafer is checked prior to scribe and break operations 3. High thermal stress on chips and fine wires during sealing

418

Epoxy package with passivated or sealed devices.

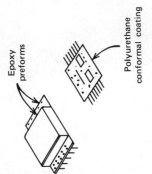

Epoxy preforms

Polyurethane conformal coating

1. Low-cost package
2. Meets military environments
3. Receiving inspection and stocking of sealed semiconductor devices is easy
4. Assembly yield climbs rapidly to high figure
5. Minimum thermal stress on chips or fine wires

1. Custom-packaged sealed semiconductors of mixed types not currently available in single packages
2. Conformal coatings and epoxy-molded packages not impervious to moisture
3. Sealed semiconductors require large overall size and weight

Epoxy package with encapsulated devices.

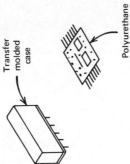

Transfer molded case

Polyurethane conformal coating

1. Lowest-cost approach
2. Receiving inspection and stocking of semiconductors is easy
3. Semiconductors readily available in wide assortment
4. Vendor reliability data readily available
5. High assembly yield
6. Minimum thermal stress

1. Will not satisfy most current military systems specifications
2. Epoxy semiconductor packages are not impervious to moisture
3. Plastic-encapsulated semiconductors require large overall size and weight

Figure 15-19 Typical hybrid IC packages: (*a*) packages not sealed and lead frames not cut and bent. This photograph allows one to see the cavity. (*b*) Sealed packages.

open, pins fall off, and package manufacturers cannot deliver enough good units to meet production demands. Despite foam-plastic shipping containers, some packages arrive with corners missing, as if mice had nibbled at them in transit. Packaged devices arrive for incoming inspection with loose or missing caps — and with damaged wire bonds or chips. Some packages look good from a distance, but users find pins attached out of registration with the contact pads on the ceramic.

It has been estimated that only about half of all packaged LSI devices survive PC board attachment, which effectively doubles their price. Although LSI devices are more reliable per function, smaller ICs may be more reliable per package — thus narrowing the LSIs' edge. Bulkier packages, more wire bonds and pins, large-area seals, and other factors seem to work against the LSI package. These packages are hard to install. Pins are frequently lost, and occasionally pins short adjoining contact pads. More packages fail in assembly as leads break off, or bend and endanger the hermetic seal. Sturdier pins do not seem to be the answer; these already are available. Either the brazed connection to the package contact breaks, or the whole contact pad lifts off the ceramic, carrying the lead with it. There is no way to repair a package when that happens.

The task of selecting a viable LSI package is further complicated by the variety of package styles needed by LSI manufacturers. The manufacturers require 40-lead packages with leads on 150- or 600-mil centers, and they are now pursuing the use of 50- and 60-lead packages. Obviously, standardization would help, but LSI and package makers alike continue to back their own designs. And for those who do not have their own designs, a compromise often is necessary. Because of these problems, many semiconductor firms are relying on their staffs to design their packages. This can only lead to further nonstandardization of LSI packages.

DIE BONDING

Die bonding means soldering, brazing, or glassing of the circuit or component, chip, or die to a package substrate. The attachment material is a mechanical contact, a thermal path, and sometimes an electrical contact, all in one.

Die attach on metal substrates is generally accomplished by a eutectic brazing alloy such as the gold–germanium alloy in the TO-5 package cross section of Figure 15-20. Note that the silicon circuit was backed with gold to enhance the bond. A eutectic alloy is two or more metals whose melting point is significantly lower than the melting point of any of the alloy constituents. In particular, the eutectic is the alloy having the minimum melting point for a given combination of metals. Certain bimetallic systems exhibit more than one eutectic point, but these generally are not important. It should be emphasized that these bimetallic alloys are mixtures, not chemical compounds. A eutectic brazing alloy is generally preferred over soft solder for die attachment because of its higher melting point.

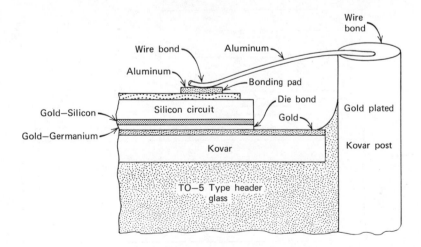

Figure 15-20   Silicon monolithic assembly.

Soft solders are used in assembling PC boards. The die bond must stay solid during board soldering, of course.

When a metal substrate is not available or if an electrical connection to the substrate is not used, as in some glass or ceramic packages, the die is usually attached with a glass frit that melts at a lower temperature than the previously formed glass–metal seals. The die and leads are held in place by a metal–glass preform. This operation is performed in an inert atmosphere at temperatures up to 525°C.

The Flip-Chip Process

A large percentage of the cost of an IC lies in the die-attaching and wire-bonding operations. An operator must perform the time-consuming duty of making 28 bonds on each 14-lead IC. Even worse, both this operation and the die attaching itself occur at elevated temperatures, with the attendant stresses applied to the dice and their protective glasses. This means also 29 possibilities of damage to the now finished and valuable die. (If tail pulling is involved, as it frequently is, another 14 possibilities are added—these 14 wires moving around are one more potential failure mechanism.)

The flip chip technique took a big step in solving some of these problems. It was the first attempt to eliminate wires and thus simplify assembly operations. There are different methods of making a flip chip, but the basic procedure involves replacing flat bonding-pad areas with raised bumps. These bumps then can be placed into contact with metallized pads similarly located on a substrate or package bottom and, by suitable means, all can be joined at one time. In usual assembly techniques, the chip with the bump is flipped upside down onto the pad pattern, thus the term "flip chip."

The chips are then mounted to the interconnecting metallization on the sub-

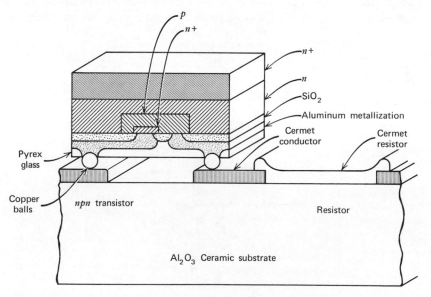

Figure 15-21   Typical cross section of flip-chip on hybrid circuit.

strate by pressure and heat or ultrasonic energy. The use of solder-coated balls is illustrated in Figure 15-21 for a hybrid circuit. In some approaches, bump contacts are formed on the die; in others, bump contacts are formed on the substrate. A soft bump is needed to firmly seat all terminals. Still another version is the leadless inverted device (LID), which is mounted on a miniature ceramic base, which is then attached to the substrate pads. This technique avoids the risks associated with handling naked chips.

The die bonding and wire bonding are performed in one operation. Thus the flip-chip technique reduces costs, achieves better quality, increases the reliability, and results in easier device handling. However, the number of device types available with bumps is still not large enough for all circuit designs.

There are several disadvantages associated with flip chips. These are listed here.

1. Bump height is critical; if equal heights are not obtained, all bonds will not uniformly contact the package bottom. In conjunction with this, is the corollary that the substrate or package must be very flat also. This raises package costs.

2. Testing, which must be done after the bump is formed, is a critical operation because excess probe pressure can cause damage and is highly likely to deform the bump, with the attendant disastrous change in height. Too light pressure gives unreliable test results.

3. Flip chips must undergo face-down bonding. Because the bond areas and pads are concealed from view, this requires expensive special aligning equipment.

4. Face-down bonding of flip chips means that the completed bonds cannot be inspected.

5. During bonding of flip chips, pressure to complete the bond must be applied to the back of the chip itself with the danger that heat, stresses, and cracks may damage the chip. The addition of a greater number of bumps means more force must be applied to achieve the minimum pressure necessary to bond.

6. Although rebonding is possible, the operation requires reapplication of heat and pressure. The effect of such procedures on reliability is questionable.

7. Because the chip is held rigidly to the substrate by the bonds, any thermal shocks to which the die may be subsequently subjected can cause stresses to build up which may damage the die or the bonds themselves.

Beam-Lead Bonding

Another method of eliminating wire bonding is beam-lead technology. Beam leading is the extension of the interconnects out over the edge of the die by the process of increasing their thickness from that of normal interconnects (ca. 1 μ) to a size that, in comparison, is a beam (ca. 12 μ — see Figure 15-22). These leads (which are rugged, yet compliant) are rigidly attached to the die because they are built right on the die. They serve as mechanical and electrical connections to the chip.

The basic beam-lead process in use today is one that was developed by the Bell Telephone Laboratories (Figure 15-23). The key features of this process are:

1. All junctions are passivated by silicon nitride, making them stable in mobile ion environments.

2. Ohmic contacts and interconnects are formed by a platinum–titanium–

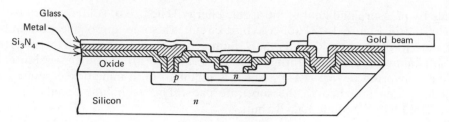

Figure 15-22   Typical cross section of beam-lead chip.

platinum–gold metal system, which is very stable even in high-humidity environments.

3. The chip-bonding medium is a cantilevered beam of gold which is easy to bond, absorbs thermal stresses, and transfers heat well from the chip to the substrate.

4. Visual inspection is easily performed.

Through most of their fabrication processes, beam-lead and packaged semiconductors are similar. As a matter of fact, until emitter diffusion is completed, identical processes are involved in creating the electrical characteristics for wafers destined to be beam-lead processed and for those which are to be package assembled. Devices to be wire bonded on packages or hybrid-circuit networks have their contact windows opened, their surfaces metallized, and intraconnects defined and made.

Beam-lead wafers go through a series of additional steps — silicon-nitride coating, intraconnect fabrication, and finally forming of the external beam-leaded wafer — which is initially more expensive than that required for a finished standard planar-processed wafer.

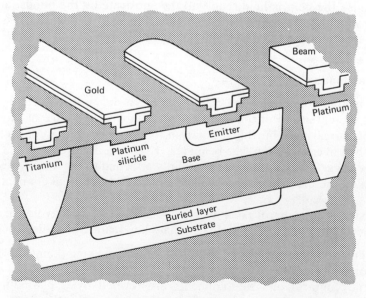

Figure 15-23   Cross section of a beam-lead chip.

In the manufacturing process of a typical beam-lead device, the conventional aluminum interconnection pattern for the device is replaced with a new metal system. First contact windows are opened in a highly protective silicon nitride layer, which has been deposited over the regular silicon dioxide passivation. The nitride forms a strongly bonded layer with high resistance to alkali ion migration. The contact areas are then made by reacting a thin sputtered platinum layer with the silicon of the device to offer a low resistance ohmic contact. The residual unreacted platinum on top of the nitride is etched off, and overall layers of titanium (for adherence to the nitride and platinum as a barrier layer) are sputtered on the wafer. The platinum is then etched to form the intraconnecting conducting paths and plated with $2\,\mu$ ($80\,\mu$in.) of gold, using photolithographic masking techniques.

Selected areas of the gold, approximately $75\,\mu$ (0.003 in.) wide and up to $200\,\mu$ (0.008 in.) long, are plated to 12-μ (0.0005-in.) thickness. The titanium layer is then etched off using the gold as a mask. The silicon under a part of the length of these so-called beams will be etched away to separate the wafer into dice and to provide a series of cantilevered gold contacts over the edges of the dice so formed. A planar wafer now is probed electrically to select potentially good devices; bad devices are identified so that they can be thrown away. The next step is to scribe the wafer and break it into separate and individual circuits.

The beam-lead wafer also is probed and the bad devices identified, but the die-separation process is different. In this case, the dice are separated by chemical means, in most cases by anisotropic etching of the silicon wafer. This in effect is the end of the manufacturing process for the beam-lead device. After electrical characterization, it is ready for use. The silicon nitride seals the device junctions, making them impervious to deleterious ambients such as moisture, and to alkali ions such as sodium and lithium. The standard device, having no alkali and moisture barrier, must be protected by means of hermetically sealed packages, which require that the silicon die be firmly attached to the package and that wires be used to connect the active device to the outside world. It is these connections, coupled with the excessive mechanical handling of devices processed by standard planar techniques, which give rise to most of the failure mechanisms present in standard devices but lacking in the beam lead devices. Figure 15-24 shows the EIA (Electronic Industry Association) beam spacings for an 18-beam die and for a 30-beam die.

The advantages of beam leads are many, but most important are the following:

1. Control of beam height or substrate flatness is not very critical, thus reducing costs.

2. Complete testing can be done while the die is still unmounted with no damage to the beams. Because no thermal or other stresses take place during assembly, it is practical to do 100% testing (ac and dc) on the die. This guarantees a chip buyer good units, and the manufacturer does not have to risk putting potentially defective die in expensive packages.

3. The die is usually bonded face down, but because the leads and bond pads are visible, the die is simple to align and utilizes no expensive specialized equipment. If it should be desired, the beam-leaded die can be just as easily bonded face up, still with no alignment problem. Face-up bonding requires a more expensive recess package, but complete visual inspection of the finished product is possible.

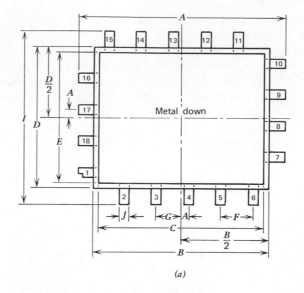

(a)

EIA Standard (in.)		
	Min	Max
A	0.0025	Typical
B	0.0550	Basic
C	–	0.0550
D	0.0450	Basic
E	–	0.0450
F	0.0100	Typical
G	0.0075	Typical
H	0.0620	0.0690
I	0.0520	0.0590
J	0.0020	0.0045

Figure 15-24a Chip outline for an 18-beam die.

4. All bonding tool force or heat is applied directly to the beam lead itself and and not to the die. Increasing the number of beams is easy because the additional bonding force required has no effect on the die itself.

5. Imperfect bonds can be easily seen, or mechanical pull tests can be performed. An imperfect bond can then be singled out and individually rebonded.

6. The chip is isolated from stresses and shock by the elasticity or ductility of the beams themselves. Indeed, they act very similar to shock mounts.

7. Due to the simplicity of assembly and the assurance of a 100% tested die, beam lead chips will be a boon to the hybrid manufacturer.

Two more very important advantages are to be found in beam-lead products. The first is the use of etching, instead of scribing and breaking, to separate the die on the wafer. Beam-lead wafers are manufactured on (100) oriented silicon, instead of the commonly used (111) material. The (100) material permits anisotropic etching, in which the etch preferentially attacks silicon in one direction or plane at an extremely accelerated rate and in a mathematically predictable manner. The

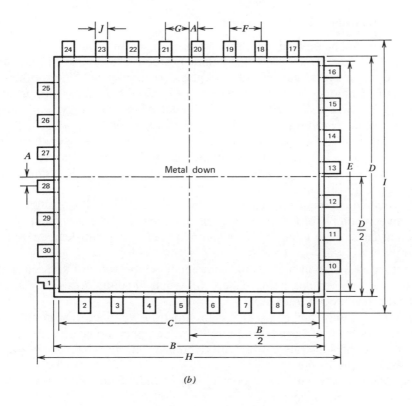

(b)

EIA Standard (in.)		
	Min	Max
A	0.0025	Typical
B	0.0850	Basic
C	–	0.0850
D	0.0750	Basic
E	–	0.0750
F	0.0100	Typical
G	0.0075	Typical
H	0.0920	0.0990
I	0.0820	0.0890
J	0.0020	0.0045

Figure 15-24b   Chip outline for a 30-beam die.

separation achieved is clean and smooth; it has no cracks, and it does not subject the die to stresses.

The second advantage of beam-lead products is the use of silicon nitride passivation to produce hermetically sealed chips. Silicon nitride is a rugged coating; it seals a die against ambients so well that the die itself is hermetic and, except for possibilities of gross damage and the problem of handling, no package is needed. This increases the reliability of a hermetically sealed die and makes inexpensive plastic packaging practical for even high-reliability usage.

Flip-Chip and Beam-Lead Comparison

The flip-chip and beam-lead devices have certain common and competing characteristics. Both are produced with integral protective coatings. Modifications in technique can make silicon nitride and glass coating applicable to either. Moreover, both techniques allow replacement of defective devices without affecting the circuit as a whole. The quality of a bond can be more easily visually inspected with beam leads, and repair of a defective bond is easier.

The greater ruggedness of flip chips can allow rougher in-process handling than is feasible with beam leads. If automatic assembly techniques are to be used, separation precision comparable to sawing or crystallographic etching is called for in order to locate contacts accurately with respect to the dice edges.

Both techniques require special device patterns either to allow sufficient separation between bumps for flip chips or to have all leads on the periphery of the chip for beam leads. It is impossible to take a wafer and beam lead or bump it at random.

In both cases, silicon real estate and, hence, chips per unit area of wafer, are competitive with regular chips except for such small-area devices as discrete low-power, high-speed transistors.

Flip chips require higher precision in substrate conductor printing, owing to the tighter location of terminal points. However, the coplanarity or flatness requirements of the substrate contact are tighter for flip chips because of the greater compliance of the beams.

Beam leads can be shaped to control device inductance, and they have lower capacitance for microwave use.

Both techniques increase reliability by reducing the number of joints per terminal to one from the three required in a packaged device.

Table 15-2 compares the salient features of the flip-chip and beam-lead technologies.

Spider Bonding

A recent die-bonding technique developed by Motorola is spider bonding. Spider bonding enhances bond integrity on the chip and combats the rising costs of standard packaging methods. Productivity and reliability are facilitated by the chip layout and spider design. Bonding reproducibility is kept high by matching or registering the positions of the aluminum bonding pads on the chip to the spider leads.

The spider frame in Figure 15-25 is stamped from a continuous coil of aluminum 2-mil (0.002 in.) thick. The tips of the spider fingers are 0.0035 in. wide on 0.008 in. centers. The fingers broaden to make the welding of the spider assembly to the external packaging frame simpler and more uniform.

In order to use spider bonding, however, the die-edge definition required is more consistent than in the standard thermocompression bonding techniques, and the die must be specifically designed for the spider. The necessary prepositioning of bonding pads imposes restrictions on circuit layout for spider bonding which result in approximately a 30% increase in die size and increase in total IC cost.

Table 15-2 Comparison of Flip-Chip and Beam-Lead Devices

Flip-Chip	Beam-Lead
• Separated from wafer by scribing and breaking, which reduces yield and may produce propagating cracks through the device	• Chips are separated by etching; no cracks
• Bump-height uniformity and flatness of substrate are critical parameters, thus resulting in expensive chips or substrates	• No such critical parameters exist
• Limited to face-down bonding	• May be mounted face up or face down for easy inspection
• Visual inspection impossible unless substrate is glass	• Ease of bonding
• Require accurate optical system for accurate positioning	• Makes possible simple and direct visual alignment of beams; with substrate conductors
• Require higher precision in substrate conductor printing because of the tighter location of terminal points	Bonding energy applied directly to each beam, not to chip itself; thus chip is not exercised
• Chip itself may be stressed during bonding because of application of heat or pressure to or through chip	• Poor heat dissipation capability
• Good thermal conductivity	• Chip isolated from stresses and shock by elasticity or ductility of beams themselves
• More rugged than beam lead	
• Requires tighter flatness for substrate contacts than beam lead	
• Allows greater packaging density	• Elements of ICs can be air dielectrically isolated by air gaps bridged by intraconnection beams
• Chips not fully tested before being mounted	
• Eliminates die-attach step and wire bonding of individual contacts	• Can be completely tested automatically in wafer form
	• Allows breathing of structure during thermal cycling without causing mechanical damage
	• Eliminates die attach, but each beam is bonded individually (bonding can be done by automatic step and repeat bonder, however)
• Has protective coating on surface of die	• Has protective coating on surface of die
• Technique allows replacement in case of defective device without affecting total circuit	• Technique allows replacement in case of defective device without affecting total circuit
• Require special device patterns to allow sufficient separation between bumps	• Requires special device patterns to have all leads on the periphery of chip
• High reliability	• High reliability

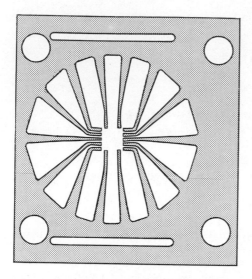

Figure 15-25 A section of the aluminum spider frame strip prior to chip-bonding operation.

Semiconductor on Thermoplastic on Dielectric (STD)

Another method using regular chips is General Electric's semiconductor on thermoplastic on dielectric (STD) microelectronic packaging technique. First the circuit pattern is developed on a substrate such as alumina, then copper mesas 50 to 75 μ (0.002 to 0.003 in.) high are built up as interconnect lands and registration checks. The entire surface is coated with fluorinated ethylene propylene (FEP) to the tops of the mesas; then the semiconductor chips, also of the same thickness, are pressed face up in the heated FEP layer. Finally, gold is evaporated over the entire surface and selectively etched to join the interconnect lands and the device terminations. The technique is of considerable interest because it employs common photolithographical and other semiconductor techniques and can utilize a wide range of standard dice.

Having barrier layers between the aluminum device lands and the gold interconnect pattern would seem to be a desirable modification if intermetallic compound formation between aluminum and gold is to be avoided. With foresight, however, this interface can be located where its potential brittleness is not mechanically disadvantageous. The pressure applied to the active face required to seat the chip is also of concern to the device engineer.

Several years ago, International Telephone and Telegraph Corporation developed a system based on a flexible Mylar PC. After conductors on the Mylar were etched to form lead fingers similar to the center of the spider lead frame, chip-sized holes were etched in the Mylar film and the fingers were bonded to chips. A version of this technique, employing etched lead frames, was also used to package monolithic ICs, but it was not commercially successful. The techniques have since been used in various modifications by aerospace manufacturers for high-density packaging.

It is worth noting that most of these techniques have been used at one time or another to package individual monolithic circuits. The ball-bumped method, for instance, was developed and is still used by IBM for computer logic, and the beam-lead technology was developed by Bell Telephone Laboratories for missile systems. Several IC manufacturers have tried flip chipping as a means of automatic packaging. However, the straightforward techniques of die brazing and wire bonding, which date back more than a decade, are still the most popular and have been refined into high-yield, high-reliability operations.

HYBRID–CIRCUIT ASSEMBLY

In hybrid-circuit technology, die attach may be accomplished by three means: by assembly of several chips face up on a substrate (multichip), by flip-chip techniques, or by beam-lead technology, as discussed previously.

In multichip assembly, the individual components are mounted on an insulating substrate and are interconnected by both metallization and wire bonding, as in Figure 15-26. Multichip assembly allows great circuit flexibility, such as the possibility of using *npn*, *pnp*, and FETs in the same circuit design. However, because of the high assembly costs imposed by the combination of wire and die bonding, it is primarily suited for low-quantity prototype production.

OHMIC CONTACTS AND WIRE–BONDING MATERIALS

Electrical contact must be established between thin-film terminals (isolated ohmic contact areas), circuit die, and package leads after the die has been attached to the package header. In hybrid circuits it is also necessary to make electrical connections between the individual circuit elements.

The metal surfaces of IC contacts are usually connected by soldering, welding, heat-pressure bonding (thermocompression bonding), or ultrasonic bonding. Some of the uses of soldering were described earlier. Package leads are generally soldered or welded. Most intraconnections are made by bonding thin wires of gold or aluminum to the thin film "pads" on the die and the gold-plated leads of the package, or to gold pads on the film conductors of a hybrid substrate.

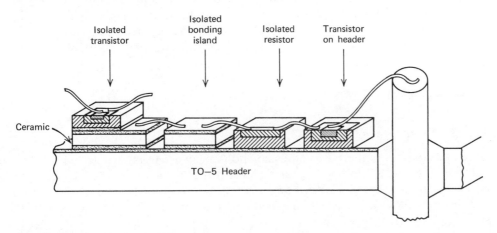

Figure 15-26   Multichip assembly technique.

Thermocompression bonding is a direct bonding method. Two metals (e.g., wire and a bonding pad) are made to seize without melting and without a third intermediate phase, such as solder. This is accomplished with high pressures and temperatures. The high temperature keeps the metals in the annealing range as they are "flowed" into atomic intimacy by the bonding pressure. These bonds are made simply, directly, and cheaply in a controlled atmosphere without melting, penetrating, contamination, or degradation of the semiconductor.

Gold and aluminum are the most common conductors for both film conductors and bonding wires. Nickel, molybdenum, and chromium are sometimes used both as plating for piece parts or for thin-film material. One application, because of the adherence of these metals to silicon dioxide, is as an interface material to permit gold metallization. All IC connection materials must be of extremely high purity, with controlled proportions when alloyed.

Gold is used in thin-film, evaporated, and plated forms, as well as in wire. Gold forms an ohmic or nonrectifying contact when alloyed with silicon. Gold combines with silicon or germanium to form a eutectic alloy widely used for bonding dice to headers or substrates. Fine gold wire (0.2 to 0.5 mil) is frequently employed for interconnection wiring. For these applications, the metal must be at least 99.99% pure. The major difficulty incurred with the use of gold is its lack of adherence to smooth silicon dioxide surfaces; absence of this property may require an underlayer as noted earlier.

The metallization of ICs is generally aluminum, because this metal adheres well to silicon and silicon dioxide. Contacts and surface interconnections are made by evaporating thin films of aluminum. Aluminum wire is then used for the other interconnection wiring. The higher eutectic temperature of aluminum and silicon (577°C) permits higher temperatures in assembly and final sealing. Aluminum wire is quite soft and deforms when squeezed or pulled. Because it is difficult to measure aluminum wire in small sizes, resistance per unit length is often specified instead of diameter. Diameters between 0.6 and 5.0 mils are available commercially.

A mixture of gold and aluminum reacts in the presence of silicon to form brittle, high-resistance compounds that will destroy bond integrity and conductivity. Termed "purple plague" or "white plague," the compounds are familiar to reliability engineers. This combination of metals should be avoided in high-reliability applications, particularly where storage or operating temperatures above 100°C might be involved. High temperature greatly accelerates plague formation.

WIRE–BONDING METHODS

Thermal Compression Bonding

Three basic types of thermocompression bonding in use today are wedge bonding, ball bonding, and stitch bonding (see Figure 15-27). Each requires local application of high pressures and elevated temperatures to produce a bond. Bonding is a molecular-adhesion process in which minute irregularities in the two metal surfaces lock them together.

In wedge bonding, the wire is located over the bonding pad by the wire-feed

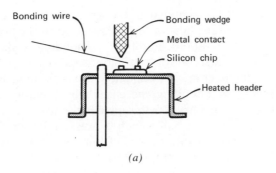

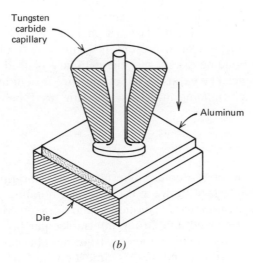

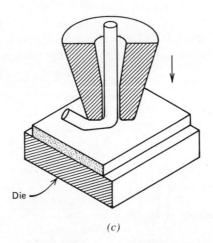

Figure 15-27 Three types of thermocompression bonding: (*a*) wedge bonding. (*b*) Ball, or nailhead bonding, in which a hydrogen flame cuts the gold wire and forms a ball. Force and heat from the substrate again flatten the gold ball for a bonding. (*c*) Stitch-wire bonding, in which the bonding wire is mechanically hooked under the lower edge of the needle. The bond is produced by the combination of heat and needle pressure.

mechanism and the wedge is located over the wire near the wire end. The heat and pressure of the wedge and the heat of the substrate form the bond. Because of the two separate alignments required, this type of bonding is slow. However, it is still the only satisfactory method for extremely small wire sizes. Wire sizes from 0.00007 to 0.002 in. diameter may be bonded to areas as small as 1 mil².

A later development, ball bonding, increases wire-bonding rates. This method locates the wire in a hot capillary which replaces the wedge, thus eliminating one alignment. The hot wire forms a ball, which is pressed flat on the die pad. Gold or aluminum wire may serve for other types of bonding, but only gold wire can be used in ball bonding. Aluminum will not form a suitable ball. Also, ball bonding usually takes a larger bond area than other methods. Because of the incompatibility of aluminum and gold (purple plague) at high sealing temperatures, ball bonding generally is not used with glass-frit-package assembly techniques.

Stitch bonding, very popular at present, is giving way to ultrasonic bonding. It is similar to wedge bonding except that the capillary is used both for feeding the wire and as the wedge. The exposed end of the wire is bent at a 90° angle instead of being balled. Stitch bonding may be used with any type of package and with aluminum or gold wire.

Ultrasonic Bonding

Ultrasonic bonding is done "cold." The ultrasonic vibration and pressure deform plastically the interface between the wire and bonding pad, forming a large contact area. Because of the wiping motion of the ultrasonic head (Figure 15-28), any oxide films are removed and strong intermolecular bonds are formed. The principal advantage is that neither the substrate nor the needle is heated. Thus surface cleaning is not critical, and there is almost no contamination of the bond or surrounding materials. Moreover, ultrasonic bonding will join dissimilar metals to produce low-resistance, low-noise joints.

Some of the differences between bonding methods are listed in Table 15-3. There is a tendency to use gold wire in thermocompression and aluminum wire in ultrasonic bonding. Thermocompression bonding with gold wire is less operator

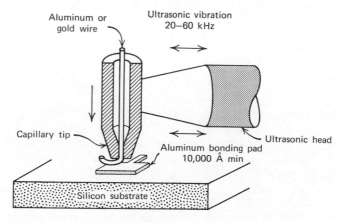

Figure 15-28 Ultrasonic bonding.

dependent. The vertical motion of the tool makes it easier to obtain uniformity from bond to bond. On the other hand, a gold bond on aluminum metallization opens the door to "purple plague." Although the presence of some intermetallics at the interface is not necessarily damaging, intermetallic formation at high temperature may be great enough to use up the aluminum around the bond, resulting in an electrical open circuit. Any time spent at elevated temperature (e.g., as during package sealing or burn-in is detrimental to gold–aluminum bonds.

Bond Inspection Criteria

Improper bonds have traditionally been the cause of most IC failures. In the past several years much time and effort has been spent in refining the bonding processes and the visual inspection criteria. For all practical purposes, this has resulted in the elimination of improper bonds as a major source of IC failures. Most IC manufacturers have adopted the bond inspection criteria of MIL-STD-883: *Test Methods and Procedures for Microelectronics*. These are listed below.

Ball Bonds (Figure 15-29a)

Ball bonds to the silicon chip that are less than two times or greater than six times the bonding wire diameter shall be rejected. Any device that exhibits evidence of rebonding on the same pad shall be rejected. Any device where the center of the bonding wire is closer to the ball bond edge than a distance equal to 1/2 the wire diameter shall be rejected.

Devices containing bonds where less than 50% of the bond is within the pad area shall be rejected. Devices with bonds in the fillet area of the bonding pad that reduce the major distance between the bond periphery and the edge of the fillet to less than 1/2 the narrowest designed width of the interconnecting metallization shall be rejected.

Ultrasonic Bonds (Figure 15-29b)

Ultrasonic bonds to the silicon chip that are less than 1.2 times and more than 3.0 times the wire diameter shall be rejected. Any device that exhibits evidence of rebonding shall be rejected.

Devices containing bonds where less than 50% of the bond is within the bonding pad area shall be rejected. Devices with bond tails longer than 3 mils in length shall be rejected. Devices with bonds in the fillet area of the bonding pad that reduce the major distance between the bond area and the fillet to less than 1/2 the narrowest designed width of the interconnecting metallization shall be rejected.

General

In addition, any device exhibiting any of the following faults shall be rejected (see Figure 15-30):

a. Wire loops greater than three times the diameter of the bonding wire when viewed from the top.

b. Nicks, cuts, crimps, or scoring of the bonding wire which reduce the wire diameter by 25%.

c. Neck down of the bonding wire caused by excessive lead tension which reduces the diameter by 25%.

d. Extra lead wires or lead tails of more than 3.0 mils in length.

e. Leads that are closer than 2.0 mils to each other at any point along their length after a distance of 10 mils from the wire to chip bond.

Table 15-3 Wire-Bonding Comparison

Wedge Bonding	Thermal Compression Bonding		Ultrasonic Bonding
	Ball Bonding	Stitch Bonding	
• Utilizes fine wedge which is brought down on the end of wire to be bonded	• Feeds wire through capillary, eliminating need for separate alignment of parts	• Uses capillary arrangement but exposed wire is bent at 90° angle instead of being balled	• Does not require heating of substrate or bonding head
• Applies high pressures at elevated temperatures locally	• Exposed end of wire is melted into ball and brought down with pressure in area of contact	• Can accommodate small-sized wires	• Bond is produced by pressing a wire against the surface to which it is to be joined; bonded by oscillator transducer and applying pressure
• Both bonds (die and package) made in direct sequence, eliminating need for two separate bonding operations	• Does not require individual registration of stripe, wire, and bonding tool	• Retains single registration feature	• Past problem has to do with mechanical resonances and total system mass
• Simple and inexpensive bonds	• Simple and inexpensive process	• Minimum bond target size lies between that of wedge and ball bonding	• Low-noise, low-resistance bond
• Requires controlled atmosphere	• Requires controlled atmosphere		

436

- Requires elaborate tooling to control movements
- Can hit much smaller contact areas
- Bond size inherently smaller
- Uses finer wires
- Higher-frequency devices can be fabricated with wedge bonding
- Can accommodate more wire types because a ball is not formed

- Fewer adjustments required
- Larger sturdier bonds, but limited to use of gold wire
- Not used with glass-package or glass-die bonding techniques
- Used mostly for hybrid circuits
- Typical pull strength: 2.7 g
- Primary problem has been purple plague

- Requires controlled atmosphere
- Smaller sized bonds
- Can be used with any package
- Can be used with either gold or aluminum wire as well as other types
- Typical pull strength: 3.7 g
- Primary problem is obtaining uniform contact areas
- Stitch bonding being phased out

- Does not require controlled atmosphere
- Surface cleaning is not critical
- Very little contamination of bonds or their surroundings
- Applicable to joining dissimilar metals
- Can be used with either gold or aluminum wire
- Used for most standard ICs; good reliability
- Typical pull strength: 1.4 g
- Primary problem is one of bond size and direction limitations

437

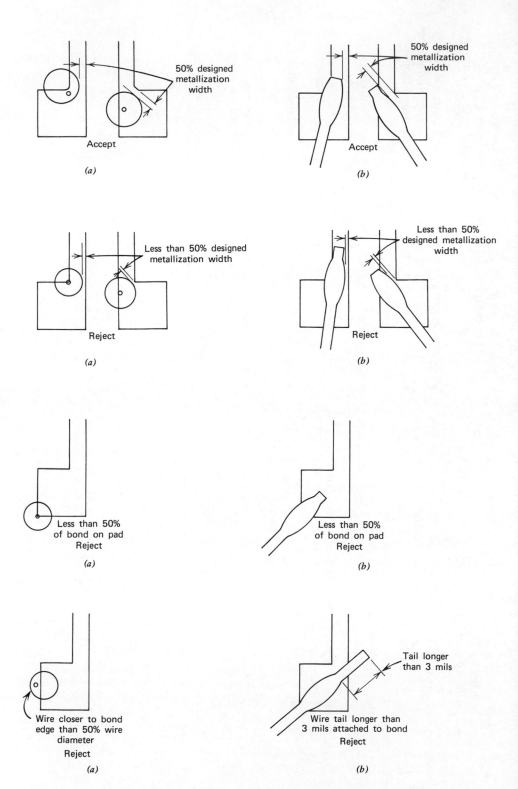

Figure 15-29 Visual inspection criteria: (*a*) ball bonds, (*b*) ultrasonic bonds.

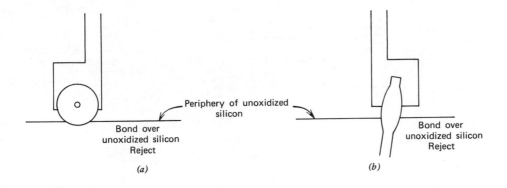

Periphery of unoxidized silicon

Bond over
unoxidized silicon
Reject

(a)

Bond over
unoxidized silicon
Reject

(b)

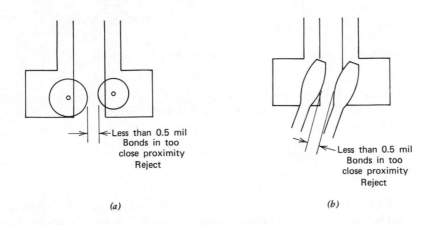

Less than 0.5 mil
Bonds in too
close proximity
Reject

(a)

Less than 0.5 mil
Bonds in too
close proximity
Reject

(b)

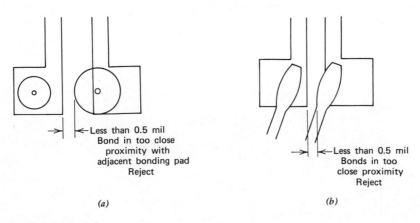

Less than 0.5 mil
Bond in too close
proximity with
adjacent bonding pad
Reject

(a)

Less than 0.5 mil
Bonds in too
close proximity
Reject

(b)

Figure 15-29 (continued).

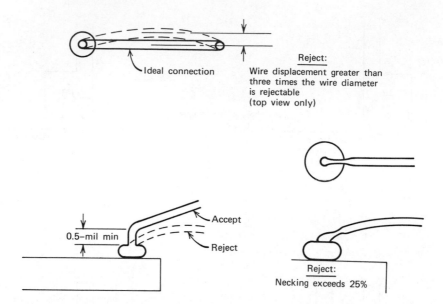

Figure 15-30   General military visual criteria, applicable to any bond.

f. For gold nail head bonded devices the lead wire shall be approximately perpendicular to the surface of the chip for a distance of greater than 0.5 mils before bending toward the package through lead.

ENCAPSULATION

After the leads of the IC have been attached by means of wire bonding, the circuit must be encapsulated. The encapsulation process used depends on the type of package contemplated. The TO-5 type of package allows the most flexible bakeout and sealing procedure. For packages other than the TO-5 type, such as the various flat packages or DIPs, the encapsulation procedure depends on the conditions required to seal the package.

BIBLIOGRAPHY

Fundamentals of Integrated Circuits, by Lothar Stern, Hayden Book Company, 1968.

Integrated Circuits — Design Principles and Fabrication, Raymond M. Warner Jr., and James N. Fordemwalt, Eds., McGraw-Hill Book Company, 1965.

Analysis and Design of Integrated Circuits, Charles S. Meyer, David K. Lynn, and Douglas J. Hamilton, Eds., McGraw-Hill Book Company, 1968.

Microelectronics: Principles, Design Techniques, Fabrication Processes, by Max Fogiel, *REA*, 1968.

Designing with Linear Integrated Circuits, Jerry Eimbinder, Ed., John Wiley & Sons, 1969.

An Up-To-Date Look At Thick Films, by John J. Cox and Donald T. Decoursey, *EDN*, September 15, 1969.

Resistors — Thick and Thin, by Thomas B. Stephenson, *EDN*, August 15, 1969.

Hybrids — Thick and Thin, by Smedley B. Ruch, *The Electronic Engineer*, October 1969.

IC Chip Interconnection Replacing the Flying Lead, by Richard J. Clark, *Electronic Packaging and Production*, December 1969.

Semiconductor Memories, by Harry T. Howard, *EDN*, February 1, 1970.

MOS on the Upswing, *Electronic Design*, **8**, April 12, 1964.

Air Gap Isolated Micro Circuits — Beam Lead Devices, by W. C. Rosvold et al., *IEEE Transactions on Electron Devices*, **ED-15**, n 9, September 1968.

Assuring Integrated Circuit Reliability During Production, by D. I. Troxel and B. Tiger, *Transactions of the 22nd Annual Technical Conference of the American Society for Quality Control*, 1968.

Avoiding IC System Design Pitfalls, by H. I. Cohen, *Electronics*, **41**, n 16, August 5, 1968.

Design of Large Scale Integrated Logic Circuits Using MOS Devices, by D. L. Critchlow et al., *Proceedings, 6th IEEE Microelectronics Symposium*, June 19–21, 1967.

EMC in Digital Integrated Circuits, by A. R. Valentine, *Proceedings of the IEEE Electromagnetic Compatibility Symposium*, July 18–20, 1967.

Electronic Packaging with Integrated Circuits, by A. Hogg, *Assembly Engineering* 11, n 4, April 1968.

Flip-Chip Bonding and Dimensional Transformation – Low Cost Microcircuit Production Technique, by J. G. Bouchard et al., *Proceedings, 6th IEEE Microelectronics Symposium*, June 19–21, 1967.

Hybrid Microcircuit Technology, by J. Goldstein, *IEEE Industrial Electronics and Control Instrumentation for Thick-Film Hybrid IC Technology, Symposium* March 22, 1968.

Hybrid Thick Film Printed Components – Materials and Processes, by A. W. Postlethwaite, *IEEE Industrial Electronics and Control Instrumentation for Thick-Film hybrid IC Technology Symposium*, March 22, 1968.

Introduction to Thick Film Hybrid Circuits, by J. J. Staller, *IEEE Industrial Electronics and Control Instrumentation for Thick-Film Hybrid IC Technology Symposium*, March 22, 1968.

Linear Integrated Circuits, IEEE Wescon Technical Paper, 11 pt 2, Session 1, 1967.

Linear Microcircuits Evaluation and Application, by M. W. Smith, *Proceedings, 6th IEEE Microelectronic Symposium*, June 19–21, 1967.

Making Integrated Electronics Technology Work, by R. N. Noyce, *IEEE Spectrum*, 5, n 5, May 1968.

Medium Scale Integration, by B. Kaplin et al., *Proceedings, 6th IEEE Microelectronics Symposium*, June 19–21, 1967.

Microcircuit Packaging, *Design and Components in Engineering*, no. 9, May 6, 1968.

Microelectronics Technology, by R. I. Walker and R. Naylor, *Electronics and Power*, 13, March 1967.

Use of MOS Transistors in Large Scale Integration, by H. W. VanBeck, *Proceedings of the IEEE Integrated Circuits Seminar*, 1967.

Plastic IC Reliability Evaluation and Analysis, by J. R. Bevington et al., *IEEE Reliability Symposium*, April 7–9, 1970.

Computer Aided LSI Design, by R. W. Ulrichson, presented at IEEE Wescon, August 19–22, 1969.

MOS Integrated Circuits – The Designer's Dilemma, by G. R. Madland, presented at IEEE Wescon, August 19–22 1969.

Noise in Integrated Circuit Transistors, by A. J. Bloderson et al., *IEEE Solid-State Circuits*, 5c, 2, April 1970.

Custom Design of Thin Film Hybrid Circuits, by K. Harrison, *Electronic Engineering*, June 1970.

Circuit Design for Thick Film Hybrid Circuits, by T. D. Tovere, *Electronic Engineering*, June 1970.

The Operational Amplifier in Linear Active Networks, by G. S. Moschytz, *IEEE Spectrum*, 7, January 1970.

A Perspective on Integrated Electronics, by J. J. Suran, *IEEE Spectrum*, 7, January 1970.

Plastic Structured Microelectronics, by J. J. Suran et al., *Digest of Technical Papers, International Solid State Circuit Conference*, 1968.

Thick Films or Thin, by R. E. Thun, *IEEE Spectrum*, 6, 1969.

On Line Graphics Applied to Layout Design of Integrated Circuits, by A. Spitalny and M. Goldberg, *Proceedings of the IEEE*, 55, November 1967.

Large Scale Integration of MOS Complex Logic, by A. Weinberger, *IEEE Transactions on Solid State Circuits*, SC-2, December 1967.

The Anatomy of Integrated Circuit Technology, by Herwick Johnson, *IEEE Spectrum*, 7, February 1970.

Beam Lead Technology, by M. P. Lepselter, *Bell Systems Technical Journal*, 45, 1966.

IEEE Transactions on Solid-State Circuits, Special Issue on Large Scale Integration, SC-2, December 1967.

IEEE Transactions on Solid-State Circuits, Special Issue on Linear Integrated Circuits, SC-3, December 1968.

Tantalum Film Technology, by D. A. McLean et al., *Proceedings of the IEEE*, 52, 1964.

Technological Advances in LSI, by H. T. Hochman and Dennis Hogan, *IEEE Spectrum*, 7, May 1970.

Technology for Design of Low Power Circuits, by C. A. Bettermann et al., *IEEE Transactions on Solid State Circuits*, SC-5, n 1, February 1970.

Interaction of Technology and Performance in Complementary Symmetry MOS Integrated Circuits, by D. W. Ahrons and P. D. Gardner, *IEEE Transactions on Solid-State Circuits*, **SC-5**, 1, February 1970.

Solid State Circuit Design and Operation, by Stanton Rust Prentiss, Tab Books, 1970.

Thermocompression Bonding of External Package Leads on Integrated Circuit Substrates, by V. P. Bolcar, *IEEE Transactions on Electron Devices*, **ED-15**, n 9, September 1968.

Practical Design with Integrated Circuits and Advanced Measurement Techniques, *IEEE Paris Seminar Proceedings*, April 1969.

The Case for Beam Lead Bonding, by J. E. Clark, *Electronic Packaging and Production*, October 1970.

IC Metallization Systems: A Comparison, by R. W. Wilson and L. E. Terry, *Electronic Packaging and Production*, October 1970.

Surveying Chip Interconnection Techniques, by Howard K. Dicken, *Electronic Packaging and Production*, October 1970.

Simplified Bipolar Technology and Its Applications to Systems, by Bernard T. Murphy, *IEEE Transactions on Solid State Circuits*, **4SC-5**, n 1, February 1970.

The Case for Emitter Coupled Logic, by Anthony A. Vacca, *Electronics*, April 26, 1971.

Electronics Special Report: The New Look in LSI Packaging, *Electronics*, April 12, and 26, 1971.

Beam Lead Semiconductor Memories, by Anthony Holbrook, *EEE*, April 1971.

Surveying Current Microelectronic Packaging, by Alfred T. Batch, *Electronic Packaging and Production*, March 1971.

STD: A New Hybrid Circuit Fabrication Process, by James P. Dietz, *Electronic Packaging and Production*, December 1971.

Thick Film Capacitor Design Nomograph, by Chester W. Young, *Electronic Packaging and Production*, July 1971.

Focus on Packaging, by George Rotsky, *Electronic Design*, **16**, August 5, 1971.

Coming Up Fast from Behind—Denser Bipolar Devices, by John A. Defale, *Electronics*, July 19, 1971.

IC Op Amps—Getting it all Together, by Thomas P. Rigoli, *EDN*, May 1, 1971.

Operation and Application of MOS Shift Registers, by Marion E. Hoff, Jr., and Stanley Major, *Computer Design*, February 1971.

Progress in Ceramic IC Packaging, by Arthur G. Cohen, *Electronic Packaging and Production*, May 1971.

Electronic Integrated Circuits and Systems, by Franklin C. Fitchen, Van Nostrand-Reinhold Company, 1970.

Operational Amplifiers, by Arpad Barna, John Wiley & Sons, 1971.

Operational Amplifiers—Design and Applications, by Jerald G. Graene, Gene E. Tobey, and Lawrence P. Huelsman, McGraw-Hill Book Company, 1971.

Thick Film Microelectronics, by Morton L. Topfer, Van Nostrand-Reinhold Company, 1971.

Linear Integrated Circuits, Theory and Applications, Jerry Eimbinder, Ed., John Wiley & Sons, 1968.

Linear Applications, National Semiconductor Corporation, January 1972.

Special Report, Improved n channel Processes Make Their Move in Memories, *Electronics*, May 8, 1972.

Implanted Depletion Loads Boost MOS Array Performance, by Bob Crawford, *Electronics*, April 24, 1972.

Ion Implantation: The Growing Technology, by Howard N. Markstein, *Electronic Packaging and Production*, April 1972.

Finding Faults in Microworlds: Electron Microscopy as a QA Tool, by Niel S. Atkinson, *Electronic Packaging and Production*, April 1972.

Tear Tabs: A New High-Frequency-Device Connection Method, by R. A. Zankoratti, *Electronic Packaging and Production*, April 1972.

Focus on CMOS, *Electronics Design*, **8**, April 13, 1972.

CMOS Unites with Silicon Gate to Yield Micropower Technology, by R. R. Burgess and R. G. Daniels, *Electronics*, August 30, 1971.

Silicon-on-Sapphire: A Technology Comes of Age, by David J. Dumin and Edward C. Ross, *The Electronic Engineer*, February 1972.

Bipolar Techniques Approach MOS in Density, Cost and Power Drain, yet Retain Speed, by Siegfried K. Wjedmann and Horst H. Berger, *Electronics*, February 14, 1972.

The Pluses and Minuses of Charge Transport Devices, by R. D. Baertsch, W. E. Engeles, and J. J. Tiemann, *Electronics*, December 6, 1971.

Epoxy B: A New IC Encapsulant, by Eugene R. Hnatek, *Electronic Packaging and Production*, May 1972.

Switching Regulator Designed for Portable Equipment, by Eugene R. Hnatek, and Larry Goldstein, *EDN*, September 15, 1971.

Low-Power TTL — New Process and Design Double Speed, by Eugene R. Hnatek, and Jerry Gray, *Electronic Products*, April 17, 1972.

Circuit Design For Integrated Electronics, by Hans R. Camenzind, Addison-Wesley Publishing Company, 1968.

Electronic Integrated Systems Design, by Hans R. Camenzind, Van Nostrand Reinhold Company, 1972.

MOS Integrated Circuits: Theory, Fabrication, Design and Systems Application of MOS LSF, The Engineering Staff of American Micro-systems Inc., Van Nostrand Reinhold, 1972.

A New phase-locked loop with high stability and accuracy, by J. A. Mattis and H. R. Camenzind, Signetics Corporation Application Note, 1970.

The Phase-locked loop — A Communication System Building Block, by J. A. Mattis, Broadcast Engineering, 1972.

Frequency Selective Integrated Circuits using Phase Lock Techniques, IEEE J. Solid State Circuits, SC-4, 1969.

A High Density Ion-Implanted CMOS Technology, by R. M. Finnila *et al.*, Electronic Packaging and Production, January 1973.

INDEX

AC coupled high input impedance amplifier, 309–310

Access time, 222, 255, 259

Active filter, 311–313

Active pull-up, 77, 79, 82, 84, 149, 223, 224, 257–258

Air gap isolation, 2, 3, 41

Alumina, 11, 12, 186, 365, 386, 430

Analog-to-digital converter ladder driver network, 319–321

Analog commutator, 320–321

ASCII code, 247, 251, 252

Aspect ratio, 376, 389, 390, 400–401

Asynchronous counter, 107

Asynchronous preset, 118

Ball bonding, 433, 434, 435, 436–437, 438–439

Bandpass filter, 142–143

BCD adders and converters, 128, 129, 130, 149

BCD counter, 112, 116, 119, 121, 128, 129, 130, 135, 137

Beam leads, 43, 46, 423–429

Bell jar vacuum system, 365, 366–367

Beryllia, 364

Bias current compensation, 295–297

Bidirectional counter, 115, 116, 117, 119

Binary counter, 107, 120, 121, 122

Bipolar digital IC's, 2–7, 14, 61–63, 70–160, 161, 171

Bipolar IC's, 2–7, 14, 18–160, 161, 171, 265–268, 270, 276–359

Bipolar IC transistors, 51, 52–65

Buffer memory, 253

Buried layer, 3, 10, 19, 27, 53, 54, 61

Bus organized system, 90–92, 148–160, 216–217, 243

Butterworth filter, 312

CAD, 260–264

CCD, 10, 189, 192, 193

Cermets, 369, 372, 373

Character generation, 247, 251, 252

Charge coupled device, 10, 189, 192, 193

Chip capacitors, 392–394

Chip resistors, 390–394

Chrome masks, 29–31

Chromium, 369, 372

Clock phase delay time, 231

Closed loop gain, 277, 278, 282, 290, 292, 294

CMOS, 9, 162, 163, 174–176, 188, 194–220, 267, 271–272

Complementary PNP transistors, 54, 55, 56

Complementary-symmetry MOS, 9, 162, 163, 174–176, 188, 194–220, 267, 271–272

Common Mode Rejection (CMRR), 284, 289, 300, 301, 318, 328

Commutating filter, 142–143

Comparison of bipolar/MOS circuits, 265–275

Composite drawings, 26, 28, 30, 31

Computer-aided-design, 260–264

Computerized artwork generation, 30–32, 261

Constant current source, 340–341

Contaminants, 17, 170, 179, 182

Coordinatograph, 28

Cost, 5, 13, 14, 84, 105, 107, 171, 176, 182, 184, 221, 265, 266, 271, 287, 288, 330, 347, 351

Counters, 14, 107–122

Current limiting, 326, 330, 343, 352, 353, 357

Current mode logic, 92–98, 267

445

D flip flop, 151, 152, 206–207, 208
D/A converter, 182
Darlington pull-up, 79, 80
David Mann pattern generator, 32, 263
DC amplifier, 298
Decade counters, 115, 116, 144
Decoders, 118, 120, 122, 123, 124, 130, 133, 135, 140, 141, 142, 148, 151, 159
Depletion mode MOS, 9, 181, 185
Die bonding, 421–431
Dielectric isolation, 2, 3, 23, 41, 42, 43, 179, 186–189, 365
Die passivation, 22, 47, 179, 423
Differential amplifier, 300–301, 327
Diffusants, 22, 33, 64
Diffused capacitors, 70, 71, 167, 189, 378–379, 380–381, 392–393
Diffused resistors, 3, 4, 21, 51, 52, 68–70
Diffusion, 33, 35–39, 40, 52, 56, 57, 64, 65, 77, 86, 95, 96, 162, 167, 173, 177–179, 180, 269–270
Diffusion furances, 33, 37, 38, 57
Digital clock, 144, 146, 147, 148
Digital filter, 142–143
Digital IC tone decoder, 144–145
Diode-Transistor Logic (DTL), 75, 76, 88, 100, 101, 102, 172, 199, 211, 212–214, 222, 265, 268
DIP, 408, 412–417
Dissipative regulator, 326, 327, 332, 334–336, 338, 339, 344, 345, 347, 348, 349
Divide-by-n counter, 114
DMOS, 185–186
Dopants, 33, 35, 39, 41, 52, 64, 66, 96, 167, 171, 181, 182
Double Diffused MOS (DMOS), 185–186
Double diffused transistor, 57
Down counter, 115, 116
DTL, 75, 76, 88, 100, 101, 102, 172, 199, 211, 212–214, 222, 265, 268
Dual-in-Line Package (DIP), 408, 412–417
Dual tracking voltage regulator, 328–329, 346–347
Dynamic (AC) Logic, 222, 223, 224–228
Dynamic RAM, 254–255
Dynamic shift register, 211, 212, 228–235, 237, 238, 239, 242–246

Eccles Jordon flip-flop, 235, 236
ECL, 93–98, 100, 101, 102, 199, 222, 265, 267, 268
Electron beam evaporation, 47, 162, 167, 366
Electronic thermometer, 323–326
Emitter clamp, 83, 84
Emitter Coupled Logic (ECL), 93–98, 100, 101, 102, 199, 222, 265, 267, 268
Encapsulation, 396, 440
Enhancement mode MOS, 9, 162–167, 181, 200, 271, 272

Epitaxy, 40, 41, 56, 319, 320
Error currents, 292, 295–299
Error function, 36, 39
Etching, 22, 32, 33, 96, 367–368
Eutectic, 12, 13

Fall time, 103, 104
Fan-in, 76, 99, 103, 151, 213
Fan-out, 75, 76, 77, 91, 92, 93, 98, 99, 151, 158, 267
FEDIS, 30–32
Feedforward compensation, 305–306, 308, 317
PETS, 2, 8, 9, 12, 50, 161, 164–166, 222, 235, 269, 313
Field effect transistor, 2, 8, 9, 12, 50, 161, 164–166, 222, 235, 269, 313
Field shield MOS, 163, 183–184
Flatpack, 15, 408, 411–412
Flip-chip bonding, 11, 16, 362, 422–423, 428, 429
Flip-flops, 61, 62, 151, 196, 197
Four phase dynamic shift registers, 233–235
Frequency bursts, 231, 233
Frequency compensation, 288–289, 301–306, 308, 356
Frequency response, 62, 64, 285, 286

Gate alignment, 173, 174, 178, 179, 181, 182, 184, 185, 273
Gaussian function, 36, 39
Gold doping, 40, 64, 66, 95
Gyrator-capacitor networks, 383

Hermeticity, 13
Hermetic packages, 407, 408, 409–415
High current voltage regulator, 351–359
High level input current, 99
High level input voltage, 98
High level output current, 99
High level output voltage, 98
High pass filter, 312
High speed TTL adders, 122, 125, 126, 127
High threshold logic, 171–174, 199
High voltage regulator, 344–345
Hybrid packages, 417, 418, 420
Hybrids IC's, 2, 6, 8, 10–13, 17, 360–383, 384–406

Ideal Op Amp, 277–279
Impurities, 39, 56, 64, 65, 163, 164
Impurity distribution profile, 36, 39
Input bias current, 282–283, 287, 292
Input impedance, 277, 283, 284, 287, 294, 318
Input offset current, 282, 287
Input offset voltage, 283, 287, 288, 291, 292, 293
Input protection, 217–219, 288

Instrumentation amplifier, 317–319
Integrator, 306, 307, 310–311
Internally compensated Op Amp, 288–289, 293, 326
Ion implantation, 10, 163, 175, 180–182, 185, 191, 192, 199
Isolation, 9, 41–44, 53, 161, 162, 184, 186–189, 267, 270, 360
Inverter, 167, 200–202, 205, 223, 224, 226, 227, 229, 272
Inverting amplifier, 280, 292–293, 299–300

J-K flip flop, 61, 62, 207, 209
Junction capacitor, 51, 52, 71–73
Junction isolation, 41–42, 53, 54

Large scale integrated circuits, 10, 15, 47, 221, 222, 260, 261, 264, 265, 267, 417–421
Laser trimming, 395–396
Lateral PNP transistor, 54, 55, 56
Light intensity regulator, 345–346
Linear IC's, 14, 26, 27, 50, 54–55, 56–61, 70–71, 72–73, 276–359
Line regulation, 331
Load regulation, 331
Low cost switching regulator, 347, 348, 349
Low frequency operation of dynamic shift registers, 231–233
Low level input current, 99
Low level input voltage, 98
Low level output current, 99
Low level output voltage, 98
Los pass filter, 310, 311, 312
Low rf noise switching regulator, 348, 350, 351
Low threshold logic, 171–174, 177, 179, 184, 199
LSI, 10, 15, 47, 221–222, 260, 261, 264, 265, 267, 417–421

Mainframe memory, 222, 257, 258
Masking, 24–27, 29, 38, 48, 177
Mechanical masks, 365, 366, 367, 368
Medium Scale Integration (MSI), 14, 15, 105, 107–160, 265, 267
Memories, 14, 44, 107, 133, 142, 149, 150, 182, 184, 186, 221, 222, 246–263, 266
Memory cells, 207, 209, 210, 222, 253–257, 260–263, 266
Metal alumina silicon MOS, 10, 186, 192, 193
Metallization, 3, 11, 15, 16, 17, 21, 22, 23, 44, 45, 46, 47, 48, 97, 167, 169, 173, 177, 181, 182, 370, 371, 386–387, 394–395, 399, 432
Metal migratron, 96
Metal Oxide Semiconductor (MOS), 2, 7, 8–10, 16–17, 70–71, 130, 151, 161–275
Minority carrier lifetimes, 65, 66

Monolithic capacitors, 70–71, 167, 189, 378–379, 380–381, 392–393
Monolithic diodes, 40, 51, 52, 65–67, 79, 81, 82, 87, 95, 266
Monolithic resistors, 3, 4, 21, 51, 52, 68–69
Monolithic transistors, 51, 52–65, 77, 86, 222
MOS, 2, 7, 8–10, 16–17, 70–71, 130, 151, 161–275
MOS bipolar interface, 275
MOS capacitors, 70–71, 167, 189, 378–379, 392–393
MOS load resistance, 166–167, 233, 269
MSI, 14, 15, 105, 107–160, 265, 267
Multiemitter transistor, 63, 77, 86, 222
Multiplexer/demultiplexer, 130, 137, 140–142, 152, 153–155, 156, 158, 159

Nand gates, 95, 102, 133, 137, 138, 139, 206, 215, 224, 225, 227
Nanoammeter amplifier circuit, 321–323, 324
Negative feedback, 277, 281
Negative voltage regulators, 347, 348, 349
Nichrome, 364, 368, 369, 370, 372, 373, 394
NMOS, 163, 170–171, 183–184, 190, 191, 194
Noise immunity, 79, 82, 91, 170, 172, 176, 202–203, 211, 213, 219, 271, 280
Noise margin, 75, 76, 93, 97, 199, 212, 213
Noninverting amplifier, 280–281, 289, 293, 294, 300
NOR gates, 102, 149, 154, 205, 206, 215, 224, 225, 227
NPN transistors, 40, 50, 54

Offset voltage compensation, 297–299, 300, 301, 321–323
Off-state output current, 99
Ohmic contacts, 169, 431–432
One-shot multivibrator, 313–314
Open loop frequency response, 281
Open loop gain, 280–281, 294, 305
Operational amplifier, 14, 26–27, 55, 56–61, 70, 71, 72, 73, 276–329
Operational amplifier design precautions, 306–308
Operational amplifier protection circuits, 288, 294, 295
Output buffer, 239, 244
Output impedance, 277, 283, 284
Overlapping clock circuits, 234–235
Oxidation, 23, 24, 167, 171
Oxide growth, 24, 162, 163, 167, 177, 178

Packaging, 12–13, 15, 221, 273, 287, 385, 386, 396, 407–421
Packaging materials, 408–409
Parasitic transistors, 68, 162, 172–173, 233
Passive pull-up, 148, 320

Phase Splitter Clamp, 83, 84
Photoetching, 25, 32–33, 167, 178, 194
Photolithography, 11, 29
Photomask preparation, 24–25, 26, 27, 28, 29, 31
Photoresist, 25, 32–33, 34, 35, 181, 368–369
Pinhoes, 167
Plastic (nonhermitic) packages, 407, 408, 409, 415–417
PMOS, 9, 163, 164–171, 190, 191, 199, 265, 267
PNP transistors, 40, 50, 53, 54
Power amplifier, 343–344
Power dissipation, 4, 5, 51, 79, 85, 86, 88, 175, 194, 195, 197, 199, 200, 201, 211, 212, 229, 233, 237, 238, 259, 266, 269, 351, 362, 363, 385–386, 389
Precision regulators, 357–359
Precharge, 239, 259
Presettable counter, 115, 118
Primary transistors, 58, 59
Processing, bipolar, 18–48
Processing, MOS, 162, 167–170, 175–176, 177–178, 183, 268, 269, 270, 272
Propagation delay, 87, 88, 104, 148, 189, 202, 215, 216, 266, 267
Punch through transistors, 56–61
Purple plague, 45, 432, 435

Quiescent current drain, 331

Radiation resistance, 187, 188
RAM, 149, 150–153, 182, 184, 186, 193, 221, 252–260
Random Access Memory (RAM), 149, 150–153, 182, 184, 186, 193, 221, 252–260
Ratio circuits, 230, 231
Ratioless circuits, 226, 227, 228, 229, 230
Read Only Memory (ROM), 14, 44, 107, 133, 142, 149, 150, 182, 184, 186, 221, 246–252, 266
Refractory MOS (RMOS), 163, 182–183
Refresh memory, 222
Reliability, 15, 16, 17, 179, 221, 271
Resistor clamp, 83, 84, 149
Resistor Transistor Logic (RTL), 75, 88, 100, 101, 102, 199
Resistor trimming, 378, 388–389, 395–396, 401
Ripple counter, 108–111
Ripple rejection, 331
Rise time, 103, 104
RMOS, 163, 182–183
ROM, 14, 44, 107, 133, 142, 149, 150, 182, 186, 221, 246–252, 266
RS flip flop, 206
RTL, 75, 88, 100, 101, 102, 199

Rubylith, 28, 29, 263, 264

Sample-and-hold circuit, 295, 310–311, 323
Sapphire, 186–188, 365
Schottky diode, 40, 66–67, 79, 81, 82, 87, 95, 266
Scratch pad memory, 252
Secondary transistor, 58, 59
Self Tuned filter, 312, 313
Semiconductor on Thermoplastic Dielectric (STD), 430–431
Series 4000A CMOS, 194, 196, 197, 205–219
Series 54C/74C CMOS, 220
Series 54/74 Standard TTL, 84, 85, 88, 89, 194
Series 54H/74H High Speed TTL, 86, 88, 89
Series 54L/74L Low Power TTL, 84, 85, 88, 89, 158, 195, 196, 197, 214, 265
Series 54S/74S Schottky Clamped TTL, 67, 87–89
Settling time, 290
Sheet resistivity, 36, 51, 52, 69, 70, 166–167, 182, 372, 373, 374, 376, 389, 395
Shift registers, 14, 48, 151, 181, 184, 189, 193, 211, 212, 228–246, 266
Short circuit output current, 99
Short circuit protection, 327
Silicon, 1–3, 19, 21, 22, 23
Silicon dioxide, 2, 3, 23, 40, 45, 47
Silicon gate, 9, 173, 174, 175, 177–179, 184, 185, 190, 191, 196, 255
Silicon nitride, 9, 163, 173–174, 179–180, 191, 192, 193, 424–425, 427
Silicon-on-Sapphire (SOS) MOS, 10, 43, 44, 186–187, 188, 192, 193, 196, 198
Silicon-on-Spinel MOS, 10, 187–189, 192, 193
Silk Screening, 384, 389, 394, 395
Sinewave oscillators, 314–317
Slew rate, 284, 289–290, 293, 305, 306
Spider bonding, 428, 430
Spinel, 10, 186, 187–189, 192, 193
Sputtering, 46, 162, 366, 370
Stable internal voltage reference, 352, 354–356
Standby (Quiescent) current brain drain, 331
Static (DC) logic, 222, 223, 224
Static RAM, 253–254
Static shift registers, 228, 229, 235–236, 237, 238
STD, 430–431
Stitch bonding, 432, 433, 434, 436–437
Storage registers, 14, 48, 151, 181, 184, 189, 193, 211, 212, 228–246, 266
Substrate PNP transistor, 54, 55, 56
Substrate preparation, 22–23, 362, 364, 365, 366
Summing amplifier, 296, 306
Super Beta transistors, 56–61

Supply current, high level output, 99
  low level output, 99
Switching regulator, 331, 332, 333, 336, 338,
  339–340, 347, 348, 349, 350, 351
Synchronous counter, 111, 112, 113, 137–140
Synchronous preset, 118, 119

Tantalum, 369, 373, 380–383
Temperature controller, 340–343
Temperature drift, 332
Testing, 17, 50, 273, 287, 288, 361, 425
Thermal shutdown, 352, 353, 354
Thermocompression bonding, 371, 381, 402, 431,
  432–434
Thick film capacitor, 392
Thick film circuit layout, 396–402
Thick film conductor design, 387–390, 399
Thick film hybrids, 2, 7, 8, 12, 362–365, 384–
  406
Thick film processing, 394–396
Thick film resistor, 363, 370, 372–378, 385,
  388–390, 400–402
Thick oxide, 67, 181, 184, 247
Thin film capacitors, 382–383
Thin film conductor design, 370–371
Thin film hybrids, 2, 6, 7, 11, 360–383, 384,
  394
Thin film materials, 364–365
Thin film processes, 365–369
Thin film resistors, 51, 363
Thin film resistor parasitics, 375
Threshold voltage, 162, 164, 165, 171–174,
  179, 181, 186, 196, 197, 198, 199, 205
TO package, 408, 409–411
Totempole output, 84
Transition time, 284

Transistor geometrics, 61–63
Transistor-transistor logic, 61, 76–89, 100, 101,
  102, 107–148, 158, 172, 189, 211, 212–
  214, 265, 268
Transmission gates, 202–205, 216, 253, 254
Tri-state logic, 79, 90–92, 148–160, 242–246
TTL, 61, 76–89, 100, 101, 102, 107–148, 158,
  172, 189, 211, 212–214, 265, 268
TTL resistors, 86
TTL transistors, 61–63, 86
Two phase clocking, 225–228
Two phase dynamic shift registors, 229–233

Ultrasonic bonding, 431, 434–435, 436–437,
  438–439
Unity gain buffer, 294, 296, 297, 300, 301, 306,
  324
Up counter, 114

Vacuum deposition, 11, 46, 47, 365
Vacuum evaporation, 366–367, 370, 379
Vertical PNP transistor, 54, 55, 56
Volatility, 186
Voltage comparators, 308–309
Voltage followers, 294, 296, 297, 300–301,
  306, 324
Voltage regulators, 4, 287, 326–329, 330–359

Wafer aligners, 33–36
Wedge bonding, 432, 433, 434, 436–437
Wire bonding, 11, 16, 370, 371, 387, 388, 402,
  431, 440

Zero crossing detector, 323, 324

COMPLETE